FRED PHILLIPS

MANUFACTURING RESEARCH AND TECHNOLOGY 17

GLOBAL MANUFACTURING PRACTICES

MANUFACTURING RESEARCH AND TECHNOLOGY

MANUFACTURING RESEARCH AND TECHNOLOGY 17

Global Manufacturing Practices

A Worldwide Survey of Practices in Production Planning and Control

Edited by

D. Clay Whybark
Gyula Vastag
University of North Carolina, Chapel Hill, NC, U.S.A.

ELSEVIER
Amsterdam – London – New York – Tokyo 1993

ELSEVIER SCIENCE PUBLISHERS B.V.
Sara Burgerhartstraat 25
P.O. Box 211, 1000 AE Amsterdam, The Netherlands

ISBN: 0 444 89978 2

This book is printed on acid-free paper.

Printed in The Netherlands

This book is dedicated to Macon Patton's family and friends,
who helped so much to make it possible.

GLOBAL MANUFACTURING PRACTICES:
A WORLDWIDE SURVEY OF PRACTICES
IN PRODUCTION PLANNING AND CONTROL

FOREWORD

Let me offer four reasons why we need to know more and more about how manufacturing is done around the world.

First, world trade in manufactured goods is increasing faster than world trade in general, and both are growing faster than the world economy. More products than ever before are being produced in one country and shipped elsewhere. We must know how our competitors do business.

Second, companies are spreading their production globally at a rapid pace. World stock of foreign direct investment grew more than 12% a year during the 1980s (outstripping the growth of the world economy) and much of that investment went to establish or acquire factories. We must know what is possible in these factories and what is not.

Third, improving manufacturing performance is the best way for most countries to improve their economies and their living standards. Many argue--and I am one of them--that this is true for almost *all* countries, rich or poor. But there is no disagreement that how well most of the world's people live depends on manufacturing performance. Because all nations are one another's suppliers, competitors, and customers, and because all share the natural resources of this planet and are fellow human beings, we must share our knowledge.

Finally, no one has a monopoly on the best manufacturing practices. World class manufacturing companies are found in many countries, and the ranking of leaders changes frequently. These companies are usually avid students of the practices of others, even of those not among the world's leaders. Comparing different practices in different settings is a rich source of ideas and insights. We all have much to learn from one another.

By providing data on manufacturing practices around the world, the Global Manufacturing Research Group (GMRG) offers an invaluable service. The GMRG is filling in big gaps in our knowledge of how manufacturing is done in the former Soviet Union, eastern Europe, and the People's Republic of China. Even in countries about whose practices we know more, the data, especially in a form that allows comparison and contrast, is of great value. The GMRG is helping us understand those countries better, as well.

Collecting such data requires not only a long-term commitment by a large group, but real devotion by the few who initiate, guide, coordinate, coach, sponsor, and manage the project. My admiration and gratitude for those who founded and are carrying on the GMRG is great. I am familiar with several other such collaborative efforts, and had the privilege of being personally involved with one--the Global Manufacturing Futures

Project (a global survey started by Professor Jeff Miller at Boston University in 1982). Among these various global surveys, however, the GMRG stands out for the nature of the research undertaken, the breadth of the database, and the number of persons involved in the effort.

As to the nature of the research, the GMRG is cleverly focused on just two industries: non-fashion textiles and small machine tools. Comparisons are easier to make and are more convincing because many variables are controlled. The industries were chosen because they are found throughout the world and because they represent two distinct forms--process industry (textiles) and batch processing (machine tools). The two industries are usually fragmented, with many firms in each, and thus provide a glimpse into the mainstream of manufacturing practices in a country.

This glimpse is quite different from that provided by the Global Manufacturing Futures Project, for example, in which there is no particular industry focus and the respondents are usually large manufacturers that dominate their markets. Other projects are attempting to discern world-class manufacturing attributes, while still others are studying the approaches to computer integrated manufacturing in advanced installations. The GMRG work therefore fills an important need by focusing on mainstream practices rather than the practices of advanced firms and market leaders exclusively. The gap between best and the mainstream is often very large indeed.

Because the GMRG focus is on two industries, and on production planning and control practices, the questions in the GMRG surveys are generally more precise than those in other questionnaire surveys. This is a welcome change for the researchers in this field, where survey questions are too often manipulated to fit many different industries, with the resulting risk of superficiality. The GMRG questions can be put more clearly and ask for more specific answers.

Another important aspect of the GMRG work is the extensive involvement of researchers from all over the world. Just a look at the table of contents of this book provides an insight into the breadth of the group. These are generous individuals who share their data not only among themselves, but also with other interested researchers. By involving an even larger group worldwide, the GMRG is creating a powerful and uniquely valuable international network of empirical researchers. For this, the GMRG deserves high praise.

The work of this dedicated group fits nicely into the tapestry of international manufacturing research. I hope that this book, as valuable as it is, will be only one of many that will report the results of the GMRG's outstanding work. I look forward to seeing more of the group's research in the future.

Kasra Ferdows
Washington, D.C., 1993

PREFACE

This volume contains research papers and the data from a worldwide survey of manufacturing practices. The project was started as the result of some questionnaire based field research on executives in Korea by Prof. Boo-Ho Rho of Sogang University in Seoul. He presented a paper on this work at the Pan-Pacific Conference in Korea in 1985. This lead to discussions with Prof. D. Clay Whybark of the University of North Carolina about starting a joint research project. The purpose would be to learn more about how manufacturing is managed in various parts of the world. The decision was made to focus on manufacturing practices (principally in the production planning and control areas), to develop a questionnaire that would be common for all data gathering activities, and to survey manufacturing practices in as many countries as possible.

The global dimension of the project meant that the industries that were to be surveyed would have be limited to those found virtually everywhere. An early decision, therefore, was to restrict the sample to two industries: non-fashion textiles (bed sheets, curtains, towels, etc.) and small machine tools (lathes, milling machines, drill presses, etc.). These industries are found in both advanced and developing countries. The choices also represent batch processing and a form of process industry. After translation of the questionnaire, the data gathering was started. The effort was called the global manufacturing practices project.

THE GLOBAL MANUFACTURING PRACTICES PROJECT

The first surveys were done in Korea with the help of the Korea Productivity Center, in the People's Republic of China with the help of the Shanghai Institute of Mechanical Engineering and in western Europe with the help of the international Institute for Management Development (IMD) in Lausanne, Switzerland. Some time after the data were gathered in these three areas, the survey was administered to firms in North America with the help of Indiana University. Shortly after the project was started there, the Indiana Center for Global Business was founded and supported the project for a period of time. In 1990, the coordination of the project was moved to the Kenan Institute at the University of North Carolina, where the Center for Manufacturing Excellence supported the installation of the data base.

To gain the support of the companies involved in providing the data, feedback on the project was provided to them. In most instances a copy of the results was sent to each of the responding companies, usually in the form of a questionnaire indicating the mean score and ranges for each response. This enabled the company to see how it was doing relative to the other companies in the industry and country.

A somewhat different approach was taken in China. In order to gain the cooperation of the participating firms, their representatives were brought together in one place to fill out the questionnaire, discuss the implications of some of the questions and share some of their management concerns. This proved to be a valuable technique for the researchers and experience for the company representatives. Other techniques have subsequently been used in other areas of the world.

The objectives of the project were much broader than gaining the cooperation and confidence of the participating companies. The first was to gain insights into the differences in manufacturing practice that exist around the world. Therefore, very little financial or other performance data was requested in the questionnaire. This also had the advantage of increasing the sample size, minimizing restrictions on data distribution and reducing the difficulties in making comparisons between countries. The concern was to develop an understanding of manufacturing practice differences would be important to improved cooperation between researchers, companies, and nations.

The second objective of the project was to make the data available to other researchers around the world, understanding that gathering such information as individuals is a very costly task. Also, involving more people would greatly improve the quality and quantity of insights drawn from the data. This required the cooperation of the participating countries and institutions. Initially, the agreement was that the data were to be made available to academic researchers in all countries where data was gathered and contributed to the project. With the publication of this book, the data can now be made generally available, opening the opportunity for substantially increased involvement in the research.

Recognizing the increasing globalization of commerce, the final objective of the project was to make the geographical scope of the activities as large as possible. This meant gathering the data in a broad sample of different cultures and economic systems, and expanding the number of people directly involved. At first this was done informally through personal contact, professional association and academic networks. In 1990 the Global Manufacturing Research Group (GMRG) was formed to provide a more formal means for including new people, sharing ideas and exchanging insights.

THE GLOBAL MANUFACTURING RESEARCH GROUP

The Global Manufacturing Research Group was formed in June of 1990. Several of the original researchers held a workshop on the project in Shanghai, China that year. The meeting was sponsored by the Shanghai Institute of Mechanical Engineering, the Korean Traders Scholarship Foundation (now the Sanhak Foundation), the Indiana Department of Commerce, and the Global Business Center at Indiana University. The workshop provided the researchers an opportunity to discuss the questionnaire, share research approaches and exchange insights gained. It also provided the group an opportunity to see some of the factories from which the Chinese data were originally gathered. There were papers presented at the meeting, but a large part of the value was in the personal contact and workshop sessions among the researchers.

Right after the workshop in Shanghai, most of the group participated in the Pan-Pacific Conference, in Seoul, Korea (where it had all started five years before). There was a manufacturing practices track in that conference, and the researchers made use of that for presenting formal papers and acquainting other people with the project. It was during the June, 1990, Pan-Pacific meeting that the group decided to create a more formal organization. Most of the participants at the meetings in Shanghai and Korea had direct experience with the manufacturing practices questionnaire and/or data base. Some brought complimentary projects, data bases or research experience to the group. All agreed that a means for continuing and expanding the collaboration was worthwhile and the Global Manufacturing Research Group (GMRG) was founded.

The group has developed a pattern of meeting internationally in conjunction with a formal group and scheduling an informal GMRG workshop closely preceding or following. For example, the group met in Budapest, Hungary, for a workshop and then, immediately afterwards, at the Decision Sciences First International Meeting in Brussels, Belgium. The Budapest workshop was supported by the Budapest University of Economic Sciences, the International Society for Inventory Research (located in Budapest), the Korea Traders Scholarship (Sanhak) Foundation, the Kenan Institute and the Kenan-Flagler School of Business at the University of North Carolina.

The 1992 workshop was held in Puebla, Mexico, followed by an association with the Pan-Pacific meeting in Calgary, Canada. These meetings were supported by the University of the Americas in Puebla, Mexico, and the Global Manufacturing Research Center at the University of North Carolina. The group is planning to meet formally with the Decision Sciences Second International Meeting in Seoul, Korea in 1993. The workshop is scheduled to follow, in Australia, with the help of the Manufacturing Center at the University of Melbourne.

During the most recent workshops, the group has worked on a revision of the questionnaire, developed means of disseminating the results of the work to date, and agreed to make the data more widely available. All of these efforts can be seen in this book and the attached data disk. Another result of these meetings was the agreement to coordinate the data gathering and information dissemination from the Global Manufacturing Research Center at the Kenan Institute on the University of North Carolina campus.

The data base now contains data from Australia, Bulgaria, Chile, Finland, Hungary, Japan, Mexico, South Korea, North America (U.S. and Canada), the People's Republic of China, western Europe (northern european countries), and the USSR (before the break-up). The group still has an objective, however, of increasing the geographical scope of the project. New researchers are being asked to use the revised questionnaire for gathering the data in their regions. It will contain questions for comparison with the first questionnaire so that changes over time can be evaluated.

THIS BOOK

This book contains a variety of papers based on the Global Manufacturing Research Project. They are included both for their content and to provide ideas for future research. As more people have become involved with the data, the quality and variety of questions, analytical techniques and insights have expanded. Much of that variety is included here. Similarly, approaches to using the data have improved as the researchers have shared ideas, provided insights, and gained experience with the data base. Thus, these papers not only provide examples of approaches to the data, but should stimulate new and better ways of using the data in the future. That this has happened already can be seen by comparing some of the early papers with those done more recently.

The papers included in this volume are organized into seven sections and an appendix. Each of the sections is preceded by a brief introduction and summary of each of the papers. The first section has a single paper, "A Worldwide Survey of Manufacturing Practices." This paper describes in detail the concepts and activities of the project and presents the questionnaire in its entirety. Additional technical details on the data gathering process, coding formats and distribution media are also presented. This paper was used to acquaint potential researchers with the original project and served as their guide for gathering, coding, and distributing the data.

The papers in most of the other sections have a clear research focus. In the second section, the papers present within country analysis. These studies look at differences between the industries, the effects of size on manufacturing practices, or the relationship between economic policies and management. Section three has papers that report comparisons between two or more regions. The first papers in the section present bilateral comparisons while the others compare three or more areas of the world. In some, explanatory theory for the observed differences is developed.

The studies contained in section four test a variety of other hypotheses. The questions they address vary from productivity and inventory relationships to the impact of economic reforms in Hungary. Section five contains papers that discuss technical issues of analysis and the data. The first two provide suggestions for analyzing the data while the last presents some potential data limitations. The sixth section contains two papers that analyze large sections of the data. The first uses multivariate statistics while the second uses a scoring method to assess differences between regions.

Section seven is a direct analog of section one. It does not contain a research paper, but has one that describes possible future activities of the group. It also contains some suggestions on conducting surveys, the revised survey, and some information on a program that supports data entry for the new survey. The new survey reflects much of what the group has learned about survey research, questionnaire design, and data sharing. Thus, the "rules" are more stringent. The intention, as before, is to get as many people as possible involved with the research and to make the data sharing as transparent a process as possible.

The appendix has information on the data base that is distributed with this book. The disk also contains a utility program for viewing and modifying the data. A brief explanation of that program is also provided. Some of the specific details on each region's data are presented and the principal researcher(s) for each region are identified. The data are in a common format and a substantial effort has gone into assuring that the data are comparable from region to region.

ACKNOWLEDGEMENTS

A large number of organizations and individuals have been involved directly and indirect-ly in the sponsorship of the Global Manufacturing Research Group and the project. They have contributed money, time, space and personnel in support of the work. These contri-butions have gone for everything from data gathering to supporting exchanges of re-searchers. Included are contributions to the GMRG workshops, the travel of the partici-pants, the conduct of the research itself, and the presentation and publication of the findings. In this section, we would like to recognize those organizations that have supported the activities. The following have provided support for the data gathering, analysis, and workshops:

> Budapest University of Economic Sciences, Budapest, Hungary
> The Center for Manufacturing Excellence, University of North Carolina-Chapel Hill, North Carolina, USA
> The Frank Hawkins Kenan Institute of Private Enterprise, University of North Carolina-Chapel Hill, North Carolina, USA
> The Indiana Center for Global Business, Indiana University, Bloomington, Indiana, USA
> International Society for Inventory Research, Budapest, Hungary
> Kenan-Flagler Business School, University of North Carolina-Chapel Hill, North Carolina, USA
> The Macon Patton Endowment, University of North Carolina-Chapel Hill, North Carolina, USA
> Sanhak (formerly, the Korea Traders Scholarship) Foundation, Seoul, Korea
> Shanghai Institute of Mechanical Engineering, Shanghai, China
> State of Indiana Department of Commerce, USA
> University of the Americas, Puebla, Mexico

Exchanges of scholars between institutions has improved the quality of the work and improved the communication among all persons involved in the project. There have been exchanges between Korea, Hungary and the United States so far and it is hoped that more can be arranged in the future. We would like to acknowledge the sponsors of the exchanges to date.

> The Fulbright Scholar Program for an exchange between Korea and the United States
> The Soros Foundation-Hungary for an exchange between Hungary and the United States
> The Citibank International Fellows Program for a visiting fellowship in the United States

Many other organizations have supported the researchers directly and indirectly. The members of the Global Manufacturing Research Group are supported by their home institutions. These institutions help with travel to the workshops and underwrite the members' research. Other organizations have supported data gathering, analysis and publication of the research by their members. Some of those findings are reported here. When that is the case, the supporting organizations are acknowledged in the appropriate paper.

IN MEMORIAM

As we were working on the final details of the book, we learned of the untimely death of Allan Lehtimäki in a car accident in January, 1993. Allan was an early contributor to the project and attended two of the workshops. He will be remembered for his visits to several of the members of the group and his knowledge of the countries in which the project has been conducted. He will be missed.

D. Clay Whybark
Gyula Vastag
Chapel Hill, 1993

TABLE OF CONTENTS

SECTION ONE

INITIATING THE PROJECT

The sole paper in this section is one of the first papers written about the project. For the new reader, this is an important introduction to the work of the Global Manufacturing Research Group. It describes some of the history of the survey, the early data gathering activities, and provides a complete copy of the questionnaire. As an introduction to some of the papers, it is useful for the new reader to become familiar with the broad categories of questions as well as to gain a sense of the general intent of the questions in each of the categories. The categories from the questionnaire are sometimes used as section headings in other articles in this book. In addition, understanding the manner in which the data were gathered is also useful, in order to give the reader some confidence in the raw material that was used for the studies reported here.

D. Clay Whybark and Boo-Ho Rho
A Worldwide Survey of Manufacturing Practices

The amount of detail reported in this paper was necessary to provide complete instructions to all researchers on the questions, the responses possible, the coding of the responses and the exact form of the dBASE III files that were used to distribute the data. The specification of dBASE III as the format for data distribution had some very important advantages. It does require an exact determination of data specification (i.e., numeric or character, length of field, and designation of variable name) and is quite easily convertible to other formats (ASCII, LOTUS 1-2-3, delimited files, etc.). On the other hand, we learned a lot about data specification, the use of character variables, and preparing the data for analysis. As the group prepares to go into a second phase of data gathering with the revised questionnaire, the lessons learned from this first phase will pay handsome dividends.

A Worldwide Survey of Manufacturing Practices

D. Clay Whybark, University of North Carolina-Chapel Hill, USA
Boo-Ho Rho, Sogang University, Korea

ABSTRACT

This paper presents general information and procedures for sharing data for a worldwide survey of actual practices in production planning and control. The purpose is to provide sufficient background information and understanding for researchers in cooperating countries to make use of the data bases for academic purposes. The form of the questionnaire, survey methodology, coding schemes, data base format, and other technical issues are all presented. The motivation for the survey, data problems and limitations are also described.

INTRODUCTION

In an effort to gain a better understanding of actual manufacturing practices in production planning and control, a project has been initiated to gather this information from companies in several countries around the world. The purposes are to help provide insights on the practices that lead to better performance within a given country or geographical region, to provide a basis for comparing practices between geographical regions, and to develop an understanding of where differences in practice might lead to problems for companies doing business or establishing joint ventures with firms in other parts of the world.

Much of the impetus for the study came from Sogang University and the Korea Productivity Center. The people there were interested in the productivity of Korean enterprises and improving the basis upon which joint ventures and international business combinations could be structured. It was in Korea, therefore, that the original questionnaire was formulated, field tested and completed. It was also in Korea where the first full data gathering program was completed.

After gaining experience in Korea, data gathering was carried out in western Europe, Hungary, the People's Republic of China and North America using the same questionnaire. Projects are currently under way or are being planned for gathering comparable data in South America, Mexico, Japan, the USSR, and Thailand. As additional countries or regions are completed they will be added to the existing data base and will be made available for performing comparative research.

SURVEY DESIGN

Recognizing that there might be large differences in manufacturing practices between different industries and knowing that collecting a large sample from many industries would be difficult, two industries were selected for study. The first of these is the small machine tool industry, the second is the nonfashion textile industry. These industries were chosen because they can be found in countries all over the world, advanced or developing. Also, the two industries represent substantially different markets, processes, technologies and degrees of integration. So, in addition to their ubiquitousness, it was felt that they would provide a rich spectrum of differences in manufacturing practices for study.

The small machine tool industry consists of companies that produce industrial products like lathes, grinders, milling machines, and other metal forming equipment. These products are used primarily by other companies for manufacturing the parts used in producing products of their own, even including small machine tools themselves. For the most part the machine tools produced by this industry are still manually controlled, although an increasing proportion have computer controls associated with them. Most companies in this industry produce their machines in a batch-oriented job shop, with some firms offering options to be determined by the customer and/or custom design flexibility.

The non-fashion textile industry consists of consumer products like towels, sheets, underwear and so forth. It was felt to be important to not include firms producing designer clothes and accessories, since the industry is so heavily market driven and such firms are not found everywhere in the world. Industrial goods produced by firms in this industry include netting, cloth wrapping material, plain woven cloth and other products used primarily by other companies either in the manufacturing process or as raw material for the products they produce. Many of the firms in this industry have manufacturing processes that are integrated from spinning through dyeing, weaving, cutting and final sewing of the products. Such integration was not a requirement for inclusion in the survey, however, so the data contains information on some firms that perform only part of the complete process.

The questionnaire was structured to gather information on the firm in general and their activities in several areas of manufacturing planning and control. There is a section in the questionnaire covering the general attributes of the firm and then sections on forecasting, production planning and scheduling, shop floor control, and purchasing and materials management. The specific questions were based on theory and reported practice, but the intent was to determine actual manufacturing practice.

To develop a questionnaire that would be effective for gathering field data, pilot tests were performed and professional feedback was obtained in Korea, China and western Europe. On the basis of that feedback and other comments from practitioners, the questionnaire was revised to the form provided in Appendix 1. Although every effort was made to keep the questionnaire consistent from country to country, and translation to translation, occasionally custom, translation difficulties, or clerical errors introduced minor differences in some of the questions. Descriptions of these differences have been included with the data.

DATA GATHERING

To provide an understanding of the data gathering process used, the experiences in Korea, China and western Europe will be presented here. In each of these areas the first step was to mail the questionnaire (in its final form) to a sample of the firms in each of the industries. In all instances the response to the mail survey was very, very poor. This had been expected to a certain degree, because of the length of the questionnaire, but the responses were even fewer than initially estimated.

After getting such a poor response to the mail-solicited questionnaires, the final sample was secured through personal interviews and field visits in Korea and China, while an intensive telephone solicitation was used in western Europe. Personal contact with firms was also used to make sure that the questions were clear and that they were correctly interpreted. A more detailed description of the activities in Korea, China and western Europe is presented in the following sections.

South Korea

Data gathering activities in South Korea were done by Sogang University in cooperation with the Korea Productivity Center. The firms to be surveyed were selected from the *Directory of Korean Companies* published by the Korean Association of Commerce and Industry in 1985. All firms with more than 300 employees and about 60% of the firms with less than 300 employees were selected. Forms were mailed to 362 firms in the machine tool industry and 327 firms in the textile industry. Only 35 firms, in total, responded. After this disappointing response, about 100 firms, randomly distributed geographically and in size, were selected for field interviews. In total, 89 usable questionnaires were available for machine tools and 33 for textiles.

While the data gathering activities were being conducted in Korea, additional feedback on the questionnaire was being received from western Europe and China. As a consequence, data for some of the questions shown in Appendix 1 are not available for the Korean firms. The question numbers and variable names for which data is not available from Korean firms is included with the data. In addition, some of the data was collected in follow-up requests to the cooperating companies, some of whom chose not to participate. Thus, the sample size is smaller for some of the questions than for others. Again, the questions affected are clear from the data.

The People's Republic of China

The data gathering in the People's Republic of China was conducted through the Shanghai Institute of Mechanical Engineering with the assistance of the Shanghai Enterprise Management Association. It started with a pilot test, which was done in Shanghai and Changzhou, of the Chinese translation of the questionnaire. In both cities a group of people from the textile and machine tool industries came to a central location, and described difficulties or problems in understanding or interpreting the questionnaire.

They also made suggestions in how to make it more closely fit the Chinese situation. As an example, the "Make-to-State-Plan" variable was added to question number 5 (Appendix 1) to accommodate the firms still having some part of their operations under state plan in China.

The questionnaire was revised once the feedback was gathered from the industry representatives, and the comments from people at the Shanghai Enterprise Management Association and the Shanghai Institute of Mechanical Engineering were collected. Since the survey was to be conducted by graduate students from the Shanghai Institute of Mechanical Engineering, budget considerations restricted the area from which firms could be selected. Consequently, a representative sample of companies in each of the two industries near Shanghai was selected to receive the questionnaire by mail.

The response to the mail questionnaire was, as in Korea, very disappointing. The students went into the field and collected information directly from the enterprises in order to increase the sample size. In total, 44 usable responses were obtained from the machine tool industry and 56 usable responses were secured from the textile industry. Data on all the questions in Appendix 1 are available for the Chinese firms.

Western Europe

The data gathering activities in western Europe also started with a mail questionnaire. The target companies were chosen by taking a random sample from the industrial listing data base maintained by COMPASS (through the Zurich office). This data base is structured along the lines of the SIC code in the United States. It provides for extracting addresses of companies by industry including names and titles of management personnel. A total of nearly 800 companies were selected to receive the questionnaire by mail using this service. To cover these companies, the questionnaire was translated into French, German and Italian.

In Europe, as elsewhere, the response rate was very low, less than one percent. In addition, the responses to the mailing indicated that the COMPASS coding did not provide as fine a discrimination as originally believed, so many of the responding companies did not fit the industry description well. This reduced the number of usable surveys even more.

After the dismal response on the mail questionnaire, a telephone solicitation was conducted using directories of trade associations in the countries where data was gathered; i.e., Scandinavia, the Benelux countries, the United Kingdom, France, Germany, Austria and Italy. Calling directly to the appropriate people in the companies selected resulted in a substantially higher response rate (still only about 25% of the firms who said they would return the questionnaire did so). Finally a total of 34 usable responses was obtained for the machine tool industry, and 24 usable responses were received for the textile industry.

Responses are available from western Europe for all the questions in Appendix 1. In some instances a line item (for example the "Other" in question 53) was left off of one of the translations. Where this occurred it is noted as part of the data. Included in the company code is a digit that identifies the country in which the company is located.

Currently under way are efforts to gather data from Hungary, Japan, North America, and South America. The experiences gained from the other parts of the world are being applied in these efforts. For example, the North American survey started with the telephone, using data on companies provided by trade associations and directories. The disappointing responses to the mail surveys in other areas made it seem better to go directly to personal contact. The target in each of these additional efforts is to get a total of 50 usable responses from each industry. As the project continues, it is expected that other parts of the world will be added.

DATA CODING

The data was coded using dBASE III to create data files for each country and industry. This choice facilitated verifying the coding and the initial data analysis. The dBASE III format also allows for fairly easy transfer to LOTUS 1-2-3, creation of ASCII files and analysis with statistical packages like ABSTAT. The coding was performed on IBM PCs and compatibles and much of the initial analysis was done on PCs as well. To facilitate analysis, the data base formats are the same for each region's data base.

The coding follows the key presented in Appendix 1. The variable name, character or numeric designation, unit of measure, and specific coding used are provided for each question. The exact dBASE III structure for an example data base is listed in Appendix 2. This provides the name, the numeric or character specification and the size of each data field.

DATA ANALYSIS

Two types of analysis have been applied to some of the data so far gathered. In China, two reports have been written on the two different industries, using just the data from China [1,2]. These two studies were filed as Master's theses at the Shanghai Institute of Mechanical Engineering. Both theses report on the factors that account for differences in the performance of companies in each of the industries. They represent examples of *within country* analysis. Other within country analyses reported at professional meetings [3,4] describe some of the differences that occur between firms in South Korea as a result of structural factors like size.

In addition to the within country studies, there are two *between country* studies that have been prepared so far. The first of these is a comparison of the practices between South Korea and Europe for both industries [5] and the second compares the results of the survey between Korea and China [6]. Both of these studies are exploratory and much more remains to be done. Additional opportunities for data analysis will present themselves as the data from other countries becomes available. For this reason, it is important to encourage many people to engage themselves in the analysis of the data.

DATA AVAILABILITY

The Indiana Center for Global Business will serve as the initial distribution facility to provide data to researchers in the cooperating countries. To qualify to have access to the data, there are a few requirements which are important in terms of the spirit of the community involved in this research. As a condition of gathering the data, a commitment was made that there would be no commercial use made of the data. Therefore, the data cannot be sold or transferred to commercial enterprises and it must be used only for academic research and teaching purposes. In addition, the results of the research must be shared with other researchers in the project and other scholars. To this end, the Center will act as a clearing house for receiving the reports that stem from the data and disseminating information on them.

Specifically the data will be made available to qualified persons who complete an application form and describe their project. To qualify, the applicant must belong to an institution in one of the countries that has contributed data to the project. The applicant must also provide a statement of purpose for using the data as a part of the application. Initially, the Indiana Center for Global Business will determine if the applicant and proposed purpose qualify for using the data.

The data will be distributed on IBM compatible format, 5 1/4 inch diskettes. The data will be in a dBASE III file, complete with notes specific to the file. A charge for each diskette is required to cover the cost of duplication and administration, and there is a postage and handling fee. Each researcher will agree to provide the Indiana Center for Global Business with a copy of any results, papers, reprints or proceedings that come from work on the data. The Center will provide information on results from such work and the availability of data from other countries as it becomes available.

REFERENCES

1. J.G. Chen, "Operations Management in Chinese Textile Enterprises," Master's Thesis, Department of Systems Engineering, Shanghai Institute of Mechanical Engineering, August 1987.

2. X.J. Yan, "Operations Management in Chinese Machine Tool Manufacture Firms, Master's Thesis, Department of Systems Engineering, Shanghai Institute of Mechanical Engineering, August 1987.

3. B.H. Rho and D.C. Whybark, "Comparative Study of Production Planning and Control Among Korea, Europe, and China," DSI Annual Meeting, Hawaii, November 1986.

4. B.H. Rho and D.C. Whybark, "Production Planning and Control in Korea," Pan-Pacific Conference IV, Taipei, Taiwan, May 1987.

5.* B.H. Rho and D.C. Whybark, "Comparing Manufacturing Practices in Europe and South Korea," Discussion Paper No. 3, Bloomington, IN, Indiana Center for Global Business, Indiana Business School, March 1988.

6.* B.H. Rho and D.C. Whybark, "Comparing Manufacturing Practices in the People's Republic of China and South Korea," Discussion Paper No. 4, Bloomington, IN, Indiana Center for Global Business, Indiana Business School, March 1988.

* This article is reproduced in this volume.

ACKNOWLEDGMENTS

The authors would like to thank the Korea Productivity Center, the Research Institute for Economics and Business at Sogang University, the Shanghai Institute of Mechanical Engineering, IMEDE (the international business school in Lausanne, Switzerland), and the Indiana Center for Global Business for supporting the project. This paper originally appeared as Working Paper No. 2 of the Indiana Center for Global Business.

APPENDIX 1
PRODUCTION PLANNING AND CONTROL QUESTIONNAIRE
KEY

Missing data:
Numeric - 99
Character blank
N = Numeric
C = Character

(Company Name):_____   CODE_____
C

I. COMPANY PROFILE

1. How many employees work for the company?   __EMPLS__   persons
N

2. How many employees are production workers?   __FCTRY__   persons
N

3. What were sales last year (in US$ or indicate the currency used)?
N   Domestic   __DOMEST ($1,000,000)__
N   Export   __EXPORT ($1,000,000)__

4. What products do you produce?
C   _____PRODUCTS_____

_____

CONVENTIONS:
1 Day = 8 Hours
1 Week = 7 Days
1 Month = 4 Weeks
1 Month = 30 Days
1 Year = 365 Days
1 Year = 52 Weeks
1 Quarter = 90 Days

5. What is the percent of make-to-order and make-to-stock products?
N   ORD   STK   PLN
N   Make-to-order (   )% Make-to-stock (   )% Make-to-State Plan (   ) %
N

6. What is your approximate average capacity utilization rate? UTL
N

    0-60% (50)   60-70% (65)   70-80% (75)   80-89% (85)   90 +% (95)

7. What is your firm's total investment in production equipment.
N   EQMT   ($1,000,000)_____

8. About how many work days per year is the average employee involved in training classes
N   (not counting on-the-job training)? __TRN (DAYS)__

II SALES FORECASTING

9. What is the position (or level) and the functional group of the person that produces the sales
C   forecast for your firm?
C

Position POSIT*		Functional Grouping FUNCT	
Managing Dir./Pres./CEO	(I)	Administration/Planning(I)	
Vice President/Director	(II)	Production/Engineering(II)	
Department/Division Head	(III)	Sales/Marketing(III)	
Group/Section Manager	(IV)	Finance/Accounting(IV)	
Other_____	(V)	Other_____(V)	

* Coded as 0 or 1 (i.e. V.P./Dir would be: 01000)

10. What are the 2 most important subjective factors in your forecast?
C

            General economic and political situation   (I) SUBJT
            General company and industry situation   (II)
            Trends in economic and political situation   (III)
            Market research and customer/vender information   (IV)
            Other _____ (V)

11. What are the 2 most important objective factors in your forecast?
C

            Past sales (demand) history   (I) OBJCT
            Current order backlog   (II)
            Economic indices (GNP, prices or investment ratios)   (III)
            Industry statistics (market shares, inventories)   (IV)
            Other_____ (V)

12. What formal techniques do you use for sales forecasting?
C

            None   (I) TECHS
            Projection (indices, moving average, exp. smoothing)   (II)
            Time series models (Box-Jenkins, spectral analysis)   (III)
            Causal models (regression)   (IV)
            Consensus (delphi)   (V)
            Other _____ (VI)

13. Which of the following ranges most closely represents the weight of the subjective factors in your forecast?
C   WT
   0-20% (1)   20-40% (2)   40-60% (3)   60-80% (4)   90 +% (5)

14. Does your firm use computers for sales forecasting? Yes (1) No (0)
C   CP

15. Is the sales forecast developed for individual products or product groups? Is it expressed in monetary or physical units?
C   FRCST
   For each product (i.e. model #) (I)   In monetary terms (i.e. $) (III)
   For product groups (i.e. lines) (II)   In physical units (i.e. #) (IV)
N
N   It covers   LGTH (MONTHS)   by   INCR (DAYS) .
    (e.g. It covers 1 year by month or It covers 2 years by quarter.)

16. About how many times a year do you modify your forecast?   MOD
N

17. Do you specify a range or a single value for the forecast?
C   RNG
           Range (1)   Single value (2)

18. What are the two most important uses of the forecast?

C   FRCSTUSE

Budget preparation	(I)	Production planning	(V)
Material/inventory planning	(II)	Sales planning	(VI)
Manpower planning	(III)	New product development	(VII)
Facilities planning	(IV)	Other _____	(VIII)

19. What is the average percentage error of your forecast?_____

N   ERR (%)

20. What is the principal measure of your forecast error?

C

No measure used	(I)	
Percentage error	(II)	ME
Average error	(III)	
Average absolute error (MAD)	(IV)	
Other _____	(V)	

III. PRODUCTION PLANNING AND SCHEDULING

21. What are the two most important factors for your production plans?

C   PRDPLAN

Actual orders/backlog	(I)	Previous sales	(V)
Production capacity	(II)	Customers' plans	(VI)
Level of inventories	(III)	The forecast	(VII)
State plan	(IV)	Other	(VIII)

22. Is the production plan developed for individual products or product groups? Is it expressed in monetary or physical units?

C   UNIT

For each product (i.e., model #)	(I)	In monetary terms (i.e., $)	(III)
For product groups (i.e., lines)	(II)	In physical units (i.e., #)	(IV)

N   It covers ___LGTHP (MONTHS)___ by ___INCRP (DAYS)___.

(e.g. It covers 1 year by month or It covers 2 years by quarter.)

23. Who prepares the production plan for your company?

C   Position   PLPOS   Functional Grouping   PLFUN

Managing Dir./Pres./CEO	(I)	Administration/Planning	(I)
Vice President/Director	(II)	Production/Engineering	(II)
Department/Division Head	(III)	Sales/Marketing	(III)
Group/Section Manager	(IV)	Finance/Accounting	(IV)
Other _____	(V)	Other _____	(IV)

24. The education of the person responsible for production planning?

C   EDUC   MAJR

C

Schooling		Field	
Primary School	(I)	Business & Commerce	(I)
High School	(II)	Liberal Arts	(II)
College Graduate	(III)	Natural Sciences	(III)
Advanced Degrees	(IV)	Engineering	(IV)

25. What are the two most important uses of the production plan?   PLANUSE
C

Budget preparation	(I)	Operations Scheduling	(V)
Subcontracting	(II)	Material/inventory planning	(VI)
Manpower planning	(III)	Purchasing/procurement	(VII)
Facilities planning	(IV)	Other _____	(VIII)

26. Who is the primary user of the production plan?
C   PUSER   FUSER
C   Position   Functional Grouping

Managing Dir./Pres./CEO	(I)	Administration/Planning	(I)
Vice President/Director	(II)	Production/Engineering	(II)
Department/Division Head	(III)	Sales/Marketing	(III)
Group/Section Manager	(IV)	Finance/Accounting	(IV)
Other _____	(V)	Other _____	(V)

27. About how many times a year do you revise the production plan?
N   REV

28. Do you revise the plan when there is a demand fluctuation?
C   FLC   LMT
N   Yes (1) No (0)   Only if the fluctuation is greater than   __%

29. Do you compare the plan to actual results?   Yes (1) No (0)
C

30. Describe (from least detailed to most detailed) the level and the units used for any other plans or schedules used in production.

1. Plan/schedule name _____
C   Made for individual project   (I) In monetary terms (III) PLN1
         Made for product groups   (II) In product units   (IV)

N   It covers (e.g. a month) LN1 (weeks) by (e.g. week) IN1 (Days)
N

2. Plan/schedule name _____
C   Made for individual product   (I) In monetary terms (III) PLN2
         Made for product groups   (II) In product units   (IV)

N   It covers (e.g. one week) LN2 (Days) by (e.g. day)   IN2 (Days)
N

3. Plan/schedule name _____
C   Made for individual products (I) In monetary terms (III) PLN3
         Made for product groups   (II) In product units   (IV)

N   It covers (e.g. one day) LN3 (Days) by (e.g. hour) IN3 (Hours)
N

31.   Is part of the production plan (or other plan or schedule) frozen or otherwise made very
       difficult to change in the near future?

C

           Yes (1)   No (0) FRZ

32.   Which of the following levels most closely describes your use of computers in produc-
       tion planning and scheduling?

C   USE
       Not at all   (1)
       A little but most of it is still done by hand   (2)
       Moderate, it's partly done by hand and partly by computer   (3)
       Extensive, it couldn't be done as it is without computers   (4)

33.   If you don't use computers, do you think they would be helpful?

C

           Yes (1)   No (0) HLP

                        IV. SHOP FLOOR CONTROL

34.   What functional grouping of people orders the factory to start production an what is the
       basis of the order?

C   Functional Grouping   STRPR   Basis of order   BASISP
C

       Administration/Planning   (I)   Actual customer order   (I)
       Production/Engineering   (II)   Production plan   (II)
       Sales/Marketing   (III)   Production schedule   (III)
       Finance/Accounting   (IV)   Shortage list   (IV)
       Other _____   (V)   Inventory position   (V)
                                           Other _____   (VI)

35.   What group orders the purchase of material and what is the basis?

C   Functional Grouping   STRPU   Basis of order   BASEPU
C

       Administration/Planning   (I)   Actual customer order   (I)
       Production/Engineering   (II)   Production plan   (II)
       Sales/Marketing   (III)   Production schedule   (III)
       Finance/Accounting   (IV)   Shortage list   (IV)
       Other _____   (V)   Inventory position   (V)
                                           Other _____   (VI)

36.   How is a purchase order transmitted to a supplier?
C   TRAN
       Orally (I)   Written (II)   Computer (III)   Other____(IV)_

37.   What percent of the fabricated parts are produced in your plant?
C   IN
       0-20% (1)   20-40% (2)   40-60% (3)   60-80% (4)   80 +% (5)

38.   What is the lead time for a typical batch from start of production until delivery to the
N   customers or inventory? _____ LT (Days) _____

39.   Does the blueprint travel with the work during production?
C   BP

         Yes (1)   No (0)   It depends on the product (3)

40.   For what portion of the operations do you have time standards?
C   TS

         Greater than 90% (1)   60-90% (2)   20-60% (3)
                             Less than 20% (4)   None (5)

41.   What portion of the operations are scheduled using time standards?
C   SC

         Greater than 90% (1)   60-90% (2)   20-60% (3)
                             Less than 20% (4)   None (5)

42.   What percent of the products have costs based on time standards?
C   CST

         Greater than 90% (1)   60-90% (2)   20-60% (3)
         Less than 20% (4)   None (5)

43.   Which of the following most closely describes your time standards?
C   QLT
                      Most are very close to actual   (1)
                      Useful, but some are accurate, some not   (2)
                      Not accurate, but consistently high or low   (3)
                      Almost useless   (4)

44.   On what portion of the orders do engineering or design changes occur after starting
         production on an order?
C   ECN
                          Greater than 60%   (1)   30-60%   (2)
                          10-30%   (3)   Less than 10%   .(4)

45.   What group establishes the sequence for releasing jobs to production and what is the
         principal basis of these priorities?
C
C   Function grouping RELSE   Basis of Priority   BASISREL

         Administration/Planning   (I)   Customer order due dates (I)
         Production/Engineering   (II)   Processing time required (II)
         Sales/Marketing   (III)   Similarity of set-ups   (III)
         Finance/Accounting   (IV)   Material availability   (IV)
                                               Marketing preferences   (V)
         Other group _____   (V)   First-come first served   (VI)
                                               Selling price of item   (VII)
         Other basis _____   (VI)   Management directive   (VIII)

46.　　　The following are factors which can change the priorities once production has started on an order. For each of them that has a heavy influence on the priorities, please enter "H" in the parenthesis, if a moderate influence "M", if little, "L".

C   CHGPRIORTY

Pressure from marketing	(I)	Manufacturing problems	(VI)
Pressure from customers	(II)	Material shortages	(VII)
Orders from management	(III)	Changes in sales plan	(VIII)
Changes in delivery dates	(IV)	Engineering problems	(IX)
Sudden surges in demand	(V)	Other "H" _____	(X)

47.　　　What percentages of the changes in the priorities are communicated to the production workers?

N　　　　　　　　　　Orally (   )%　　By document (   )%

N　　　　　　　　　　　　　　　　　　ORAL　　　　　　　　　　　　　　　　DOC

48.　　　How is the delivery date for customer orders determined?

C   NEG

　　　　Mostly by customer (1)　　　Negotiation (2)　　　　　　　Most by us (3)

49.　　　What is the usual delivery time promised to customers?

N

　　　　A fixed time period of FIX (Days) (e.g., 2 weeks or 4 months)

N

N　　　　or the period varies from LOW (Days) to HI (Days) depending upon:

C　　　　Product complexity　　　　(I)　　　Importance of customer　(III)

　　　　Production load　　　　　　(II)　　　Material availability　　(IV)

　　　　　　CAUS　　　　　　　　　　　　Other_____·_____(V)

50.　　　What percentage of the orders are late?　LAT

C

　　　　Less than 5% (1)　　5-10% (2)　　10-20% (3)　　　　　　20-30% (4)

　　　　　　30-40% (5)　　40-50% (6)　　　>50% (7)

51.　　　What is the average lateness for those orders that are late?

C   AVG

　　　　Less than 1 week (1)　　1-3 weeks (2)　　3-6 weeks (3)

　　　　6-8 weeks (4)　　　　　2-4 months (5)　　4+ months (6)

52.　　　What are the two most frequent causes of lateness?　WHYLATE

C

Lack of machine capacity	(I)	Material shortages	(V)
Lack of labor capacity	(II)	Quality problems	(VI)
Production bottlenecks	(III)	Due date changes	(VII)
Transportation Problems	(IV)	Other _____	(VIII)

53. Several alternatives for changing the capacity required to meet a peak schedule are shown below. For each alternative that is highly useful to you, please enter "H" in the parenthesis, if only moderate, "M", if not useful, "N".

C

CAPACITY

Increases		Decreases	
Hire additional workers	(I)	Lay off the extra workers	(VI)
Use overtime	(II)	Use undertime (idleness)	(VII)
Subcontract production	(III)	Reduce the work time	(VIII)
Backorder the production	(IV)	Build inventory	(IX)
Lease temporary capacity	(V)	Lease capacity to others	(X)
		Other_____	(XI)

54. What has been the annual growth rate in the <u>physical</u> volume of production over the last two years?

C

GRW

Declining (1)   0-10% (2)   10-20% (3)   20-40% (4)   40+% (5)

V. PURCHASING AND MATERIALS MANAGEMENTS

55. In which of the following activities is purchasing involved?

C

PURCHACT

Materials shortages	(I)	Capacity changes	(V)
Engineering changes	(II)	Quality programs	(VI)
Make or buy decisions	(III)	Cost reductions	(VII)
Product design	(IV)	Inventory programs	(VIII)

56. What is the breakdown of the inventories and the total amount?

	Approximate %	Total amount of inventory
N		(end of year)
N Purchased materials and parts	RAW (%)	
N Work-in-process inventories	WIP (%)	
N Finished goods inventories	FIN (%)	TOTAL ($1,000,000)

57. Do you use the "Economic Order Quantity" (EOQ) concept to specify purchase or production amounts?

C   Yes (1)   No (0)   EOQ

58. What is your exposure to "MRP" (Material Requirements Planning)?

C

MRP

Never heard of it	(1)
Using it and benefiting from it	(2)
Using it but not benefiting from it	(3)
Understanding it, but feeling no necessity to introduce it	(4)
Just starting to introduce it	(5)
Trying to introduce it, but having difficulty doing so	(6)

59.   What is your most frequent method for purchasing raw materials?

C   BUY

Purchase on a periodic basis (e.g., weekly, monthly) to bring inventories up to the
desired level   (I)
Purchase fixed amounts whenever needed to build inventories up to desired levels (II)
Purchase the necessary amounts based on experience   (III)
Purchase only for specific customers' orders   (IV)
Purchase according to a production plan or schedule   (V)
Raw material distributed by the State   (VI)

60.   What are the two most important actions you take to assure the timely supply of materials and purchased parts?

C   SUPPLY

Long-term contract	(I)	Large quantity purchases	(IV)
Hedging	(II)	Multiple sourcing	(V)
Single sourcing	(III)	Using sister plants	(VI)
Other _____		(VII)	

61.   On the average, how many vendors do you have per purchased part?

C   NUM

One (1)   Two or three (2)   Four or more (3)

62.   Check the two most important reasons for subcontracting outside?

C   SUBCON

Production load	(I)	Cost are lower	(IV)
Production difficulty	(II)	Quality is higher	(V)
Company policy	(III)	Production is faster	(VI)
Other _____		(VII)	

63.   Do you have someone titled Materials Manager (i.e., responsible <u>only</u> for most or all of
these activities: transportation, warehousing, production planning, inventories and
purchasing)?

C   MMG

Yes (1)   No (0)

64.   What is your exposure to the JIT (just-in-time) system?

C   JIT

Never heard of it   (1)
Using it and benefiting from it   (2)
Using it but not benefiting from it   (3)
Understanding it, but feeling no necessity to introduce it   (4)
Just starting to introduce it   (5)
Trying to introduce it, but having difficulty doing so   (6)

65.   How many weeks of finished goods inventory do you hold for protection against unexpected demand fluctuations?

N   __SFY__ (Weeks)

APPENDIX 2
DATABASE STRUCTURE
(dBASE III Listing)

Field	Field Name	Type	Width	Dec
1	CODE	Character	4	
2	EMPLS	Numeric	5	
3	FCTRY	Numeric	5	
4	DOMEST	Numeric	8	3
5	XPORTS	Numeric	8	3
6	PRODUCTS	Character	10	
7	ORD	Numeric	3	
8	STK	Numeric	3	
9	PLN	Numeric	3	
10	UTL	Numeric	3	
11	EQPMT	Numeric	8	3
12	TRN	Numeric	3	
13	POSIT	Character	5	
14	FUNCT	Character	5	
15	SUBJT	Character	5	
16	OBJCT	Character	5	
17	TECHS	Character	6	
18	WT	Character	1	
19	CP	Character	1	
20	FRCST	Character	4	
21	LGTH	Numeric	3	
22	INCR	Numeric	3	
23	MOD	Numeric	3	
24	RNG	Character	1	
25	FRCSTUSE	Character	8	
26	ERR	Numeric	3	
27	ME	Character	5	
28	PRDPLAN	Character	8	
29	UNIT	Character	4	
30	LGTHP	Numeric	3	
31	INCRP	Numeric	3	
32	PLPOS	Character	5	
33	PLFUN	Character	5	
34	EDUC	Character	4	
35	MAJR	Character	4	
36	PLANUSE	Character	8	
37	PUSER	Character	5	
38	FUSER	Character	5	
39	REV	Numeric	3	
40	FLC	Character	1	
41	LMT	Numeric	3	
42	CMP	Character	1	
43	PLN1	Character	4	
44	LN1	Numeric	3	

DATABASE STRUCTURE (cont.)
(dBASE III Listing)

Field	Field Name	Type	Width	Dec
45	IN1	Numeric	3	
46	PLN2	Character	4	
47	LN2	Numeric	3	
48	IN2	Numeric	3	
49	PLN3	Character	4	
50	LN3	Numeric	3	
51	IN3	Numeric	3	
52	FRZ	Character	1	
53	USE	Character	1	
54	HLP	Character	1	
55	STRPR	Character	5	
56	BASISP	Character	6	
57	STRPU	Character	5	
58	BASEPU	Character	6	
59	TRAN	Character	4	
60	IN	Character	1	
61	LT	Numeric	3	
62	BP	Character	1	
63	TS	Character	1	
64	SC	Character	1	
65	CST	Character	1	
66	QLT	Character	1	
67	ECN	Character	1	
68	RELSE	Character	5	
69	BASISREL	Character	9	
70	CHGPRIORTY	Character	10	
71	ORAL	Numeric	3	
72	DOC	Numeric	3	
73	NEG	Character	1	
74	FIX	Numeric	3	
75	LOW	Numeric	3	
76	HI	Numeric	3	
77	CAUS	Character	5	
78	LAT	Character	1	
79	AVG	Character	1	
80	WHYLATE	Character	8	
81	CAPACITY	Character	11	
82	GRW	Character	1	
83	PURCHACT	Character	8	
84	RAW	Numeric	3	
85	WIP	Numeric	3	
86	FIN	Numeric	3	
87	TOTAL	Numeric	8	3
88	EOQ	Character	1	
89	MRP	Character	1	

DATABASE STRUCTURE (cont.)
(dBASE III Listing)

Field	Field Name	Type	Width	Dec
90	BUY	Character	6	
91	SUPPLY	Character	7	
92	NUM	Character	1	
93	SUBCON	Character	7	
94	MMG	Character	1	
95	JIT	Character	1	
96	SFY	Numeric	3	
97	NOTES	Memo	10	
Total **			383	

SECTION TWO

WITHIN COUNTRY STUDIES

The papers in this section all share one attribute. Each presents a study of the data from a single country. That means that regions (like North America and western Europe) are not included and the paper on the USSR was based on data before the breakup. The first few papers all contain analyses of the firms within the country, often comparing practices between industries or between firms of different sizes. In some cases recommendations are made for improvements based on the comparisons. In the latter papers, all dealing with former centrally planned economies (CPEs), a different perspective is included. In these papers, the influence of central planning is traced in the manufacturing practices of those countries. There is some evidence that the reforms are already affecting manufacturing practices, particularly in Hungary.

Amrik S. Sohal and Danny Samson
Textile Industry Practices in Australia

The Australian data was gathered only for the textile industry. There was not a large enough population of machine tool firms in Australia to provide a viable sample size. The analysis focuses on the effectiveness of the practices of the firms in the industry. The sector is an important one in the economy. Planned reductions in tariffs are a threat to the health of the sector and changes are needed to assure viability. The conclusions provide suggestions for changes in the firms' practices to help meet the pending competition.

Antonio E. Kovacevic, J. Claudio Lopez and D. Clay Whybark
Manufacturing Practices in Chile

The Chilean data comprised both textile and machine tool firms. Included in the machine tool category are also some firms that make other metal parts and/or products using machine tools. After it was verified that the differences in practices between the makers and users of machine tools was minimal, the data was pooled for the comparisons reported in the paper. Several differences between the industries are found, fewer than expected overall, but some that were not expected. The analytical tools used were parametric and non-parametric statistical tests, like the Student's t test of significant differences in population means or the Kolmogorov-Smirnov test of significant differences between distribution (of replies, for example). The conclusions provide some suggestions for improving the practices in both industries in the country.

Benito E. Flores, Felipe Burgos and Arturo Macias
Manufacturing Practices in Mexico: An Example of the Non-Fashion Textile and Machine Tool Industries

The area around Puebla, Mexico, is a major center of textile production in Mexico. It was where much of the data was gathered for this article. In the paper, the manufacturing practices in the textile industry are compared with those in the machine tool industry. Substantial differences in the practices of the two industries are shown, using techniques similar to those in the preceding papers. The firms in this detailed study are relatively small, but they are not immune from the pressures of the North American Free Trade Agreement.

Allan Lehtimäki
Materials Management Practices of Finnish Manufacturing Companies

This paper on the manufacturing practices in Finland is based on the GMRG questionnaire and on some additional questions, designed to gather information on the techniques used in materials management and purchasing. Moreover, the companies in the Finnish survey were the larger firms in the machine tool and electrical industries (as opposed to the textile industry). The focus of the analysis is on materials management and purchasing practices. Topics covered include the means of controlling material flows; the use of techniques (i.e., MRP, JIT and OPT), computers and expert systems; and the size and distribution of inventories.

Boo-Ho Rho and D. Clay Whybark
Manufacturing Practices in Korea

The data gathered in South Korea were among the first for the project. Perhaps because of this, there is a high percentage of non-responses for some of the financial variables. Nevertheless, the overall sample size was large enough that meaningful comparisons of practices could be performed as a function of size. As with the Australian analysis, differences were found between different size firms and, as with the Chilean work, between industries. The basis for these conclusions was analysis based on the same techniques as the earlier papers. Some commentary on why the differences exist is offered as well as suggestions for the improvement of practice in some of the companies.

Krisztina Demeter
Changes in Hungarian Manufacturing Strategies

This article is the first of several on manufacturing practices in CPEs. It reports on the changes in the Hungarian economy and their influence on manufacturing strategy and practice. It uses the data from a survey administered in 1986 and re-administered in 1991. The focus of the article is on the link between manufacturing strategy and practices. It traces the development of manufacturing strategy in response to the economic reforms and how those changes in strategy are reflected in manufacturing practices. The impact of the economic reforms can already be seen in manufacturing activities.

Pavel Dimitrov and D. Clay Whybark
Manufacturing Management Practices in Bulgaria Under Conditions of a Centrally Planned Economy

The survey data on Bulgarian companies is first compared to national statistics for the machine tool and textile industries. This clearly establishes the importance of these industries in the Bulgarian economy. The manufacturing practices exposed by the survey are then related to the economic polices of the country. The data were gathered before the reform movements had time to substantially affect manufacturing. Thus the effect of the polices of the central economy on manufacturing practices are clearly reflected. Implications are drawn for changes in manufacturing management necessary to facilitate the economic reforms.

Alexander Ardishvili and Arthur V. Hill
Manufacturing Practices in the Soviet Union

This paper, using data from the former USSR, describes comparisons of the two industries and among different size firms in that part of the world. The analysis was done with Student's t and (fittingly) the Kolmogorov-Smirnov tests. The findings with respect to industry and size differences are very similar to those of the other papers. Notably, however, this paper gives some insight into the transition underway at the time of the data gathering and speculates on the conditions that may exist as the market reforms take place. This can hav
e considerable value for firms interested in joint ventures or other collaborative work with firms in the USSR. The conclusions also indicate important directions of change for management practices in the country.

Textile Industry Practices in Australia

Amrik S. Sohal, Monash University, Australia
Danny Samson, University of Melbourne, Australia

ABSTRACT

This paper presents an analysis of the textile industry in Australia. Industry statistics and government policies for this segment of Australian economy (Textile, Clothing and Footwear (TCF) firms) are presented as a background for the survey on manufacturing practices. The survey indicates that there is room for improvement in many of the practices in the industry. In particular, the areas of production planning and control, forecasting, inventory control, quality control, and the manufacturing-marketing relationship are mentioned. As tariffs are lifted on imported products, there is concern about the ability of many of the firms in the industry to survive unless some of the changes suggested are made.

INTRODUCTION

The Textile, Clothing and Footwear (TCF) Industry in Australia makes a significant contribution to the Australian economy [1]. Around 120,000 people are employed in these industries accounting for nearly 10 percent of the total manufacturing employment. In 1984-85, this industry accounted for $A2.7 billion in value added which was around seven per cent of manufacturing value added. The TCF industry is an important employer in a number of regional areas. A significant number of the workers in these industries are women many of whom were born overseas.

Firms within the TCF industry range from small single-owner enterprises to very large corporations employing 1000's of people having several divisions and operating companies. An example of the latter is the Linter group of companies.

Most of the firms in the TCF industry have in the past focused on the domestic market only. The nation has a substantial resource of raw materials (wool, cotton, hides) however, only a small proportion of these raw materials are processed into intermediate or finished products. In 1986-87, exports across the spectrum of TCF products accounted for $A1.4 billion, including early stage processing. In terms of intermediate and finished products, exports were relatively low.

Over the past few years major changes have taken place in the TCF industry. Restructuring and rationalizing is currently taking place in response to the phasing out of tariffs on clothing imports by 1995, at which time tariff only assistance of 55% (50% developing countries) will apply. These changes are the result of the TCF Industries Development Strategies announced by the Government at the end of 1986.

The government's aim is to increase the exposure of TCF industry to international competition and make them more competitive. The government's program for restructuring and revitalizing these industries has two facets [1]:

Table 1. Trends in

Description	Establishments			Employment		
	1984-85	1986-87	Change %	1984-85	1986-87	Change %
	(No.)			('000)		
• cotton ginning	17	17	0.0	0.4	0.6	50.0
• wool scowing/top making	21	22	4.8	1.5	1.5	0.0
• man made fibers & yarns	16	16	0.0	3.4	3.6	5.9
• manufactures woven fabric	33	34	3.0	4.2	3.6	-11.9
• cotton yarns & woven fabrics	39	38	-2.6	4.3	4.0	-7.0
• worsted yarns & woven fabrics	12	12	0.0	2.2	2.3	4.5
• woollen yarns & woven fabrics	26	28	7.7	2.0	2.3	15.0
• narrow wovens & elastics	27	24	-11.1	1.3	1.2	-7.7
• textile finishing	61	76	24.6	2.4	3.3	37.5
• household textiles	76	65	-14.6	2.0	1.4	-30.0
• textile floor coverings	41	49	19.5	3.5	4.2	20.0
• felt & felt products	11	8	-27.3	0.7	0.7	0.0
• canvas & associated products	195	188	-3.6	2.5	2.7	8.0
• other	13	17	30.8	0.5	0.4	-20.0
	66	67	1.5	2.2	2.6	18.2
Total textiles	654	661	1.1	33.0	34.4	4.2
Clothing & footwear Knitting Mills						
• hosiery	31	30	3.2	3.0	3.3	10.0
• cardigans & pullovers	142	118	-16.9	4.9	4.7	-4.1
• other knitted goods	93	109	17.2	5.4	5.3	-1.9
Total knitting mills	266	257	-3.4	13.3	13.2	-0.8
• mens trousers, shorts and work clothing	124	113	-8.9	9.4	8.3	-11.7
• mens suits, coats and water proof clothing	90	92	2.2	4.8	5.0	4.2
• womens outerwear	748	655	-12.4	15.6	14.1	-9.6
• foundation garments	15	15	0.0	2.0	2.1	5.0
• underwear & infant clothing	206	187	-9.2	10.6	10.3	-2.8
• head wear & other clothing	361	551	52.6	5.9	8.9	50.8
Total clothing	1544	1613	4.5	48.4	48.6	0.4
Footwear	202	210	4.0	12.5	12.3	-1.6
Total TCF	2666	2741	2.8	107.2	108.5	1.2
Total manufacturing	27647	28795	4.2	1018.7	1022.7	0.4
TCF share of total (8)	9.6	9.5		10.5	10.6	

Source: ABS No. 8707.0 Manufacturing Establishment

the TCF industry

Wage & salaries ($M)			Turnover ($M)			Value added ($M)		
1984-85	1986-87	Change %	1984-85	1986-87	Change %	1985-84	1986-87	Change %
37.4	44.6	19.3	370.4	454.8	22.8	66.6		
75.5	93.6	24.0	165.6	228.4	37.9	67.4		
71.6	71.1	-0.7	304.4	37.9	22.8	144.9		
78.6	78.4	-0.3	269.1	329.7	22.5	111.6	372.1	44.0
34.7	43.2	24.5	121.7	144.1	18.4	48.0		
32.4	37.9	17	162.0	165.4	2.1	52.0		
19.5	20.4	4.6	81.3	91.8	12.9	37.0		
45.3	71.1	57.0	159.6	324.4	103.3	77.9		
26.9	18.9	-29.7	146.1	105.7	27.7	52.8		
58.1	78.7	35.5	416.9	558.8	34.0	130.2		
13.8	17.3	25.4	62.0	73.8	19.0	34.3		
32.6	40.2	23.3	153.7	190.9	24.2	62.8	536.9	31.2
9.2	8.0	-13.0	36.7	46.7	27.2	16.2		
45.8	51.0	11.4	230.4	317.0	37.6	112.8		
581.4	674.5	16.0	2680.0	3405.2	27.1	1014.5	1409.0	38.9
46.2	59.3	28.4	155.7	208.5	33.9	78.8	435.5	
68.0	72.1	6.0	225.7	275.9	22.2	116.2		
82.3	97.0	17.9	402.7	505.7	25.6	142.1		
196.5	228.5	16.3	784.0	990.1	26.3	337.1	435.5	29.1
128.4	121.8	-5.1	437.0	442.1	1.2	212.0		
66.2	73.1	10.4	155.3	203.9	31.3	89.3		
191.4	190.9	-0.3	770.1	834.0	8.3	324.6	1197.0	
27.4	35.6	29.9	81.2	101.6	25.1	51.9		
146.1	155.0	6.1	563.5	529.0	14.1	237.9		
72.0	120.4	67.2	257.9	410.1	59.0			
631.5	696.9	10.4	2165.1	2520.8	16.4	1040.3	1199.0	15.2
175.4	201.1	14.7	584.2	723.3	23.8	289.5	347.5	20.0
1584.8	1801.0	13.6	6213.3	7639.4	22.9	2681.7	3391.0	26.4
18779.9	21415.7	14.0	98024.3	115839.5	18.2	38253.5	45420.7	18.7
8.4	8.4		6.3	6.6		7.0	7.5	

Summary of Operations, Australia 1984-85 and 1986-87

(i) Lower the level of Government assistance to these industries over a period of time. Phase down quotas and tariffs. Rationalize all other assistance measures.

(ii) Through the implementation of the Industries Development Strategy, encourage and facilitate the development of more efficient, competitive and internationally oriented TCF industries.

Both the Government and the TCF industry recognize that with appropriate restructuring and rationalizing the industry can make a more effective contribution to the Australian economy [1]. Many large corporations in the TCF industry are now taking steps to meet the challenges of domestic and international competition. For example, the National Textile Operation of the Linter group have a $A20 million investment planned for 1990 to increase its capacity [2].

RECENT INDUSTRY TRENDS

Employment and Output

Table 1 shows recent trends in the TCF industry. The data was derived from the 1984-85 and 1986-87 census of manufacturing establishments which was carried out by the Australian Bureau of Statistics. Between this period, for the textile sector, employment increased by 4.2% to 34,400 employees, turnover increased by 27.1% to $A3405.2 million and value added increased by 38.9% to $A1409 million.

Table 2 shows the output figures for Australian Textiles for the year ending August 1988. It is clear that production has declined when compared to the corresponding period for the previous year.

The following quote is taken from *Communique* (13 December 1988), the official newsletter published by TCFCA [3].

> While wool yarn production remained relatively stable, this was in the main due to a 13% increase in the production of woollen carpet yarn. Production of other woollen yarn declined by 27% and worsted yarn output also fell, by 21% for machine knitting yarn and 11% for other worsted yarn.
>
> Cotton yarn production also fell significantly (by 16%) over this period, as did total discontinuous synthetic fibre yarn (down 17%), including a fall of 25% in the production of polyester/cotton blend yarn.
>
> Woven wool fabric production fell by 5%. The decline in woollen fabric output of 18% was largely responsible for this fall, although the increased production of worsted fabric of 15% offset this to a large extent.

Woven cotton fabric also recorded a fall in production for the eight months to August of

Table 2. Australian textiles production: year to date August

	YTD Aug 1987	YTD Aug 1988	% change
Yarns (tons)			
Total wool yarn	15186	15164	-0.1
• woollen - carpet	9257	10471	13.1
- other	2255	1646	-27.0
• worsted - machine knitted	2274	1802	-20.8
- other	1401	1245	-11.1
Total cotton yarn	14593	12260	-16.0
Total discontinuous synthetic yarn	7103	5874	-17.3
• polyester/cotton mixes	3751	2804	-25.2
Broadwoven fabrics ('000 square meters)			
Total wool fabric	7127	6759	-5.2
• woollen	4382	3609	-17.6
• worsted	2745	3150	14.8
Total cotton fabric	26553	25032	-5.7
Total man-made fiber fabric	6000	111686	-2.5
• terry toweling and similar	6000	6021	0.4
• dcf polyester/viscose	13302	11685	-12.2
• dcf polyester/wool	2797	2107	-24.7
• cf polyester	2392	3295	37.8
• cf polyolefin	65145	63811	-2.0

6%. However, production of terry toweling fabrics managed to remain relatively stable compared to the same period a year earlier.

Investment in Capital

In 1987-88, the TCF sector's expenditure on new plant and equipment amounted to $A308 million representing an increase of 32% on previous year. This was expected to increase to $A334 million for the 1988-89 fiscal year. The 1987-88 increase in capital investment by the TCF sector exceeded the growth rate for total manufacturing sector expenditure, which only increased by 11% over this period [3].

The TCF industry in Australia is currently operating in a volatile environment. Imports are increasing both within quota and total clearances. "Total textile quota category clearances for within quota imports increased by 8% in 1988, although total clearances of the quota assisted textile products rose only by 3%" [3].

The nervousness in the industry is perhaps reflected in the response to the survey discussed below.

THE SAMPLE AND THE RESPONSE RATE

From its total membership of over 530 firms, the Textile, Clothing and Footwear Council of Australia (TCFCA) provided us with a sample of 466 firms which were considered to be suitable for inclusion in this study. The questionnaire provided by the Center for Global Business at Indiana University was used without any alternations. However, two additional questions were attached to the questionnaire at the request of TCFCA. One of these questions was concerned with incentive bonus schemes and the other was concerned with employee contributions.

The questionnaire with a covering letter was mailed to 466 firms on 30 March 1989. By the end of April (within a month) only 32 completed questionnaires had been returned. By the end of May another ten completed questionnaires had been received, bringing the total number of usable responses to 42. A targeted follow-up undertaken by the TCFCA at the end of May only resulted in four more responses.

The 10% response rate achieved from the Australian survey is somewhat better than the questionnaire survey response rates achieved in South Korea, the People's Republic of China and western Europe [4].

THE RESPONDENTS

Table 3 presents the respondents analyzed by industry sector (i.e. textiles, clothing or footwear) and size of the company as measured by the number of employees. Overall, 32 companies represented the textiles sector, 12 represented clothing and two companies represented the footwear sector. The average company in the sample employed nearly 200 people. The largest company in the textiles sector employed 385 persons and in the clothing sector, the largest company had 2,000 employees.

Table 3. Sample classified by sector

	Total people	Under 20	20-49	50-99	100-199	200-499	500-1000	Over 1000
Total sample	44	13	7	6	8	6	2	2
Textiles	11	2	1	3	2	3	0	0
Clothing	31	11	6	2	6	3	2	1
Others	2	0	0	1	0	0	0	1

* Two firms provided no information on the number of employees

Sales turnover figures for the previous financial year are presented in Table 4. Of the 38 companies providing this data, 24 did not export any of their output. The average domestic sales and exports for companies in the textiles sector was $12.7 million and $3 million, with a maximum of $39.8 million and $5 million, respectively. In comparison, the average domestic sales and exports for companies in the clothing sector was $15.5 million and $345,000, with a maximum of $92 million and $5 million, respectively.

Table 4. Sample classified by $ sales

	Domestic Sales	Export Sales
Less Than $500,000	6	9
$500,000 - $1M	2	0
$1M - $2M	6	1
$2M - $5M	6	3
$5M - $10M	6	0
10M - $20M	3	1
Over $20M	7	0
No Export	0	24
Total	38	38

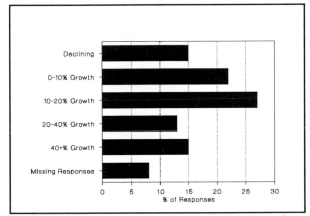

Figure 1. Annual growth rate in the physical volume of production

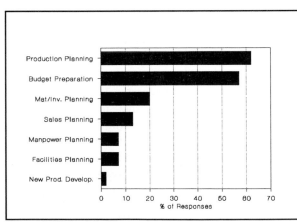

Figure 2. Use of forecasts

Figure 1 shows the annual growth rate in the physical volume of production over the last two years in the TCF companies. On average around three-quarters (73.8%) of the products were made-to-order (71.5% in the textile industry and 74.4% in the clothing industry) with the remainder being made-to-stock. The average investment in production equipment was $3.2 million with an average utilization of 82.7%. The average investment in the textile industry ($4.4 million) was three and a half time higher than the average for the clothing industry ($1.2 million).

FORECASTING PROCEDURES

Figure 2 shows the various uses of sales forecasts in the TCF industry. The three most important uses were for production planning (61% of firms), budget preparation (57% of firms) and material/inventory planning (20% of

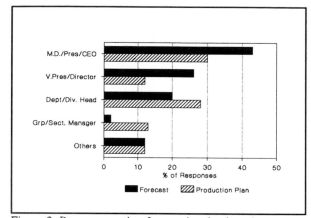

Figure 3. Person preparing forecast/production plan

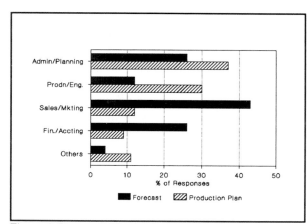

Figure 4. Responsibility for forecast/planning

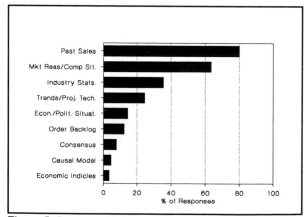

Figure 5. Sources of information

firms). Little use was made of forecasts for new product development, manpower planning and facilities planning (3%, 7% and 7% of firms respectively), although a slightly higher usage was made for sales planning (13% of firms).

Generally, the forecasts were prepared by top managers (around 70% of firms) from the sales/marketing, finance/accounting and administration/planning departments (43%, 26% and 26% of firms respectively). Group/section managers and department/division heads, representing middle and lower management respectively, had a limited role in producing the forecasts, as did personnel from the production and engineering departments (see Figures 3 and 4). The fact that a large proportion of the TCF companies and the sample companies are small-sized explains the high level of involvement of senior management in preparing forecasts and plans.

Figure 5 shows the sources of information used by respondents in preparing the sales forecast. TCF firms make extensive use of past sales history (80% of firms), the general company and industry situation (63 % of firms) and market research and customer/vendor information (63 % of firms). Overall, little use was made of the formal techniques of sales forecasting, with 59% of the firms indicating no usage at all. Projection techniques such as indices, moving average and exponential smoothing were most commonly used (20% of firms) whilst regression and the Delphi method were being used by less than 7% of the companies.

The application of computers for preparing forecasts was found to be a low key activity in the TCF industry. Only a third of the firms in the sample were using computers for forecasting. Generally, forecasts were developed for product groups (70% of firms) rather than individual products, covering on average a period of one year with an increment of around six weeks. About half the firms provided a range for the forecast rather than a single value.

Almost a third (30%) of the firms expressed the forecast both in monetary terms and in physical units, but overall a slightly higher proportion of firms expressed the forecast in monetary terms (61%) than in physical terms (54%). Nearly two-thirds (63%) of the firms modify the forecast between two and four times per year, 15% of the firms modify the forecast each month and one clothing manufacturer indicated that it modified the forecast 16 times per year. The average for the sample was five times per year.

PRODUCTION PLANNING AND SCHEDULING

More middle and junior managers were responsible for preparing production plans than preparing forecasts (43% compared with 22%). However, it was primarily top management which carried the responsibility for preparing the production plans (43% of firms) and these personnel were mostly from administration/planning (37%) and production/engineering (30%). Sales/marketing (13%) and finance/accounting (9%) personnel played a minimal role in this respect (see Figures 3 and 4). The users of the production plans were no different than the preparers in terms of management level, however a higher proportion of users, than preparers, were from administration/planning and production/engineering (80% compared with 67%) (see Figure 6).

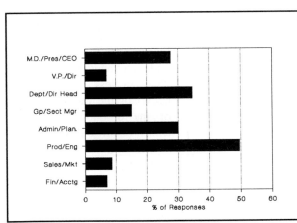

Figure 6. Primary users of production plan

Since three-quarters of the firms manufacture-to-order, it was not surprising to find that actual orders/backlog (59%) and production capacity (33%) were the two most important factors used for preparing production plans (see Figure 7). However, it was surprising to find that production plans were little used for budget preparation (15%), manpower planning (20%), facilities planning (13%) and purchasing/procurement (26%). As shown in Figure 8, operations scheduling (50%) and material/inventory planning (52%) were the two most common uses of the production plan.

The production plan was equally likely to be developed for individual products or product groups. On average, it covered nine months in increments of 35 days. In 41% of the firms the production plan covered 12 months and in one company in the clothing sector it covered three years. In almost one half of the companies the production plan was revised three or four times per year, 17% of the firms revised their plan each month and two clothing manufacturers modified the production plan weekly. Ninety-one per cent of the firms revised their production plan when there was a demand fluctuation and a fifth (nine companies) set a limit and the plan was revised only if fluctuation exceeded this number. This limit ranged from 5% to 100% with an average of around 25%. In less than a quarter (22%) of the firms some part of the production plan was frozen or difficult to change.

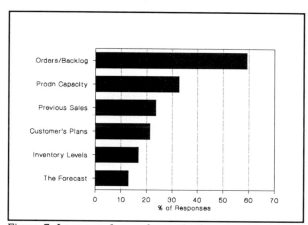

Figure 7. Important factors for production planning

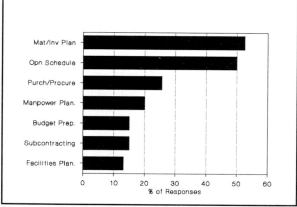

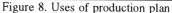

Figure 8. Uses of production plan

The use of computers was more common for production planning and control than was the case for forecasting (72 % compared with 33 %), with more than half making extensive (39%) or moderate (20%) usage. Furthermore, a third of the non-users said that computers would be helpful for production planning and control.

SHOP FLOOR CONTROL

Administration/planning personnel rather than production/engineering personnel ordered

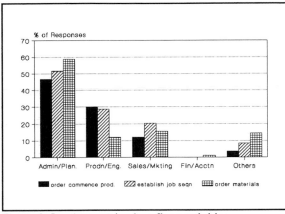

Figure 9. Involvement in shop-floor activities

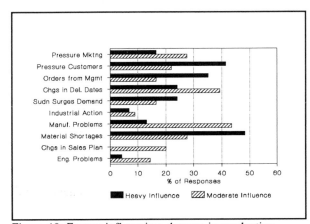

Figure 10. Factors influencing changes in production priority

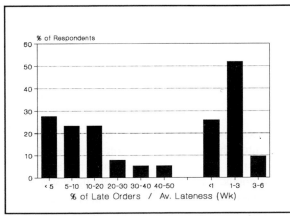

Figure 11. Proportion of late orders and average lateness

the factory to start production and established job sequence in around one half of the companies (see Figure 9). They also ordered the purchase of material in 59% of the firms. These authorizations were generally based on actual customer orders, the production plan/schedule and material availability. A somewhat surprising finding was that in a fifth of the companies the sales/marketing personnel established the sequence for releasing the jobs to production. Generally, this activity is the responsibility of the production or planning personnel who are more familiar with what goes on on the factory floor.

Material shortages (48%) and pressure from customers (41%) had a heavy influence on changing the priorities once production had started on an order. Internal factors such as engineering problems and industrial action had little influence (see Figure 10). Generally, changes in the priorities were orally communicated (67%) to the production workers rather than by document.

The average lead time for a typical batch from start of production until delivery to the customer or inventory was 32 days whilst the usual delivery time promised to customers averaged at 42 days. These figures indicate that delivery promises to customers can be improved if communication between design, marketing and production (i.e. product complexity and production load) and material availability can be improved.

In one quarter of the firms, 5-10% of the orders were late and in another quarter of the firms, 10-20% of the orders were late. In over one half (54%) of the companies the average lateness was two weeks and the most common causes of lateness were quality problems (50%), production bottlenecks (33%) and lack of machine and labor capacity (17% each) (see Figures 11 and 12).

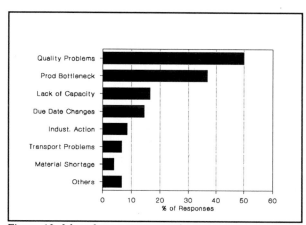

Figure 12. Most frequent causes of order lateness

Companies used a variety of alternatives to change the capacity required. Overtime working (52% of firms) and subcontracting (35% of firms) were highly used to increase production and during times of reduced demand companies build-up inventory (highly used, 20%; moderately used, 26%) or lay-off the extra workers (highly used, 15%; moderately used, 28%). Leasing production facilities to increase or decrease capacity was little or never used in majority of the companies (see Figure 13). Standard time for operations were reasonably commonly established and used for scheduling and costing products.

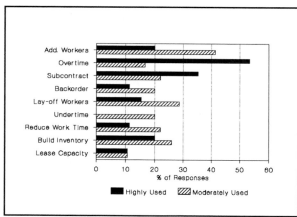

Figure 13. Means of changing capacity

PURCHASING AND MATERIALS MANAGEMENT

Around two-thirds of the firms in the sample provided data on the amount of inventories in stock. The average total amount of inventory held was $2.3 million (maximum figure was $12.7 million) and this was broken down as follows: purchased materials and parts, 37%; work-in-process, 30%; and finished goods, 33%. Companies held three weeks of finished goods on average as protection against unexpected demand fluctuations.

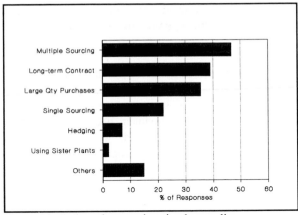

Figure 14. Actions for assuring timely supplies

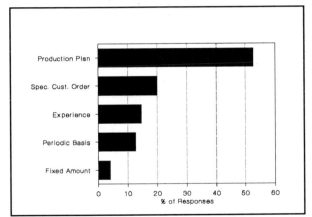

Figure 15. Methods of purchasing raw materials

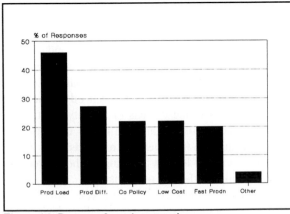

Figure 16. Reasons for subcontracting

Figure 14 shows the actions taken by companies to assure the timely supply of materials and purchased parts. The most important actions taken were multiple sourcing (46%), long-term contracts (39%) and purchasing larger quantities (35%). A third of the companies had two or three vendors for purchased parts and nearly a quarter had four or more vendors. Companies which purchased larger quantities than necessary had obviously not become aware of the benefits of keeping minimum inventory in stock. Also single sourcing was not commonly practiced, only 22% of the firms were using this practice to assure timely supply of materials and purchased parts.

The most frequent method of purchasing raw materials was the production plan or schedule (52%). Some companies (20%) purchased raw materials for specific customer's orders (see Figure 15). This is obviously the case for companies in the clothing sector, especially the high fashion clothing companies.

The two most important reasons identified for subcontracting outside were production load (48%) and production difficulty (28%). It is also interesting to note that 22% of the firms were subcontracting outside because costs were lower and 20% of the firms were doing the same because quality of the subcontracted work was higher (see Figure 16). This is a clear indication of in-house manufacturing ineffectiveness and inefficiency of many Australian companies.

It was interesting to note that purchasing personnel were involved in a wide range of activities apart from their regular responsibilities such as materials shortages (78%) and inventory programs (43%). As shown in Figure 17 they were involved in product design (39%), capacity changes (28%) and quality programs (24%).

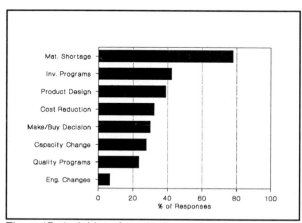

Figure 17. Activities of purchasing personnel

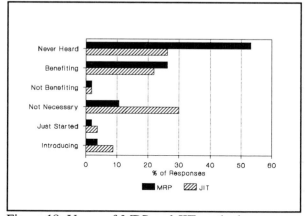

Figure 18. Usage of MRP and JIT methods

Finally, only a quarter (24%) of the firms had a person titled Materials Managers; the Economic Order Quantity (EOQ) concept was being used in only 28% of the firms; more than half (52%) of the respondents had not heard of Material Requirements Planning (MRP). Only 26% of the companies were using it and benefiting from it. Just-in-Time (JIT) systems were more common with 22% of the firms using it and benefiting from it, 9% trying to introduce it but having difficulty, and a further 4% just starting to introduce the JIT concepts (see Figure 18).

CONCLUSION

Australia's textile and clothing industries have not generally been internationally competitive, as evidenced by the flood of imports into Australia and the lack of exports in these areas of activity. Part of the reason for this is clearly related to high unit labor costs relative to newly industrialized countries and part is due to sub-optimal production management methods. Our survey found that opportunities for improvement exist in production planning methods, forecasting, inventory control and in the manufacturing-marketing relationship.

Some firms in the industry were clearly efficient operators, while others have many opportunities for improving their production control techniques, quality assurance and hence their competitive performance in terms of lead times, reliability, de-bottlenecking, inventory levels and customer responsiveness.

Although this survey did not explicitly measure variables associated with human relations issues, the high degree of sub-contracting used in the industry and the predominant 'hire-fire' strategy indicate that many firms do not achieve a high degree of employee commitment in the workforce. The lack of quality control in 'piecework' sub-contract garment assembly is a problem for many firms. Relative to other industries, the general lack of sophistication in planning and control systems, the cost structure and the low status of the production function and workers explain many of the difficulties of this industry.

However, this does not mean that the industry as a whole is doomed. As protection (tariffs) are decreased, those firms which restructure appropriately and improve their operational effectiveness and performance should be able to achieve sustainable competitiveness levels.

REFERENCES

1.   *Australian Industry: New Directions*, Australian Government Publishing Service, Canberra, 1987.

2.   *Business Review Weekly* (Australia), April 14, 1989, p. 59.

3.   *Communique*, Textile, Clothing and Footwear Council of Australia Newsletter, December 13, 1988.

4.*   D.C. Whybark and B.H. Rho, "A Worldwide Survey of Manufacturing Practices," Discussion Paper No. 2, Bloomington, IN, Indiana Center for Global Business, Indiana Business School, May 1988.

∗ This article is reproduced in this volume.

ACKNOWLEDGEMENTS

A version of this paper appeared in the Proceedings of the Pan-Pacific Conference, Seoul, Korea, June 1990.

Manufacturing Practices In Chile

Antonio E. Kovacevic, Pontificia Universidad Católica, Chile
J. Claudio Lopez, Pontificia Universidad Católica, Chile
D. Clay Whybark, University of North Carolina-Chapel Hill, USA

ABSTRACT

The purpose of this paper is to identify, catalog and compare the production planning and control practices of the small machine tool and non-fashion textile industries in Chile. Specifically, this paper presents the key differences in the practices between the two industries. The results of these comparisons provide a base for subsequent comparisons of Chilean practices with those of other countries; studies which can be of incalculable value for companies considering "joint ventures" or other forms of business with firms of other countries.

INTRODUCTION

This paper presents an analysis of a particular set of data on the manufacturing planning and control practices gathered from a sample of Chilean companies. Several machine tool and nonfashion textile firms were surveyed between December 1988 and May 1989 as part of a world-wide project in both developing and advanced nations. Data has been collected from North America, Western and Eastern Europe, South Korea, the People's Republic of China, Australia, Japan and Chile. This paper concentrates, however, on the data from Chile. Detailed information on the data gathering methodology for the total study, with details on the questionnaire, format, and data availability, can be found in Whybark and Rho [1]. The benefits of the study come in three broad areas. They are: comparisons of firms within a country, comparisons between countries, and guidelines for improving joint manufacturing programs. For each country in the study, the questionnaire has been designed to provide information for comparing activities between firms. Differences between firms having better or worse performance can be developed by the researchers and the firms themselves can see how they compare with other companies within their country.

Perhaps of more academic interest, but important anyway, is the comparison between countries [2,3]. The industries selected for study (small machine tools and non-fashion textile) are common to all the countries being surveyed. By looking at the data on a country-to-country basis, macro differences in the industries approaches can be determined. Significant differences in views of inventory, planning and scheduling, purchasing and so on, could have significant implications for industrial policy in some countries.

An absolutely essential ingredient in international joint ventures is understanding the approaches of the other firms. The results of this study can provide to guidelines for firms that are considering supply relationships or joint manufacturing ventures. Knowing how production planning and control activities are done in another country, for example, will enable a firm to determine how much adjustment is necessary to successfully match their systems with those of a prospective partner.

The importance of gathering empirical data from the field is very clear when the objective is to compare practices between different parts of the world. It is even more important when part of the interest concerns doing business with firms in these areas. Although theoretical models were used for developing the questions to be asked, it is the firm's actual practices that are of interest in this paper.

BACKGROUND

This international project has been developed through the collaboration of several institutions and individuals. The Korea survey has been supported by Sogang University and the Korea Productivity Center (KPC). In Europe, the activities have been supported by IMEDE (the management institute in Switzerland), where the questionnaire was translated into French, German and Italian. In China the project has been carried out at the Shanghai Institute of Mechanical Engineering (SIME), whereas, in Japan, Waseda University supported the data gathering. In North America, support came from Indiana University's Global Business Center and the work in Chile has been supported by the School of Management of the Catholic University in Santiago.

The study has been designed to achieve the objectives in three steps. The first step is to collect the data from each country individually and build the country database. The survey form has been designed to have maximum overlap of questions from country to country, but individual tailoring has been done where necessary to take into account local needs. The individual country databases will provide enough data to generate the comparisons between firms. Cooperating firms have been provided summary results of the survey, but individual firms' responses are held in strict confidence.

The key requirement for between-country comparisons is the willingness of a country to share the data. All countries that do so will be provided all data from the other countries and the results of the analysis of the data. Questions of common interest between countries will be addressed in cross-country analysis. Requests for data from specific countries could be addressed if funding can be provided.

The development of guidelines for joint ventures will come from the analysis of between-country results, and a knowledge of the within-country-between-firm results. These guidelines would be published for fairly wide dissemination, although individual requests by firms could be entertained if data confidentiality can be assured.

DATA GATHERING

The Chilean data gathering activity was supported by the School of Management of the Pontificia Universidad Católica de Chile. The effort was one that involved personal interviews and a representative sample of around 15 companies in each industry. The

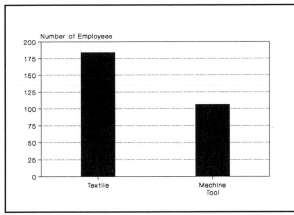

Figure 1. Number of employees

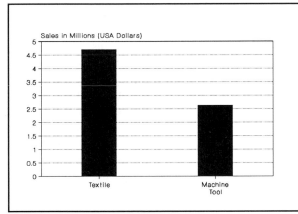

Figure 2. Sales

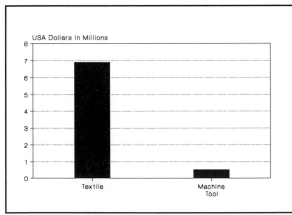

Figure 3. Investment in equipment

sample of small tool manufacturers includes firms that produce valves, gears, spare parts, hinges, dies, etc. The textile group consists of companies that produce articles such as curtains, sheets, blankets, towels, mattress covers, etc.

The questionnaire included sections for general company information, forecasting, production planning, scheduling and shop floor procedures, materials management and purchasing. This paper describes the findings for each of these categories.

GENERAL COMPARISONS OR COMPANY PROFILE

The average sales and employment level for Chilean small machine tool companies are less than those of the Chilean textile companies. Investment in equipment in the textile industry is higher than that for machine tools. (These comparisons are shown in Figures 1, 2 and 3). It is important to mention that the textile industry in Chile is actually a new industry, having developed substantially after the economy was opened and adjusted in 1975 - 1976. Nevertheless, export sales have already reached more than 10 percent of the total sales of these firms. Make-to-order and make-to-stock are not exclusive categories in either industry. Both industries have a mixture of these categories, but firms in both industries produce basically on a make-to-order basis.

SALES FORECASTING

Sales forecasts are produced mainly by top executives of each of the industries. This activity is conducted substantially by administration/planning personnel with the involvement of marketing. In small machine tool firms production/engineering people also play an important role, as seen in Figure 4.

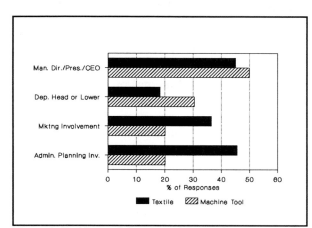

Figure 4. Position and function of forecasting

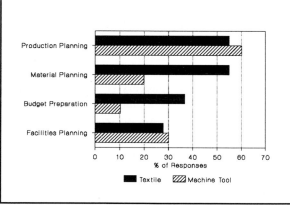

Figure 5. Use of forecast

There are no substantial differences between industries on how forecasts are used. Only a few firms use forecasts for manpower planning or new product development. The machine tool firms, especially, do not make much use of forecasts for budget preparation. The textile industry mostly uses forecasts for production and materials planning. The machine tool industry basically uses forecasts for production planning. See Figure 5.

Figures 6 and 7 summarize the results concerning the information used to prepare the forecasts. No significant differences are found between industries. In terms of objective factors, the textile industry uses past sales and backlog as the most important. The machine tool industry also considers past sales as most important, where backlog and company/industry situation (a subjective factor) are tied for next most important.

Only 20% of the machine tool firms and 37% of the textile firms use computers for forecasting. It is important to mention, basically, that neither industry uses formal techniques for forecasting either. It can be noticed, however, that subjective factors such as company and industry situation, economic and political situation trends, and market information all have importance in forecasting for 40% - 60% of the firms.

Overall, the use of formal techniques and error measurements are much higher in the textile industry. Differences are also found in the way the products are aggregated for forecasting. In the machine tool industry, about 70% of the firms forecast on a product group basis while some 55% of the textile firms forecast on an individual product basis.

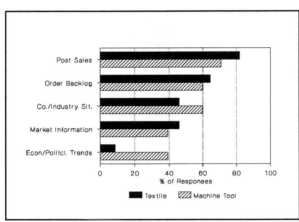

Figure 6. Forecasting inputs

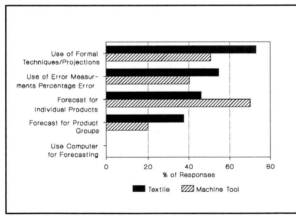

Figure 7. Forecasting practices

There are many interesting similarities for the industries. For example, the forecasting period for the textile industry averages about 7 months, and the increment is about 2 months; while the average for the machine tool industry is about 9 months, and the increment is about a quarter. Forecast revisions are performed on average 3 to 4 times a year in both industries. Some 70% to 80% of the firms provide a range for their forecasts, as opposed to a single point. Also between 60% and 80% of all firms forecasts are in physical units rather than in monetary value.

PRODUCTION PLANNING AND SCHEDULING

The primary factors used by both industries in developing production plans are the production capacity and actual orders (about 45% to 50% of the firms). The textile industry also considers customers' plans.

Looking at the units used for production planning, again Chilean machine tool firms use product groupings as opposed to the more detailed individual product units used by textile firms. Figure 8 shows this. In contrast to the results in the section on forecasting,

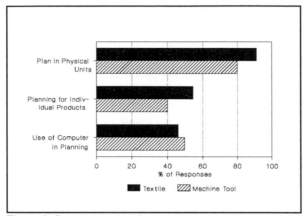

Figure 8. Production planning practices

the use of computers in production planning is slightly lower in the textile than in the machine tool industry.

There is a small difference between industries with respect to the function of the people who prepare the production plans (see Figure 9). In general it can be said that production planning is centralized since the administration/planning group is heavily involved. In the textile industry, production planning is done at a higher level while the machine tool industry involves the production group more.

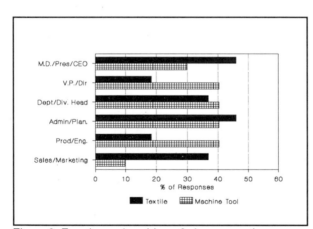

Figure 9. Function and position of plan preparation

Administration/planning, production personnel, and sales/marketing are the main users of production plans in both industries (see Figure 10). The production plan is used by firms in both industries primarily for operational activities such as scheduling or material planning rather than in planning activities like budgeting or facilities investment.

The modal production plan for the textile industry is for 6 months. It is divided into increments of about 2 months. For the machine tool firms it covers about 3 months in increments of a month. In both industries the plan is revised every 50 days on average. Around 70% to 90% of the firms in both industries revise plans when there is a demand fluctuation.

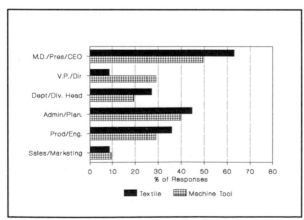

Figure 10. Function and position of plan users

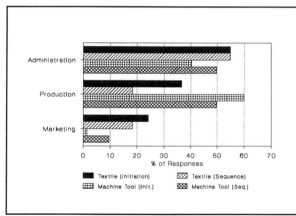

Figure 11. Involvement in shop floor activities

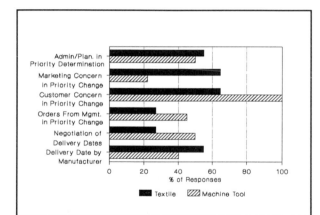

Figure 12. Influences on the shop floor

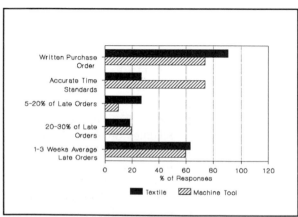

Figure 13. Shop floor practices

SHOP FLOOR CONTROL

In both industries, administration/planning or production starts, or gives the order, to start production. These same groups also set the sequence that establishes priorities. The degree of their participation is shown in Figure 11.

For the textile industry, marketing and customer concerns dominate in establishing and changing shop floor priorities, but orders from management are also considered. In determining delivery dates, however, it is the manufacturer's opinion that counts (see Figure 12). In the machine tools industry, shop floor priorities are also influenced by customers' concerns, but delivery dates are negotiated.

Figure 13 summarizes several shop floor practices. Data is transmitted to suppliers in written form in both industries, but there is more written communication of priority changes to the shop floor in the textile than in the machine tool industry. More than 90% of the operations have time standards in both industries, but in the machine tool industry, time standards are mostly used for scheduling and determining product costs. Although time standards are quite accurate in the case of machine tool firms, the percentage of late orders is higher than in the textile firms.

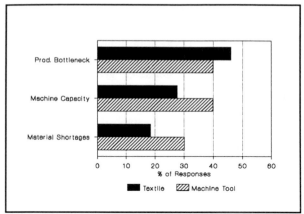

Figure 14. Causes of lateness of orders

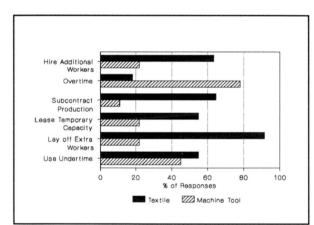

Figure 15. Means of capacity change

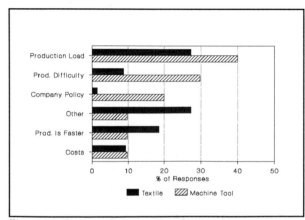

Figure 16. Reasons for subcontracting outside

However both industries have about the same average lateness for those orders that are late (between 1 to 3 weeks).

The causes of lateness, as seen in Figure 14, are similar in both industries. In both cases production bottlenecks appear to be the primary cause, with machine capacity and material shortages involved as well.

When capacity changes are necessary, there are a variety of responses in both industries: leasing, subcontracting and hiring are all used for capacity increases in the textile industry. In machine tool firms the means for capacity increases are overtime, leasing and hiring. During times of reduced demand; the extra workers are laid off in the machine tool industry and undertime is used in the textile industry. Subcontracting is used for augmenting production and cost reduction in both industries. In the textile industry this is company policy. These findings are summarized in Figures 15 and 16.

PURCHASING AND MATERIALS MANAGEMENT

About 50% of the inventory holdings of textile firms is in the raw materials or purchased parts. In the machine tool industry this percentage is around 30%. The purchasing of these raw materials is largely based on production plans or schedules for textile firms and on customers' orders in the machine tool industry. Neither industry uses single sourcing, both have more than 2 vendors per purchased part. There appears to be slightly more use of EOQ, MRP and JIT techniques in the machine tool industry as seen in Figure 17.

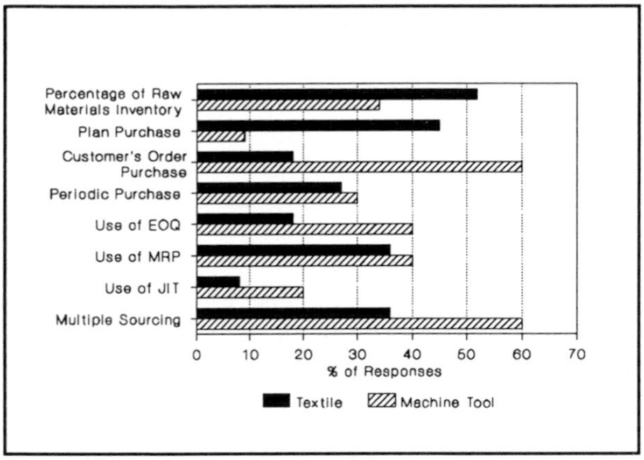

Figure 17. Purchasing and materials management

SUMMARY AND CONCLUSIONS

This study has disclosed that there are very few significant differences between the textile and machine tool industries in Chile. There are differences in the investment in equipment, where the machine tool industry's investment represents only about 7% of the textile companies' amount (see Figure 3) and in sales, where the textile companies' sales are almost 50% than those of the machine tool firms' (see Figure 2).

In sales forecasting, also, there are few significant differences between the industries. One difference is in the way products are aggregated for forecasting. In the machine tool industry, forecasts are made on a product group basis whereas textile firms forecast on an individual product basis (see Figure 6). This difference is also reflected in production planning (see Figure 8). Also because of the type of products offered by the machine tool industry, machine tool firms involve their production personnel more heavily in production planning and scheduling.

The Chilean economic scene, and the way all industries reacted during the recession period, may explain the similarities between the industries. The Chilean economy has gone through an interesting process during the last 15 years. Before 1973 the economy was closed and the national textile and machine tool industries supplied the demand with normal quality products. When the economy was opened (after 1973), imported goods such as cloth and fabrics of all types arrived in huge quantities competing strongly with lower prices against Chilean textile companies. The tools and parts that were imported were of inferior quality, but they were inexpensive.

Firms in both the textile and machine tool industries, in order to survive against this strong competition, started to obtain investment loans. These loans were mainly in U.S. dollars since the interest rates were lower than those for Chilean peso loans and the exchange rate was frozen at 39 pesos per US$. In June of 1982 a devaluation was executed and debts in dollars were immediately raised. These high debts and strong competition, caused a great number of companies in both industries to go bankrupt or on public sale. Those that survived now continue operating and offering competitive products of high quality. Surviving firms in both industries were affected the same way during this period and reacted the same way to the pressures. This may explain why both industries have similar systems for forecasting, production planning and control.

REFERENCES

1.* D.C. Whybark and B.H. Rho, "A Worldwide Survey of Manufacturing Practices," Discussion Paper No. 2, Bloomington, IN, Indiana Center for Global Business, Indiana Business School, May 1988.

2.* B.H. Rho and D.C. Whybark, "Comparing Manufacturing Practices in the People's Republic of China and South Korea," Discussion Paper No. 4, Bloomington, IN, Indiana Center for Global Business, Indiana Business School, May 1988.

3.* B.H. Rho and D.C. Whybark, "Comparing Manufacturing Practices in Europe and South Korea," Discussion Paper No. 3, Bloomington, IN, Indiana Center for Global Business, Indiana Business School, April 1988.

* This article is reproduced in this volume.

ACKNOWLEDGEMENTS

A version of this paper previously appeared as Discussion Paper No. 40, Indiana Center for Global Business, April 1990, and in the Proceedings of the Pan-Pacific Conference VII, Seoul, June 1990.

Manufacturing Practices In Mexico: An Example of the Non-Fashion Textile and Machine Tool Industries

Benito E. Flores, Texas A&M University, USA
Felipe Burgos, University of the Americas, Mexico
Arturo Macias, University of the Americas, Mexico

ABSTRACT

The purpose of this paper is to identify and compare the production planning and control practices of the small capital goods companies such as machine tools and non-fashion textile industries in Mexico. Specifically, this paper presents the key differences in the practices between the two industries. The results of these comparisons can provide a base for subsequent comparisons of Mexican practices with those of other countries. Also, this study can be of value for companies considering joint ventures or other forms of business with firms of other countries.

INTRODUCTION

This paper presents an analysis of a data set on the manufacturing planning and control practices gathered from a sample of Mexican companies. Several machine tool and non-fashion textile firms were surveyed between September 1990 and August 1991 as part of a world-wide project being carried out by the Global Manufacturing Research Group in both developing and advanced nations. To date, information has been collected from North America, Western and Eastern Europe, Finland, South Korea, the People's Republic of China, Australia, Japan and Chile. This paper is devoted to the data from Mexico. Detailed information on the data gathering methodology for the total study, with details on the questionnaire, format, and data availability, can be found in Whybark and Rho [1].

The benefits of the study come in three broad areas. They are: comparisons of firms within a country, comparisons between countries, and guidelines for improving joint manufacturing programs. For each country in the study, the questionnaire has been designed to provide information for comparing activities between firms. Differences between firms having better or worse performance can be contrasted by the researchers and the firms themselves can see how they compare with other companies within their country.

Perhaps of more academic interest is the comparison between countries [2,3]. The industries selected for study (small machine tools and non-fashion textile) are common to all the countries being surveyed. By looking at the data on a country-to-country basis, macro differences in the industries' approaches can be determined. Significant differences in views of inventory, planning and scheduling, purchasing and so on, could have significant implications for industrial policy in some countries. This is especially true in the case of Mexico as negotiations on the Free Trade Agreement with the U.S. and other countries advance.

An absolutely essential ingredient in international joint ventures is understanding the approaches of the other companies. The results of this study can provide helpful guidelines for firms that are considering supply relationships or joint manufacturing ventures. Knowing how production planning and control activities are done in another country, for example, will enable a firm to determine how much adjustment is necessary to successfully match its systems with those of a prospective partner.

The importance of gathering empirical data from the field is very clear when the objective is to compare practices between different parts of the world. It is even more important when part of the interest concerns doing business with firms in these areas. Although theoretical models were used for developing the questions to be asked, it is the firm's actual practices that are of interest to this paper.

BACKGROUND

This international project has been developed through the collaboration of several institutions and individuals. The Korea survey has been supported by Sogang University and the Korea Productivity Center (KPC). In Europe, the activities have been supported by IMD (the management institute in Lausanne, Switzerland), where the questionnaire was translated into French, German and Italian. In China, the project has been carried out at the Shanghai Institute of Mechanical Engineering (SIME), while in Japan, Waseda University supported the data gathering. In North America, support came from the Indiana Center for Global Business at Indiana University and the Global Manufacturing Research Center at the Kenan Institute, University of North Carolina-Chapel Hill. The work in Mexico has been supported by the College of Business of the University of the Americas and the Center for International Business Studies at Texas A&M University College of Business Administration.

The study has been designed to achieve the objectives in three steps. The first step is to collect the data from each country individually and build the country database. The survey form has been designed to have maximum overlap of questions from country to country, but individual tailoring has been done where necessary to take into account local needs. The individual country databases will provide enough data to generate the comparisons between firms.

Cooperating firms have been provided summary results of the survey, but individual firms' responses are held in strict confidence. The key requirement for between-country comparisons is the willingness of a country to share the data. all countries that do so will be provided all data from the other countries and the results of the analysis of the data. Questions of common interest between countries will be addressed in cross-country analysis. Requests for data from specific countries could be addressed if funding can be provided.

The development of guidelines for joint ventures will come from the analysis of between-country results, and knowledge of the within-country-between-firm results. These guidelines would be published for fairly wide dissemination, although individual requests by firms could be entertained if data confidentiality can be assured.

DATA GATHERING

The Mexican data gathering activity was supported by the College of Business of the University of the Americas in Puebla, Mexico. The effort was one that involved personal interviews with 14 samples from the machine tool industry and 22 firms representative of the textile industry.

The sample of small tool manufacturers includes firms that produce stamping machines, ball crushers, farm implements, shears, bending presses, band saws, lathes, presses, milling machines, hydraulic equipment, drill presses, grinders, etc. The textile group consists of companies that produce articles such as cotton socks and stockings, cotton and polyester thread and cloth, sheets, etc.

The questionnaire included sections for general company information, forecasting, production planning, scheduling and shop floor procedures, materials management and purchasing. This paper describes the findings for each of these categories.

GENERAL COMPARISONS OR COMPANY PROFILE

The employment level of the Mexican small machine tool companies is less than for that of the textile firms as shown in Figure 1. As the values show, the companies are small. The ratio of total employees to factory workers is 67% and 74% for the industries. Thus, overhead is higher for the textile companies. On the other hand, sales of the companies averaged .8 and 1.7 million dollars respectively as shown in Figure 2. Export sales were

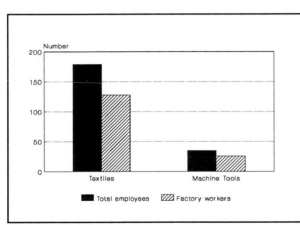

Figure 1. Work force

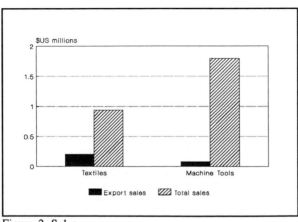

Figure 2. Sales

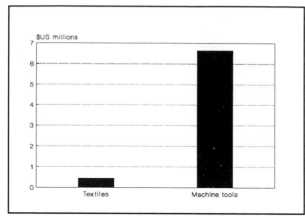

Figure 3. Equipment investment

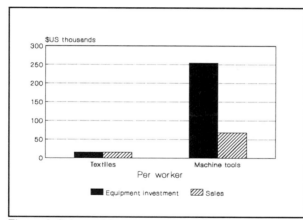

Figure 4. Sales and investment per worker

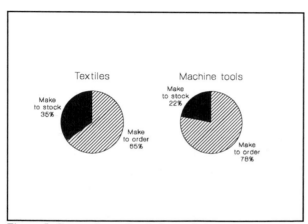

Figure 5. Comparison of ordering techniques

small in the textile (18% of the total sales) and the machine tool industries (5%). Worker training is small for both averaging 9 and 5 days respectively.

To simplify the paper, the following convention will hold. Unless specified, data are averages and will be referred first to the textile industry and then to the machine tool industry.

Investment in equipment in the machine tool industry (US$6.6 million) is higher than that for textiles (average near US$500,000). A probable explanation may be the age of some of the textile equipment, which tended to be old (Figure 3).

Sales per capita of productive workers shows that for the industries the values are: $6,239 and $64,615 respectively. The investment per capita is $8,349 and $253,846. The difference in the values points out that the better equipped worker in the machine tool industry generates more sales. This is shown in Figure 4.

Make-to-order and make-to-stock are the usual categories in either industry. Both industries produce basically on a make-to-order basis. This is shown as Figure 5.

SALES FORECASTING

Sales forecasts are produced mainly by the top executives in each of the industries. This activity does not seem to be delegated. This activity is conducted substantially by administration/planning personnel with the involvement of sales/marketing executives (Figure 6).

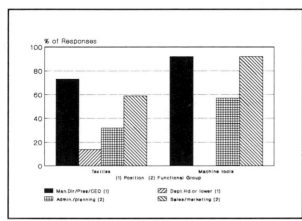

Figure 6. Forecasting responsibility

The use of the forecast is directed towards the short and intermediate horizon. Activities such as budget preparation, materials planning, and sales planning, are stressed more in the machine tool industry. In the case of the textile industry, production planning is influenced more by the forecast. Longer term forecast horizon activities such as new product planning or facilities planning are done infrequently if at all (Figure 7).

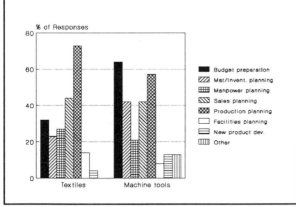

Figure 7. Use of the forecast

It is clear that many of the forecasting methodologies available are not being used. Extrapolation of past history seems to be the prevalent means of generating a forecast, specially in the machine tool industry.

The textile industry uses subjective factors much more than the machine tool industry. In the textile industry, the sample did not show any company that used a statistical forecasting technique in the forecasting process. In the case of the machine tool industry, there were a few cases. These facts are illustrated in Figures 8 through 10.

About 50 percent of both the machine tool firms and the textile firms use computers for forecasting. As was noted before, neither industry uses formal time series techniques for forecasting when they could be using the computer to do so. Subjective factors such as company and industry situation, economic and political situation trends, and market

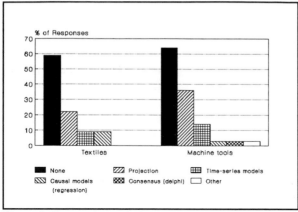

Figure 8. Forecasting techniques

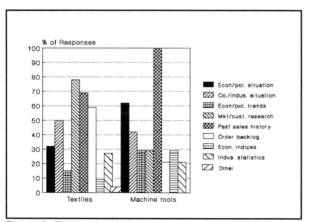

Figure 9. Forecasting input factors

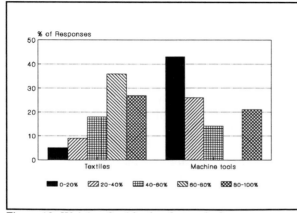

Figure 10. Weight of subjective factors in forecast

information all have importance in forecasting for most of the firms, especially in the textile industry.

Differences are also found in the way the products are aggregated for forecasting. While only 27% of the textile firms forecast on an individual product basis, about 57% of the machine tool firms forecast on a product group basis. The large majority of the companies forecast in physical units (73% and 93% respectively).

There are some interesting similarities between the industries. The average percentage error for both industries is 17%. This accuracy measure is the most prevalent (45% and 79% respectively). Also, the average forecasting horizon for the textile industry averages about 10 months, and the increment is about 1 month; while the average for the machine tool industry is about 9 months, and the increment is about six weeks. In both cases, the sum is close to 10 months. There seems to be little confidence in the stability of the forecast (or the economy); or it may be that the subjective weighting of the developers of the forecast has a frequent impact. Forecast revisions are performed on average about once a week in both industries. Some 73% and 43% of the firms provide a range for their forecasts respectively, as opposed to a single point value.

PRODUCTION PLANNING AND SCHEDULING

There is a difference between industries with respect to the function of the people who prepare the production plans (see Figure 11). Perhaps because the companies are smaller, the level of the people that do the production planning in the machine tool industry is higher than in the textile indus-

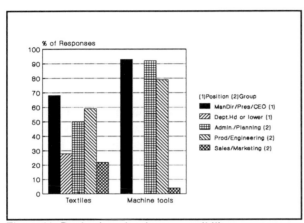

Figure 11. Production planning responsibility

try. As is to be expected, the personnel involved is either on the administrative side of the company or the production/engineering side. In general, it can be said that production planning is centralized since the administration/planning group is heavily involved, especially in the machine tool industry. The role of marketing/sales in the preparation is rather small.

The primary factors used by both industries in developing production plans are the production capacity (in the machine tool industry) and the actual orders (in the textile industry). The textile industry considers customers' plans far more heavily than the machine tool industry (55% vs. 43%). Previous sales, which in the forecast process were a major factor, are not as important in the production planning process (see Figure 12).

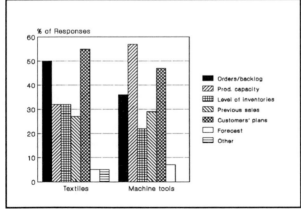

Figure 12. Production planning input factors

Examining the units used for production planning, both textile and machine tool Mexican firms use individual products as opposed to product groupings to do the production plan (73% and 64% respectively). The plan is made in physical units (86% and 100% respectively).

The majority of the people involved in production planning have college educations (68% and 71%), with some executives having graduate degrees (18% and 14%).

The production plan for the textile firms is used mostly for material and inventory planning. Other major uses are scheduling and manpower planning. For the machine tool industry, the largest use is operations scheduling (79%) followed by materials/inventory (57%) and budget preparation (50%). The results are shown in Figure 13.

The users of the production plan are at the top levels of the organization (68% and 100%) in the administration/planning, and production/engineering areas of both industries. Marketing use of the information is more limited.

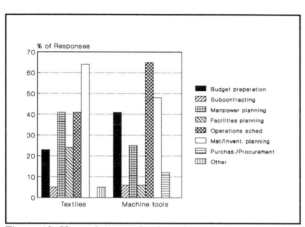

Figure 13. Uses of the production plan

The model production plan for the textile industry is for 9 months, divided in increments of about 1 month. For the machine tool firms it covers about 7 months in increments of about a month. In both industries the plan is revised every week on average. This revision coincides with the forecast revision process and reflects large instability in the production process. Around 90% of the textile firms revise plans when there is an average demand fluctuation of about 16%. Machine tool firms revise plans when there is an average demand fluctuation of 8%.

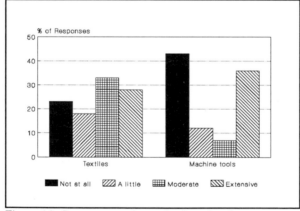

Figure 14. Computer use in production planning

The percentage of firms that consider the production plans to be very rigid is about 32% and 14% respectively. The fact that the percentages are not high matches the many changes in the production plan. The results seem to indicate that machine tool firms require a more flexible manufacturing environment than the textile firms.

Computer use in the development of the production plan is low in both industries (Figure 14). The firms that were not using computers thought (40% and 83%) that they could benefit from them, indicating an area of opportunity for the firms.

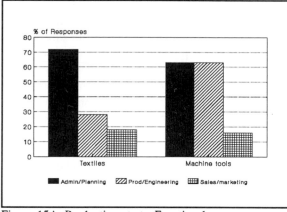

Figure 15A. Production start: Functional group

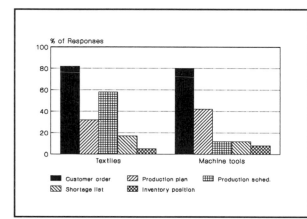

Figure 15B. Production start: Basis of order

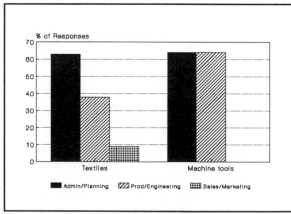

Figure 16A. Material purchasing: Functional group

SHOP FLOOR CONTROL

In the textile industry, administration/planning starts production, or issues the order to start. In the machine tool industry, the production/engineering group initiates the activities along with the administration group. The degree of their participation is shown in Figure 15A. Both industries base the start on two factors: actual customer orders (which has the most influence) and the production plan. This is consistent with the results presented earlier on make-to-order and make-to-stock (Figure 15B).

Part of the shop floor control activities include the purchase of the necessary materials. Their acquisition in the textile industry is handled principally by the administrative group and secondarily by the production group. In machine tool firms, the responsibility is shared equally (Figure 16A). As before, customer orders and the production plan are the basis for these purchases in both industries (Figure 16B). In a surprising number of cases, the purchase order is placed orally (50% and 43% respectively), but in many cases a written order is also placed (50% and 71%). The computer is not used in any significant way (9% and 7%).

The percentage of the com-

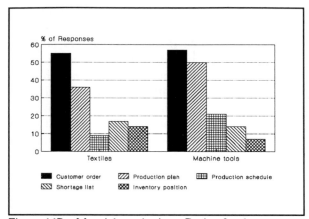

Figure 16B. Material purchasing: Basis of order

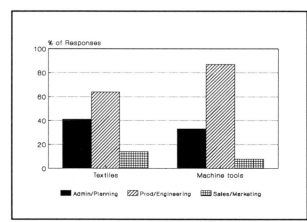

Figure 17A. Job sequencing: Functional group

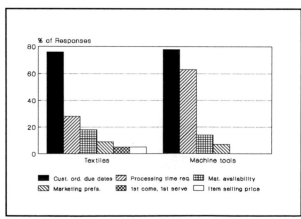

Figure 17B. Job sequencing: Basis of priority

ponents of the final product made in-house is quite high. This implies little subcontracting for the components. Lead times for the products made in-house range from 20 weeks for the textile industry to 34 weeks in the machine tool industry.

There seems to exist a prevalence of time standards in both industries that are used both to schedule and to estimate production costs. There seems to exist a degree of confidence that the time standards are close to the actual times.

The design process seems to be adequate since once the order is given to manufacture, there are very few engineering or design changes.

Production seems to be given the upper hand in determining the sequence the products should have when they are being manufactured (64% and 86%). Marketing does not seem to have high influence (Figures 17A and 17B). The basis for the priorities, though, is the due date given to the customer. In the case of the machine tool industry, another very high priority is technical (processing time required).

Thus, the delivery date determination is quite important. For both industries, the date is negotiated (50% and 64% respectively) as shown in Figure 18.

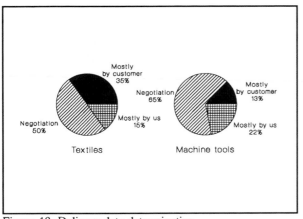

Figure 18. Delivery date determination

Once the date is established, there may be reasons for modifying the production priorities for the order. The reasons for modifying the priorities are shown in Table 1.

The changes in the scheduling of the production orders are usually sent to the shop in verbal form (71% and 82%). The main reason is probably that the size of the plants does not require a formal procedure.

Table 1. Importance of various reasons
for modifying production priorities

Industry	Weight	Reason
Textile	High	Customers and marketing
	Medium	Manufacturing problem; sales plan change
	Low	Delivery date change; materials shortages
Machine Tool	High	Customers and changes in delivery dates
	Medium	Orders from management; manufacturing problems
	Low	Marketing; material shortages

The average delivery dates are given to the customers either as a fixed time period or as range. The fixed values of the time period are 17 and 25 weeks respectively. The range of values are from 21 to 58 weeks in the case of the textile industry. For the machine tool industry the values are 38 to 68 weeks in length. The length of the delivery date in the textile industry is influenced by the shop load, the importance of the customer, and the availability of material. For the machine tool industry, product complexity and production load are by far the most important variables.

The weighted averages of the number of shop orders that are late are about 11% and 13% respectively. This is shown as Figure 19. The weighted average values of the duration of the lateness for each order are two weeks for both industries.

The causes of lateness are not much different for both industries. In both, machine capacity is important (27% and 29%), as are material shortages (32% and 29%) and labor capacity (50% and 29%).

When capacity changes are necessary, there are a variety of responses. Overtime and hiring are used for capacity increases in the textile industry. In machine tool firms, the means for capacity increases are overtime, hiring and some subcontracting. During times of reduced demand, the textile industry builds inventory and reduces work time. The machine tool industry lays off extra workers and reduces time.

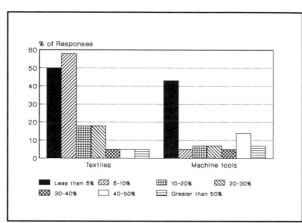

Figure 19. Late orders

The inventories of the industries have been increasing in weighted percentages in the following rates: 11% and 16% respectively.

MATERIALS MANAGEMENT

The values of the inventories are broken down as follows in the usual categories: raw materials, work in process and final goods. The graphical description is shown as Figure 20.

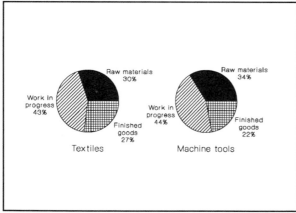

Figure 20. Inventory levels

One of the better known production planning systems is MRP (Materials Requirement Planning). The textile industry is less knowledgeable about the system than the machine tool industry, as the percentage of people that never heard of MRP was 59% to 36%.

The purchasing of the raw materials is done largely on a periodic basis in textile firms and on customers' orders in the machine tool industry (Figure 21). Both industries use multiple sourcing, with a weighted average of 2.4 to 3.1 sources respectively.

Subcontracting is used sparingly by both industries. When it is used in the textile industry, the reasons why are lower costs (27%) or faster production (27%). Production load (86%) is the reason for subcontracting in the machine tool industry.

EOQ is used by only 23% and 21% of the firms. JIT is not known by 45% and 43% of

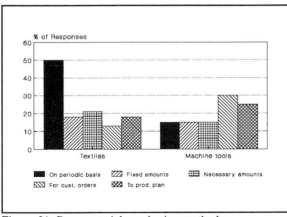

Figure 21. Raw material purchasing methods

the firms. Of textile firms that know of JIT, 23% use it but do not benefit from it, while 21% do not use it.

The safety stock inventory level of the firms in the two industries is 4.7 and 3.3 weeks.

SUMMARY AND CONCLUSIONS

This study has disclosed that there are significant differences between the textile and machine tool industries in Mexico. The textile companies are larger in terms of personnel, though not in investment terms. The machine tool industries tend to be newer and thus smaller. For both industries, the absolute size is small, however.

The firms are too small to have much impact on the export scene. Between 1982 and 1989, many of these companies started to attempt exports. This does show up in the export sales values. For the textile industry, export sales are more significant.

The demand for the products is not stable, as reflected in the percentage of sales that are made to order. The equipment investment reveals this instability in a more significant manner. The many changes to forecasts and production plans undoubtedly also affect productivity.

Because of the size of the companies, there are few levels in the organization. This means that the top levels of the firms carry out many of the functions that in other companies would be delegated.

There is little interaction between marketing and production. This is not to say that marketing should dominate the companies. There is a need to integrate the marketing and the production strategy in order to have a better managed and more successful firm.

The need to use more advanced technologies in many areas--time series forecasting, material requirements planning, etc.--shows up in the data. The fact that the planning staff is well qualified should in time show up in increments of the technology that will be used in the manufacturing practices.

The Free Trade Agreement being negotiated between the U.S. and Mexico is going to put pressure on these companies to increase productivity and quality. There is an awareness of this. Imports of machine tools and other products at perhaps lower prices and better quality will force the small companies to improve. It will create a compete-or-else environment.

REFERENCES

1.[*]   D.C. Whybark and B.H. Rho, "A Worldwide Survey of Manufacturing Practices," Discussion Paper No. 2, Bloomington, IN, Indiana Center for Global Business, Indiana Business School, May 1988.

2.[*]   B.H. Rho and D.C. Whybark, "Comparing Manufacturing Practices in the People's Republic of China and South Korea," Discussion Paper No. 4, Bloomington, IN, Indiana Center for Global Business, Indiana Business School, May 1988.

3.[*]   B.H. Rho and D.C. Whybark, "Comparing Manufacturing Practices in Europe and South Korea," Discussion Paper No. 3, Bloomington, IN, Indiana Center for Global Business, Indiana Business School, April 1988.

[*] This article is reproduced in this volume.

ACKNOWLEDGEMENTS

The support of the College of Business of the University of the Americas in Puebla, Mexico, and the Center for International Business Studies at Texas A&M University, College Station, Texas, USA, is gratefully acknowledged. A version of this paper was prepared for presentation at the Global Manufacturing Research Group Workshop at the University of the Americas in Puebla, Mexico, June 1992.

Materials Management Practices of Finnish Manufacturing Companies

Allan Lehtimäki, University of Oulu, Finland

ABSTRACT

In this paper materials management practices of the Finnish plants in machine and electrical industry are reported. The results are based on the survey realized in Finland in 1990. The topic of the survey was production planning and materials management in Finnish manufacturing companies. The survey is a part the activities of the Global Manufacturing Research Group. Materials management and purchasing are analyzed from several points of view: Functions controlling material flows, used techniques like MRP, JIT and OPT, sending purchase orders, assuring timely supply of materials, subcontracting, organization and coordination, use of computers, utilizing expert systems, and size of inventories and their distribution between raw materials, WIP and finished goods. The survey also gave knowledge of production, lot sizes and capacity utilization rate.

INTRODUCTION

This survey is a part of research activities of the Global Manufacturing Research Group (GMRG). Founders of the Group are from the USA, Korea, Australia, Hungary, China, Switzerland, Bulgaria, Chile and Japan.

The first project of the GMRG has been a worldwide survey of actual practices in production planning and control. Information has been gathered from companies in several countries around the world. The purposes are to help provide insights on the practices that lead to better performance within a given country or geographical region, to provide a basis for comparing practices between geographical regions, and to develop an understanding of where differences in practice might lead to problems for companies doing business or establishing joint ventures with firms in other parts of the world. A more detailed description of the project is presented by Whybark and Rho [1].

This survey realized in Finland is partly based on the questionnaire used in the worldwide survey. However, some adaptations have been made, and several questions have been added. These questions were particularly related with used methods, techniques, and application of computers.

GOALS OF THE RESEARCH

Some goals of the Finnish survey were:
- to know how well various production planning and control, and materials management and inventory control methods are known by Finnish manufacturing units.
- to know in which extent these methods are applied.
- to know which kind of organization is used in materials management and purchasing, and what is their relationship with production planning and control.

- to know in which extent computers are utilized in production planning and control, and materials management and inventory control.
- to know which kind of computer systems (hardware) and software are used in production planning and control, and materials management and inventory control.
- to know are expert systems used in production planning and control, and materials management.
- to know how satisfied manufacturing units are in applying computers.

In this report we particularly handle the results from the point of view of materials management. In the next report practices of production control will be reported.

SAMPLE AND SOME GENERAL DATA

We supposed that bigger companies apply more sophisticated production planning and control methods than smaller ones. That's why we decided to base our sample on bigger companies and their units. However in the case of Finland the size of bigger companies is relatively small, and also the number of these companies. (About the structure of the Finnish industry, see [2].)

In the surveys realized in the other countries the industries included were textile and machine tool. However, the author was more interested in production planning and control in electrical industry than in textile. So in this survey the industries are machine industry and electrical industry.

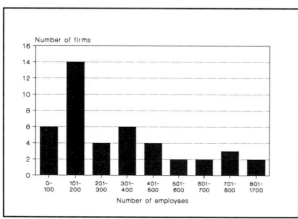

Figure 1. Number of employees

In forming the sample we used the lists of companies including their separate units. Because Finland is a small country also the number of companies is limited. That's why all bigger companies and their units in the two industries are included in the sample.

In the mail interviews one hundred questionnaires were sent out and the number of non-respondents was half of that. Some of the questionnaires we rejected because of missing information. Finally we could analyze 43 questionnaires. The interviews were conducted in the spring of 1990.

The questionnaires were sent to production managers of the units. When answering the questions they were asked to cooperate with other people like materials managers and sales managers, if necessary.

Only 3 of the units are independent companies. 23 are profit centers of companies. The number of employees of the company or parent company is varying between 160 and 42,000. The number of employees of interviewed units is varying between 27 and 1700. 51% of these units have less than 225 employees. The distribution of number of employees is presented in Figure 1.

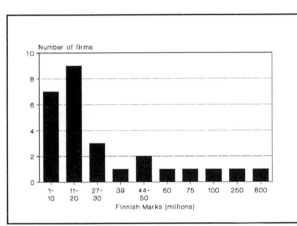

Figure 2. Value of manufacturing equipment

The number of manufacturing workers is varying between 16 and 1000 and some 51% of the units have less than 135 workers. Domestic sales of the companies or parent companies was varying between 32.4 and 6.000 million Finnish marks (FIM), and export between 0 and 86.000 million FIM in the year 1989. Domestic sales of interviewed units was varying between 3 and 1215 million FIM, and export between 0 and 1280 million FIM. The value of manufacturing equipment in balance sheet was at least 20 million FIM for half of the companies. The variation was between 1 and 800 million FIM. In Figure 2 there are presented distribution of value of manufacturing equipment.

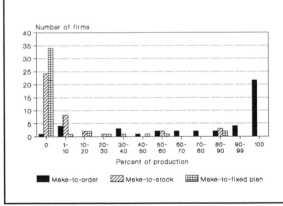

Figure 3. Type of production

In Figure 3 there are presented distribution of type of production. In 22 of the interviewed plants the whole production is manufactured to customer order, and at least 50% in 35 units. There is only one unit which does not manufacture to customer order at all and 24 plants do not manufacture to stock at all.

In Figure 4 there are presented typical lot sizes of orders and lot size in manufacturing. Half of the companies have one as typical lot size of orders. There are 10 units for which lot size is higher than 13. One unit has 10.000 as typical lot size of orders. There are 6 companies for which typical lot size in manufacturing is one. Half of the companies have lot size 13 or less. When looking at Figure 4 we see that lot sizes in manufacturing tend to be larger than lot sizes of orders.

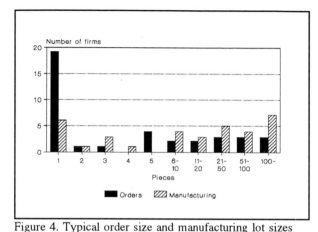

Figure 4. Typical order size and manufacturing lot sizes

The approximate average capacity utilization rate was 80-90% in 30% of the plants, and over 90% in 51% of the plants.

The results of the survey will be presented under the headings of the sections of the questionnaire.

SALES FORECASTING

There were several questions about sales forecasting. We handle here only the most important ones.

The priorities for the uses of the forecast were as presented in Table 1. We see that sales forecasts are primarily used for budget preparation. However, they have important role also in purchasing and materials management, as well. For 20.5% it is the primary purpose of forecasts.

Table 1. Frequencies(%) of the priorities for the uses of the forecast

Purpose	Priority 1	Priority 2	Priority 3
Budget preparation	43.6	23.1	10.5
Planning finished stocks inventories	0	2.6	2.6
Purchasing, materials management	20.5	10.3	31.6
Facilities investment planning	0	7.7	5.3
Production planning	7.7	35.9	39.5
Sales planning	25.6	20.5	7.9
Product development	0	0	0

In the most cases (46.7%) no formal techniques are used in forecasting. The most popular used technique was the group: moving average, exponential smoothing, indices (used primarily in 40.0% of cases). Computers for sales forecasting are used in 57.1% of cases.

PRODUCTION PLANNING, SCHEDULING AND SHOP FLOOR CONTROL

Next we handle some issues of production planning which are related with materials management. The most important uses of production plan are presented in Table 2. In the most cases(41.5%) operations scheduling is the primary use of the production plan. However, material/inventory planning has been the primary use in 34.1%, and the secondary use in 29.3% of the cases. Purchasing/procurement has been the primary use only in 4.9%, but the secondary use in 17.1% of the cases. Priority 3 was given by 26.8% of the respondents.

Table 2. The most important uses of production plan

Use	Priority 1 Frequency %	Priority 1 Frequency n	Priority 2 Frequency n	Priority 3 Frequency n
Budget preparation	17.1	7	2	-
Subcontracting		-	4	8
Manpower planning	2.4	1	6	9
Facilities planning		-	3	-
Operations scheduling	41.5	17	6	5
Material/inventory planning	34.1	14	12	5
Finished stocks planning		-	1	3
Purchasing/procurement	4.9	2	7	11
Other		-	-	-

The principal basis of establishing the sequence for releasing jobs to production is customer order due dates in 90.5% of the cases, and material availability in 2.4% of the cases. In 23.5% of the cases material availability has priority 2 and in 25.0% of the cases priority 3. It seems to be at the third place when establishing the sequence.

The most important reason for changing the sequence of jobs after releasing is material shortages. Of the respondents, 41.9% evaluate that this reason has very strong effect, and 30.2% a moderate effect.

The most important criteria for determining delivery time is the capacity situation (67.5%). However complexity of the product and materials availability are almost as important criteria.

Of the manufacturing units, 80.9% report that 5%-30% of the orders are late. All the units have some lateness. The average lateness is 1-3 week in the most cases or less than 1 week. Material shortages are the most frequent causes of lateness. Of the respondents

42.9% have given priority 1 to this cause, and 36.8% priority 2. Related with these causes we also asked the corresponding quality problems. The most frequent quality problems were related with materials and components. Of the respondents 31.6% have placed this at the first place and 20.6% at the second place.

MATERIALS MANAGEMENT

Next we go to the basic subject of this report: materials management. We handle the topic from several points of view.

Functions controlling material flows and purchasing

In Table 3 there are presented frequencies for the question: Which function is controlling material flows in your unite? We see that production planning and shop floor control is the most frequent function for controlling material flows.

Table 3. Frequencies for the functions controlling material flows

Function	Frequency (%)
Production planning and shop floor control	67.4
Materials management	23.3
Purchasing	7.0
Other	2.3

In the most cases requirements of purchased materials are determined by the production department. Of the respondents 62.8% have given the first priority to this alternative (see Table 4). The marketing department is mainly determining requirements in 11.6% of the cases, and the purchasing department in 16.3% of the cases.

Table 4. Frequency of the departments determining
requirements of purchased materials.

Department	Frequency for "mainly"		Frequency for "also"
	%	n	n
Administration	2.3	1	-
Production	62.8	28	3
Sales	11.6	5	5
Accounting		-	-
Purchasing	16.3	7	6
Other	7.0	3	1

Sales and purchasing departments also have a secondary function in determining requirements.

Determination of requirements of purchased materials is mainly based on real customer orders (60%, see Table 5). Also production plan (22.5%) and shop floor control plan (17.5%) are used. Production plan and inventory are priority 2 and 3.

Table 5. Determining of purchase requirements of materials/components

Based on	Priority 1 Frequency		Priority 2 Frequency	Priority 3 Frequency
	%	n	n	n
Real customer orders	60.0	24	2	-
Production plan	22.5	9	10	3
Shop floor control plan	17.5	7	6	1
List of shortages	-		2	1
Inventory level of material	-		4	9
Other	-		-	-

Techniques used in materials management

Next we handle various techniques which could be utilized in materials management. At first we asked the use of MRP or MRP II. In Table 6 there are presented roles of MRP or MRP II in materials management. We see that 66% of the manufacturing units use MRP or MRP II and benefit from it, 13% understand it but feel no necessity to introduce it, 15.8% have never heard of it. One of the companies was using it but not benefiting from it.

Table 6. Role of MRP or MRP II in materials management

Role	Frequency	
	%	n
Never heard of it	15.8	6
Using it and benefiting from it	65.8	25
Using it but not benefiting from it	2.6	1
Understanding it but feeling no necessity to introduce it	13.2	5
Just starting to introduce it	2.6	1
Trying to introduce it, but having difficulty in doing so		-

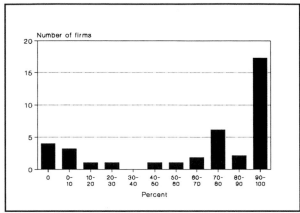

Figure 5. Items included in MRP

In Figure 5 there are presented frequencies for percentages of items included in MRP. We see that in the most cases 90-100% of items are included. This is quite high percentage, because usually computing times are quite long if all the items are included. Of the respondents 65.7% reported that requirements planning is realized on the net basis.

In Table 7 there are presented role of JIT in materials management. Frequency for using it and benefiting form it is very high: 83.7%. Three of the plants understand it, but feel no necessity to introduce it.

Table 7. Role of JIT in materials management

| | Frequency | |
Role	%	n
Never heard of it		-
Using it and benefiting from it	83.7	36
Using it but not benefiting from it	2.3	1
Understanding it but feeling no necessity to introduce it	7.0	3
Just starting to introduce it		-
Trying to introduce it, but having difficulty in doing so	7.0	3

In Table 8 there are presented role of OPT in materials management. Of the plants 32.4% are using it and benefiting from it. Of the respondents 29.7% have never heard of it. The frequency is the same for understanding it, but feeling no necessity to introduce it.

Next we were interested to know, on which plans purchases are primarily based on. The results are presented in Table 9. We see that in the most cases purchases are primarily based on production planning. Only in 4 cases are they based on plans of materials management function.

Next we look at the methods on which purchases are based. We also asked the respondents to prioritize used methods. The frequencies are presented in Table 10.

Table 8. Role of OPT in materials management

	Frequency	
Role	%	n
Never heard of it	29.7	11
Using it and benefiting from it	32.4	12
Using it but not benefiting from it	5.4	2
Understanding it but feeling no necessity to introduce it	29.7	11
Just starting to introduce it		-
Trying to introduce it, but having difficulty in doing so	2.7	1

Table 9. Frequencies of the plans on which purchases are based

	Frequency	
Plan	%	n
Production planning	88.6	31
Own plans of purchasing function		-
Plans of the own materials management function of the unit	11.4	4
Plans of the materials management function above your unit in the organization		-

Table 10. Frequencies of the methods on which purchases on based

	Priority 1 Frequency		Priority 2 Frequency	Priority 3 Frequency
Method	%	n	n	n
Purchasing budget, annual purchasing plan	120.8	5	4	1
MRP	41.0	16	8	1
Fixed amounts whenever needed	2.6	1	6	4
On a periodic basis		-	1	1
For specific customers' orders	17.9	7	4	3
According to production plan or operations schedule	15.4	6	3	7

(continued)

Table 10, continued

	Priority 1 Frequency		Priority 2 Frequency	Priority 3 Frequency
Method	%	n	n	n
Based on materials purchasing of the whole company	-		-	-
Based on trade policy and compensation trade	-		-	-
Annual contracts	2.6	1	4	1
Other	-		-	5

We see that in almost half of the cases purchases are primarily based on MRP. Although the frequency of using JIT and benefiting from it is higher than that of MRP (see Tables 7 and 8), MRP is the more frequent method on which purchases are based. Purchasing budget, for specific customers' orders, and according to production plan or operations schedule have also been frequently used methods.

Of the interviewed manufacturing units 72% use ABC-analysis in material management and benefit from it (Table 11). Two of the respondents have never heard of it, and 14% understand it, but feel no necessity to introduce it.

Table 11. Role of ABC-analysis in materials management and inventory control

	Frequency	
Role	%	n
Never heard of it	4.7	2
Using it and benefiting from it	72.1	31
Using if but not benefiting from it	4.2	2
Understanding it but feeling no necessity to introduce it	14.0	6
Just starting to introduce it	2.3	1
Trying to introduce it, but having difficulty in doing so	2.3	1

Scheduling and lot sizing of purchases are determined by purchasing itself in the most cases (37.8%, see Table 12). However, the frequency for production is almost as high. In 21.6% of the cases it is done by materials management function above the manufacturing unit in the organization. Of the interviewed units 29.3% use EOQ for specifying purchase amounts. It is mainly done by purchasing function (Table 13). Of the units 22.6% use EOQ for specifying production amounts. Typical purchasing lot sizes are bigger than production lot sizes in 56.1% of the cases. Purchasing lot sizes are the same in 34.1% of the cases, and smaller in 4.9% of the cases. Determining levels of the safety stocks is based on statistical methods in 29.2% of the cases.

Table 12. Frequencies for functions which determine scheduling and lot sizing of purchases

Function	Frequency	
	%	n
Production	32.4	12
Purchasing	37.8	14
Materials management of your own unit	8.1	3
Materials management above the manufacturing unit in the organization	21.6	8

Table 13. Use of the Economic Order Quantity

For specifying	Used by	Production Frequency		Purchasing Frequency	
		%	n	%	n
Purchase amounts		4.9	2	24.4	10
Production amounts		20.0	8	2.6	1

Means of sending purchase orders

Mail (36.6%) and telefax (39.0%) are the most frequent ways to send purchase orders to suppliers (see Table 14). Also telephone is quite important. Only four companies use e-mail, but only as secondary means. No one uses kanbans.

Table 14. Means to send purchase order to suppliers

Means	Priority 1 Frequency		Priority 2 Frequency	Priority 3 Frequency
	%	n	n	n
Orally by visiting supplier	2.4	1	1	1
Orally when supplier visits company		1	1	2
Telephone	19.5	8	8	10
Mail	36.6	15	7	7
Telefax	2.4	1	2	5
Telegram		-	-	-

(continued)

Table 14, continued

Means	Priority 1 Frequency %	n	Priority 2 Frequency n	Priority 3 Frequency n
E-mail	-		2	2
Card as kanban	-		-	-
Component as kanban	-		-	-
Other	-		-	-

The documents needed (for mail or telefax) in ordering are mainly produced by computers (in 71.4% of the cases). In 19.0% of the cases they are prepared manually, and in 9.5% of the cases they are not needed at all.

Assuring timely supply of materials

Next we were interested to know by which means companies assure material and components deliveries to arrive in time. The results are presented in Table 15. Long-term contracts seem to be absolutely the most frequent action. After that come multiple sourcing, and keeping in touch with the suppliers. The frequencies show that many of the manufacturing units use several actions.

Table 15. Actions for assuring timely supply of materials and purchased parts.

Action	Priority 1 Frequency %	n	Priority 2 Frequency n	Priority 3 Frequency n
Long-term contracts	78.6	33	5	-
Single sourcing	2.4	1	5	3
Large quantity purchases	2.4	1	-	3
Multiple sourcing	7.1	3	15	3
Using sister plants		-	1	-
Keeping in touch with the suppliers	7.1	3	9	12
Keeping in touch with transportation/forwarding companies		-	-	3
Utilizing reserve suppliers	2.4	1	1	9
Other		-	1	-

In the most cases (65%, see Table 16) the plants have 2-3 vendors per purchased part, 33% have only one vendor, and only one plant 4 or more vendors.

Table 16. Frequency for number of vendors per purchased part

Number of vendors	Frequency %	n
One	32.6	14
2-3	65.1	28
4 or more	2.3	1

Subcontracting

Next we handle some issues of subcontracting. In the most cases (60.5%, see Table 17) production is primarily responsible for subcontracting. Purchasing is primarily responsible in 20.9% of the cases, and secondary in 14 plants. Materials management is primarily responsible in 16.3% of the cases.

Table 17. Frequency of the functions responsible for subcontracting

Function	Priority 1 Frequency %	n	Priority 2 Frequency n
Production	60.5	26	2
Materials management	16.3	7	2
Purchasing	20.9	9	14
Other	2.3	1	-

In Table 18 there are presented reasons for using subcontractors. The most frequent reasons are lack of capacity, or simply the policy of the plant. Lower costs are the most frequent secondary and third reason.

Table 18. Frequency of reasons for using subcontractors

Reason	Priority 1 Frequency %	n	Priority 2 Frequency n	Priority 3 Frequency n
Lack of capacity	40.5	17	5	1
Difficult to manufacture	4.8	2	3	3
It belongs to the policy of the unit	40.5	17	7	1

(continued)

Table 18, continued

Reason	Priority 1 Frequency		Priority 2 Frequency	Priority 3 Frequency
	%	n	n	n
Better quality	-		-	2
Quicker delivery	-		1	-
Other	7.1	3	-	1

Organization of materials management and coordination with other functions

Next we handle some organizational issues of materials management. In 67.4% of the plants there is someone titled Materials Manager. In Table 19 there are presented division of work in production control and materials management. We see that in the most cases the same team/person is responsible for production control and materials management including purchasing. It is most probably the easiest way to synchronize these functions.

Table 19. Division of work in production control and materials management.

Type of division of work	Frequency	
	%	n
The same team/person is responsible for production control and materials management including purchasing	47.6	20
The same team/person is responsible for production control and materials management excluding purchasing	23.8	10
There is special team/person both for materials management (including purchasing) and production control	28.6	12

Materials management is centralized at the upper level of the organization only in 11.6% of the plants. Frequency is the same for the centralization of purchasing, as well. In Table 20 there are presented data of the degree of controlling purchasing from the upper level. We see that it is completely controlled only in one case, and a little in 39.5% of the plants.

We can summarize that the plants are quite independent in materials management and purchasing. We have also earlier seen (Table 9) that in none of the companies are purchases primarily based on the plans of materials management above the own organization. The case is the same when analyzing methods used in purchasing (see Table 10).

Table 20. Degree of controlling purchasing activities
from the upper level of the organization

Degree of controlling	Frequency	
	%	n
Completely	2.3	1
A little	39.5	17
Not at all	55.8	24

Next we were interested in the activities in which purchasing is involved. The results are presented in Table 21. We see that more actively purchasing is participating in materials shortages, in engineering changes, make or buy decisions, cost reductions and inventory programs.

Table 21. Purchasing's involvement in some activities.

Activity	Priority 1 Frequency		Priority 2 Frequency		Priority 3 Frequency	Priority 4 Frequency
	%	n	%	n	n	n
Materials shortages	96.2	25	-	-	-	-
Engineering changes	3.8	1	42.3	11	-	-
Make or Buy decisions		-	26.9	7	3	-
Product design		-	7.7	2	-	-
Capacity changes		-	3.8	1	4	1
Quality programs		-	3.8	1	4	1
Cost reductions		-	7.7	2	10	4
Inventory programs		-	7.7	2	2	6

We also surveyed the use of computers in materials management and purchasing. Here we present some of the results. We asked, what is the role of computers. The results are presented in Table 22.

We see that in 78% or 75% of interviewed plants the use of computers is extensive in materials management and in purchasing respectively. Extensive use means that without computers it would not be possible to organize materials management as it is realized now. Only two of the plants in purchasing and four in materials management don't use computers at all.

Table 22. Role of utilizing computers in materials management and purchasing

	Materials management Frequency		Purchasing Frequency	
Role	%	n	%	n
Don't use at all	9.8	4	5.0	2
Are used, but operations are mainly done manually	9.8	4	17.5	7
Extensive use	78.0	32	75.0	30

The used software is tailored software in half of the cases (Table 23). The proportion of standard software is about a third. Also self-made tailored software is used in some cases.

Table 23. Type of software used in computer applications
of materials management and purchasing

	Materials management Frequency		Purchasing Frequency	
Type of software	%	n	%	n
Standard software	31.3	10	33.3	9
Tailored software	56.3	18	55.6	15
Self-made tailored software	12.5	4	11.1	3

Finally we asked the application of expert systems. The results are presented in Table 24. We see that the use of expert systems is very unusual until now. Only in two or three cases are they used in materials management or purchasing. Additionally six manufacturing units plan to introduce them in the future. But 17 respondents think that it would be useful to introduce expert systems for materials management and purchasing.

Table 24. Use of expert systems in production planning, operations scheduling,
materials management and purchasing

	Production Frequency	Operations Frequency	Materials Frequency	Purchasing Frequency
Phase of utilization	n	n	n	n
Are used currently	4	2	3	2
Will be used in the future	8	7	6	6
Think it would be useful to introduce	18	13	17	17

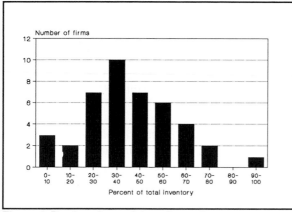

Figure 6. Purchased materials and parts inventory

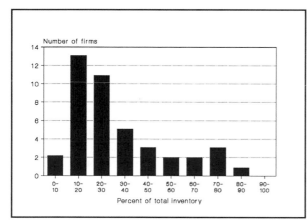

Figure 7. Work in process inventory

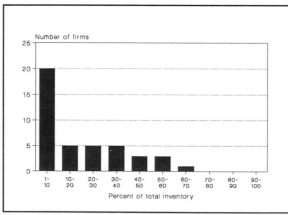

Figure 8. Finished goods inventory

Size of inventories

We asked the production managers to give data about how the value of inventories is divided between purchased materials and components, work in process and finished goods inventories. The results are presented in Figures 6, 7 and 8. In most cases the highest proportion is in purchased materials and components inventories. Minimum is 10% and maximum 100% (Figure 6). The next highest proportion is in WIP with minimum of 0% and maximum of 90% (Figure 7). Finished goods inventories represent the lowest proportion varying between 0 and 67% (Figure 8). Of the plants 23.8% have 0% in finished goods stocks. The total amount of inventory is varying between 2 and 500 million FIM.

In Figure 9 there is presented size of finished goods inventories against fluctuations in demand. We see that almost half of the companies have inventories for zero weeks, 18.4% for 2 weeks, and 15.8% for 4 weeks.

SUMMARY AND CONCLUSIONS

In this paper materials management practices of the Finnish manufacturing units are reported. Materials management was analyzed from several point of view: functions controlling are

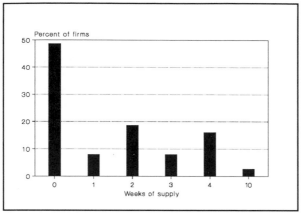

Figure 9. Finished goods inventories held against fluctuations in demand

material flows, used techniques, sending purchase orders, assuring timely supply of materials, subcontracting, organization and coordination, use of computers and size of inventories.

In the sample we have included manufacturing units of bigger companies in machine and electrical industries. Half of the units returned the questionnaire. The sample represents well the current machine and electrical industry in Finland. The manufacturing units internationally quite small. This is partly due to current trend to decentralize manufacturing. Typically these manufacturing units manufacture to order and have no finished goods inventories. Typical lot size of orders is one. However lot size in manufacturing is larger.

The most important findings of the survey are:
- Material flows are in the most cases controlled by production planning and shop floor control.
- Determining of purchase requirements is mainly based on real customer orders.
- Some 66% of the plants are using MRP and benefiting form it.
- The percentage is even higher (84%) for JIT. It means that many of the plants use both of these methods.
- Of respondents 32% report that OPT is used and beneficial.
- Purchases are primarily based on MRP.
- In the most cases purchasing or production determine scheduling and lot sizing of purchases.
- About 30% use economic order quantity (EOQ) to specify purchase amounts.
- Telefax is now the most important means to send purchase orders to suppliers. However mail is almost as important.
- Long-term contracts are the most frequent action for assuring timely supply of materials and purchased parts.
- Usually production is responsible for subcontracting.
- In 29% of the cases there is special team/person both for materials management and production control.
- When looking at the coordination, purchasing involvement is primarily in materials shortages, engineering changes and make or buy decisions.

- Materials management and purchasing are centralized at the higher level of the organization only in 11.6% of the plants. The plants are quite independent in materials management and purchasing.
- In the most cases the highest proportion of inventories (around 40%) is in purchased materials and components inventories. It indicates that the manufacturing units have been successful to make material flow thinner in process itself. WIP's proportion is around 20%.
- In 75% of the plants computers are in extensive use in materials management and purchasing. Expert systems are used only in two cases.

In this paper only materials management practices in Finland were analyzed. In further papers production planning and control practices will be reported. Other topics will be the international comparisons.

REFERENCES

1.[*] D.C. Whybark and B.H. Rho, "A Worldwide Survey of Manufacturing Practices," Discussion Paper No. 2, Bloomington, IN, Indiana Center for Global Business, Indiana Business School, May 1988.

2. A. Lehtimäki, "The Structure of Small and Medium-sized Industrial Firms in Finland," Institute of Industrial Economics, Tampere University of Technology, 1981/3

[*] This article is reproduced in this volume.

ACKNOWLEDGEMENTS

An earlier version of this paper was published as Industrial Engineering and Management Laboratory Working Paper 4/1991, University of Oulu, Finland, 1991.

Manufacturing Practices in Korea

Boo-Ho Rho, Sogang University, Korea
D. Clay Whybark, University of North Carolina-Chapel Hill, USA

ABSTRACT

A comparison of manufacturing practices as a function of size is presented in this paper. Smaller size firms, quite naturally, have higher levels of management involved in the details of manufacturing. In addition, they are less exposed to technology and are less formal in their manufacturing practices. Despite the differences resulting from size, however, several common suggestions for improving the practices among Korean firms are provided in the paper.

INTRODUCTION

As demonstrated by Japan recently, effective management of production operations is recognized to be important in strengthening corporate competitiveness. However, there has not been much study on how production management is done in Korea and how it might be improved. We have placed too much emphasis on macro policy variables such as tax, exchange rates and interest rates, which have only short-term effects on competitiveness, while neglecting the improvement of production management which is essential to long-term competitive strength.

The study is the first attempt to identify the status of production management in Korea and has the purpose of describing the current practices of production planning and control in Korea and identifying differences due to size. The firms are divided into three size categories. "Small" firms have less than 50 employees, "medium" firms have 50 to 300, and "large" firms have 300 or more. The study is based on a questionnaire and in-depth interviews.

Machine tool and non-fashion textile companies were selected for the survey as they have diverse companies where production planning and control is one of the most important aspects in their production management. Production planning and control for this study is divided into four parts: forecasting, production planning, shop floor control, and materials management. The complete questionnaire and background for the study can be found in Whybark and Rho [1].

The appendix provides the detailed results of the survey including the statistical analysis in terms of size. Problems and suggested improvements are also presented as a part of this report.

DATA GATHERING

The data gathering activity was supported by Sogang University and the Korea Productivity Center. The effort involved both a mail survey and personal interviews. The responses to the mail survey were disappointingly small, and the decision was made to send teams of interviewers into the field to directly gather the data from the companies. The data, therefore, is from a substantial portion of the population of firms producing small machine tools and non-fashion textiles. In the textile industry, some integrated firms (from spinning and weaving through sewing) are included as well as some smaller less integrated companies. Small tool manufacturers include firms that produce small drill presses, mills, other forming equipment and some computer-controlled equipment. The sample is representative of the Korean firms doing business in these industries, with 89 machine tool and 33 textile companies included. The distribution of companies surveyed is shown in Table 1.

Table 1. Distribution of firms surveyed

Number of employees	Machine tool	Textile
Over 299	20	11
50-299	47	15
Under 50	22	7
Total	89	33

Averages for the number of total employees and production workers, the annual sales, percentage of export sales, and the amount of investment in production equipment in each industry are shown in Table 2.

Table 2. Company profiles

	Total	Machine tool	Textile
Employees in total	472	468	484
Production workers	361	317	410
Total sales ($mil)	14.6	14.4	15.3
Export ratio (%)	0.17	0.14	0.28
Equipment investment (%)	9.5	10.1	8.1

FORECASTING

Of the surveyed firms, 70.5% reported practicing demand forecasting. The smaller the size of firms, the lower the percentage that practice forecasting. (See Figure 1.) Forecasting is produced mostly by the section chief or department head in the sales/marketing group. However, the percentage of sales/marketing group participation is lower in smaller firms in favor of the administration and planning groups. In smaller and medium firms, Vice Presidents and Presidents do participate in producing the forecasts. In terms of input data to forecasting, important subjective factors are market research and customer/vendor information and the most important objective factor is past sales history. Smaller firms use more market/customer information in their forecasting. Most of sampled corporations do not use formal techniques or computers in forecasting, although about half of sampled corporations said that they need computers. There is, however, a significant difference in the need and use of computers in forecasting and different sizes of firms. Smaller firms need and use computers less than larger firms.

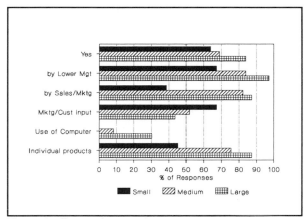

Figure 1. Forecasting practices

Forecasts are mostly in the form of individual products and physical units rather than product groups and monetary terms. For all firms the planning horizon is mostly one year with an increment of either a month or a quarter (an average increment of two months).

Forecasts are modified mostly less than five times a year. There is a significant difference in the number of modifications depending on the size of forms, with the larger firms doing more frequent modifications. More than half the sampled corporations give ranges to the forecast, with ranges given usually to the higher side of the forecast. Most firms do not measure forecast error. Forecasts are used mostly in production planning, materials planning and sales planning. However, forecasts are of less importance for production planning in smaller firms.

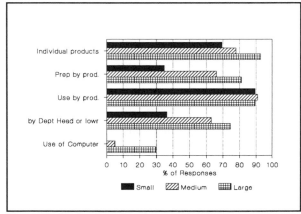

Figure 2. Production planning practices

PRODUCTION PLANNING AND SCHEDULING

The most important information items used to produce production plans are the forecast, actual orders and customer plans. The production plan is produced mostly in individual product and physical units. (See Figure 2.) In smaller companies, production planning by product groups gets higher percentage values. Planning horizons are mostly one year with a modal increment of one month. The functional group and position of people who produce production plans are mostly section chiefs or department heads in the production/engineering group.

Production plans are used by the same functional groups and positions that prepare them. However, in preparing production plans, there is more involvement from administration/planning groups and sales/marketing groups in smaller firms. In a high percentage of the small firms, there is involvement by the CEO. There is a significant difference in the position of people using production plans among different sizes of firms. Unlike the preparation of production plans, there is more involvement from lower level management in the use of the production plan in smaller firms. There is also a higher percentage of involvement by Presidents and Vice Presidents in smaller firms. The survey indicates that there is an evenly spread involvement from all levels of management in smaller firms.

Production plans are used mostly for operation activities such as material/inventory planning and operations scheduling rather than managerial activities such as budget preparation and facilities planning. Also, production plans are revised almost every month, more frequently than forecasts and larger firms revise more frequently. Only about ten percent of the firms use computers in producing production plans, although about half of the sampled corporations said they need computers. The percentages of need and actual use of computers are significantly lower in smaller countries.

SHOP FLOOR CONTROL

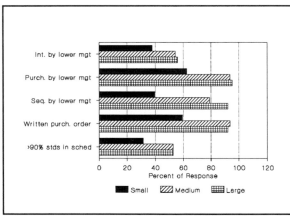

Figure 3. Involvement in shop floor practices

Production is initiated mostly by section chiefs or department heads in the production/engineering group. In "small" firms, there is relatively high percentage involvement by Presidents and Vice Presidents in the initiation of production. (See Figure 3.) The sales/marketing group has some role in starting production, especially in the "medium" size firms.

Purchasing is also initiated mostly by lower management such as section chief or department head in the production/engineering group. The "small" firms have relatively larger roles of higher management including the President and Vice President levels. Purchase orders are transmitted to the supplier by written documents, although smaller firms have fairly large percentages of oral transmissions. Computer transmission is not utilized in Korea.

On the shop floor blueprints are not always attached to the work, but more than half of the work travels with a blueprint. Larger firms tend to have more operations that have time standards, but the differences aren't statistically significant. However, larger firms do more operations scheduling using time standards. It also turned out that, in general, the more operations schedules are made using time standards, the better the delivery performance (i.e. the higher the percentage of use, the more the firms are in the less than 10 percent of orders late category as shown in Table 3). The exception is the "under 20%" standards category where delivery performance is fairly good. Most of the firms view time standards as accurate enough to be useful.

Table 3. Schedule using standards and delivery performance

Use of standards for schedules	<10% of orders late
Over 90%	56.0%
60-90%	47.4%
20-60%	17.6%
Under 20%	46.7%

Engineering change orders are not very frequent in Korean firms. Most firms have engineering change orders on less than 10 percent of their orders.

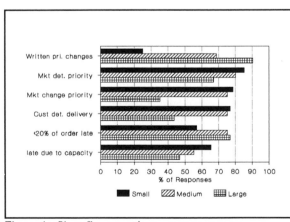

Figure 4. Shop floor practices

Production sequences are determined mostly by section chiefs or department heads in production/engineering groups. In smaller firms they are determined significantly more by higher level management. Priority changes are communicated to workers mostly by written documents. (See Figure 4.) The larger the company, the higher the percentage or written communication. Customer and marketing concerns dominate in establishing and changing shop floor priorities, and in determining delivery dates. Market and customer influence in shop floor activities is even more significant in smaller firms.

Orders are late mostly under 20% of the time. There is no significant difference between "medium" and "large" firms but "small" firms show a higher percentage of orders that are late. When orders are late, they are mostly late by less than 1 week. They are seldom late more than 3 weeks (4.9%). The cause of lateness are capacity problems, material shortages and quality problems. There seems to be some difference in terms of size. The smaller the firms, the higher the percentage of capacity problems and the lower the percentage of material and quality problems.

When there is a need to increase capacity, a variety of means including overtime, hiring and subcontracting are used. In times of capacity decrease, inventory buildup is usually used.

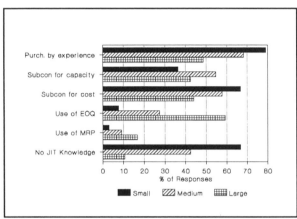

Figure 5. Materials management practices

PURCHASING AND MATERIALS MANAGEMENT

About 50% of the firms' inventory is in raw materials and purchase parts, and the rest is about equally distributed between work in process and finished goods. Purchasing of raw materials is largely based on experience. (See Figure 5.) Systematic management of inventory such as periodic purchase or ROP purchase is not much utilized. Periodic purchase is least utilized. This is especially true in small firms.

More than 50% of the firms have only one vendor per part. This single sourcing is, however, not a common mechanism to guarantee supply. Large quantity purchase (and, thus, holding inventory) is the most common means of supply assurance. Subcontracting is used mainly for capacity expansion and cost reduction. A fairly high percentage of "small" firms indicated no subcontracting. Quality improvement is not important factor for subcontracting. Cost reduction and quality improvement are more important in "large" firms.

As the size of firms gets bigger, there is more exposure to EOQ, MRP and JIT, although the level of exposure is not very high. About 30% of the firms use EOQ, 10% of the firms use MRP and less than 1% use JIT. About 50% of the firms indicated that they have never heard of JIT and MRP.

SUMMARY AND CONCLUSION

This study is based on an analysis of the mail responses and interview data concerning manufacturing practices in Korean machine tool and textile firms. The sample size is believed to be large enough to represent the general characteristics of production management practice in Korean industries. Many in-depth interviews were also conducted to grasp other important aspects that may not be identified through the survey. The results indicate differences between firms of different size and provide the basis for several suggestions.

Differences Due to Size

Several elements of manufacturing practice that show significant differences among different sizes of firms have been identified. The differences lie in organizational aspects, the level of management sophistication, the detail and flexibility of plans and the exposure to technology.

Organizationally, there is more involvement from lower level production and engineering groups in forecasting, production planning and shop floor control in larger firms. In smaller firms, there is relatively high involvement from higher level administration/planning people. There is also relatively high influence from marketing in the initiation of production and in the management of shop flow activities in smaller firms. As the size of a firm gets bigger, priority changes and purchase orders are communicated more by written documents. This indicates that larger firms are more formalized. More use of market/customer information in forecasting by the smaller firms also indicates their informality.

Larger firms are more sophisticated in their management. There are utilizing time standards more in establishing operations schedules and, as a result, show a lower percentage of orders that are late. They are also more systematic in purchasing materials and more strategic in subcontracting. In larger firms, ROP systems are more utilized frequently in purchasing and strategic factors, such as cost and quality, are given more importance in subcontracting.

Forecasting and production plans are more detailed and flexible in larger firms. Their forecasting and production plans are based more on individual product and revised more frequently than those of smaller firms.

Larger firms are exposed more to technology. A higher percentage of them utilize formal techniques in forecasting, need and use computers in preparing forecasts and production plans, and are exposed to material management techniques such as EOQ, MRP and JIT.

Observations and Suggestions

The following general observations and suggestions are made to improve the level of production management in Korea.

Performance measurement is not done well. Performance measurement is basic to management as it tells where we are and makes it possible to set objectives. Improvements can only be possible through the process of setting and achieving objectives. For example, most firms do not measure forecast error. Also, performance measurement is mainly based on simple labor productivity such as volume per man-hr. Measurements related to delivery and cycle time are not available.

Uncertainty and change management capability is lacking. This problem was found in production, but also seems to extend to management as a whole in Korea. For example, many firms do not establish long range plans, do not give ranges to forecast for contingency planning and do not modify forecasts or production plans early or frequently enough to adapt to a changing environment. Because of this, Korean firms place orders without the necessary lead time or change their orders suddenly, making it difficult to meet delivery dates. However, Korean firms meet delivery dates satisfactorily by utilizing inventory and overtime extensively. The interviews indicated that Korean firms tend to take inventory lightly. This practice results in raising manufacturing costs.

Introduction of management techniques and technology is lacking. There are few companies which utilize formal forecasting techniques or computers and there is a lack of understanding of EOQ, MRP or JIT. This may be an indication that production is neglected as compared to other functions in Korean firms. It should be recognized that production is a very important factor to competitive strength, and the most important in these industries. Education to raise the level of understanding of technology and management techniques of production managers has to be strengthened.

Management perspectives to see production as an integrated system are lacking. Forecasting is not well utilized in production planning. Delivery dates are determined mostly by customer without giving sufficient consideration to the production situation. It is not recognized that inventory is correlated to quality and cycle time. Materials management is not integrated in the sense that raw materials, work-in-process, and finished goods are managed in separate departments. As forecasting, production planning, shop floor control and materials management are parts of a system, smooth information flow is required among these parts and the interrelationship among these parts has to be understood.

REFERENCES

1.*   D.C. Whybark and B.H. Rho, "A Worldwide Survey of Manufacturing Practices," Discussion Paper No. 2, Bloomington, IN, Indiana Center for Global Business, Indiana Business School, May 1988.

* This article is reproduced in this volume.

APPENDIX

(Values in percent unless otherwise indicated)

	Total	Small	Medium	Large	significance[*]
FORECASTING					
Yes	70.5	62.1	67.7	83.9	<0.20
By lower mgmt	83.9	66.6	82.6	96.6	<0.05
Sales/Mktg	74.2	38.9	80.4	86.2	<0.05
Input: Mkt/Cust info	51.1	66.7	51.5	42.6	<0.20
Past sales	87.5	91.7	83.3	91.7	No
Tech-Yes	8.6	0.0	6.5	17.2	<0.10
Need computer	49.5	27.8	41.3	75.9	<0.05
Use computer	14.0	0.0	8.7	31.0	<0.05
Indiv. products	72.0	44.4	73.9	86.2	<0.05
Phys. units	58.0	63.6	56.5	57.5	No
Horizon-1 yr.	96.8	100.0	95.7	96.5	No
Increment-month	44.1	44.1	34.8	58.6	No
quarter	52.7	55.6	60.9	37.9	
Revision/yr.	3.5x	2.1x	2.8x	5.3x	<0.05
Range	55.1	64.7	56.8	46.4	No
Error measure	22.8	16.7	21.7	28.6	No
Use: Prod. plan	32.7	26.7	37.0	29.6	No
Material	20.6	30.0	19.8	16.7	
Sales					
PRODUCTION PLAN					
Input: Forecast	44.4	39.5	42.1	51.9	No
Actual order	45.5	50.0	45.5	42.3	
Indiv. product	80.5	70.4	78.7	93.5	<0.10
Phys. units	65.1	74.3	62.8	62.2	No
Increment: 1 mo.	92.5	85.7	93.5	96.7	No
Prep. prod.	64.3	35.7	66.6	73.4	<0.05
Used: prod.	90.2	88.8	91.0	89.3	No
Dept. head/Sect chief	60.6	36.3	63.0	75.1	<0.10
Budget	15.3	17.1	11.1	23.7	No
Facilities	11.04	1.8	4.9	4.3	
Material	31.3	37.1	34.4	18.4	
Scheduling	25.8	25.7	24.4	29.0	
Revisions/yr	10.0x	8.9x	9.8x	11.2x	<0.20
Need computer	47.9	22.2	44.3	77.4	<0.05

(continued)

Appendix, continued

	Total	Small	Medium	Large	significance
(PRODUCTION PLAN, cont.)					
Used computer	10.1	0.0	4.9	29.0	<0.05
SHOP FLOOR					
Initia. prod.	701.6	64.3	65.5	83.3	No
by Mktg	14.3	20.7	6.7		
Pres./VP	11.5	37.5	6.8	0.0	<0.05
Dept/Sect chief	51.3	37.5	54.2	56.7	
Purch. prod.	73.6	46.1	81.1	72.0	<0.05
by Adm.	14.3	46.2	7.6	12.0	
by Lower mgmt	89.5	62.7	94.3	96.2	<0.05
Write Purch.	86.4	60.0	94.8	92.6	<0.05
Oper std>90%	40.8	31.0	43.3	45.2	No
Sched. std>90%	41.7	31.0	45.0	45.2	<0.05
Accurate std	83.5	82.8	82.0	87.1	No
ECO.<10%	76.9	77.8	78.0	74.2	No
Sequence, prod	94.1	93.3	93.2	96.4	No
Adm./Planning	5.9	6.7	6.8	3.6	
by Lower mgmt	73.2	39.3	79.0	93.0	<0.05
Write shop	63.1	24.1	67.7	90.3	<0.05
Mkt/Cust pri	76.8	83.3	79.5	66.7	<0.20
Mkt change	67.2	78.6	75.8	35.7	<0.05
Cust. set del	67.6	76.9	75.0	42.3	<0.05
<20% late ord	71.3	55.2	75.8	77.4	<0.10
1 wk late	57.4	51.7	54.8	67.7	No
Due to cap	54.6	63.4	54.6	46.9	<0.15
Material	28.7	22.0	30.9	29.8	
Quality	9.7	2.4	10.3	14.9	
Cap. incr: hire	28.1	29.6	31.0	20.0	No
OT	50.9	54.6	46.0	57.5	
subcon	21.1	15.9	23.0	22.5	
Cap. decr.: Inv.	73.0	87.5	61.9	87.5	No
MATERIALS MANAGEMENT					
Raw mat'l	50.6	52.3	50.5	50.4	No
Purch. experience	66.9	79.3	69.4	50.5	<0.10
Rop	26.6	20.7	24.2	33.3	
Single vendor	55.7	66.7	58.3	44.0	No
Supply ass. quant.	66.4	79.4	69.4	48.4	No
					(continued)

Appendix, continued

	Total	Small	Medium	Large	significance
(MAT. MGT, cont.)					
Single sourcing	20.5	17.2	17.7	29.0	
Subcon. capacity	46.9	37.5	54.8	42.1	<0.05
Cost	19.7	12.5	6.1	31.6	<0.05
Quality	7.6	0.0	3.2	21.1	
None	13.2	37.5	6.5	0.0	
EOQ use	7.1	27.6	51.7	<0.05	
MRP use	9.3	3.6	8.5	16.7	<0.05
No MRP Know.	48.3	75.0	54.2	12.9	
JIT use	.9	0.0	1.7	0.0	<0.05
No JIT know.	40.6	65.5	43.1	10.3	

* The level of significance is for the differences among different sized firms. χ^2 tests are used. The indication "No" represents a significance level exceeding 0.20.

Changes in Hungarian Manufacturing Strategies

Krisztina Demeter, Budapest University of Economic Sciences, Hungary

ABSTRACT

The relationship between the macro level reforms in Hungary and the micro level reactions at the firm are presented in this paper. The reforms have already had an influence on the manufacturing strategy of companies as they seek to replace lost markets and shifting resources. The details of some of the changes can be seen in the results of the survey on manufacturing practices administered in 1986 and 1990. The conditions are ripe for the continued evolution of manufacturing strategy to meet the changing conditions.

INTRODUCTION

The fundamental changes in the economies in Eastern Europe are widely discussed. There are less words about the managerial foundations of these changes, though it is well understood that macroprocesses are the results of many microlevel decisions.

In this paper, I concentrate on a single field of company operations, manufacturing, and give some indications of the changes which can be experienced in the manufacturing strategy at state-owned or previously state-owned Hungarian companies. These changes have their roots in the environment: the collapse of the COMECON market, together with the declining internal market, the disappearance of state support, high inflation and taxes, and privatization. The emerging of new types of customers, partners, and competitors also made it necessary for companies to redefine their strategies.

The characteristics of manufacturing strategies matching the redefined corporate strategies are described and analyzed in the paper. The analysis is based on a two-step survey of production management in 1986 and 1990 at Hungarian companies and on personal observation of company experience.

I will discuss the connection of macroprocesses and microlevel decisions, especially from a strategic change point of view first. These highly influence the second and main topic, the situation of strategic decisions in manufacturing. Trends are promising, but results show that Hungarian companies have just started to learn the techniques and the way of thinking of the companies in market economies.

CONNECTION OF MACROPROCESSES AND MICROLEVEL DECISIONS

Here are some facts that affect company operations. The COMECON collapsed in 1990, the internal market is continuously declining, the inflation rate is more than 30%, interest rates follow the movements of the inflation, and taxes are also very high. The state gave up subsidizing company operations and started to privatize, partly with the aim to decrease the high budget deficit.

The market squeeze and monetary processes put companies in a very difficult financial situation from one day to another. Their primary strategic goal of growth has been exchanged for survival. They have basically the tools described below to carry out this strategy.

Passive reaction

Hungarian companies have huge amounts of excess resources--employees, capacities, inventories--but no money. Since capacities cannot be reduced very fast, and inventories cannot be sold in the short term because of the market situation, they have no other possibility in the short term than to lay off people or go into bankruptcy. This logical but painful reaction is shown by the unemployment rate, which is now slightly over 10%.

New markets

To avoid the painful alternatives, companies have started very fast to look for new markets. However, since internal markets are declining in almost every segment and COMECON markets are practically eliminated, the only direction is towards market economies. How large are the changes this will cause? Here we look at two groups of companies--machine tool and textile. In 1990, machine tool companies still exported one-fifth of their products to COMECON markets, thus they have to find new markets for this portion of production, plus for the portion they cannot sell in the internal market. (From 1986 to 1990 sales to the internal market declined in value by 40%.) Textile industry exported only 5% of the production to COMECON countries in 1990, but the reduction of the internal market is at the same level as the machine tool. What about the average profits companies reached in 1990? Machine tool: -74 million Hungarian forints (HUF), or about -US$925,000. The textile industry made 21 million HUF (US$262,500). That is, the machine tool industry made losses even though it could sell products in COMECON markets. Since this market is surely the least demanding market (e.g., in quality), the machine tool industry must face an extremely difficult situation if it wants to sell products into market economies.

Machine tool is not the only industry facing this situation, and all Hungarian industries surely realize the need for new markets. The question is how many of them try to look for new markets with a strategic approach for the long run as against dumping into markets using low, unprofitable prices with the only aim to get rid of goods. Those who make attempts at finding long term markets and learning their competitive requirements realize that they have to change both their production technology and their products. But for this reason, they need a lot of money.

Privatization

Privatization came in at the right time. It gives hope of finding a deep pocket. What kind of approaches can be observed concerning privatization? (i) Some companies will look for customers who invest and are satisfied with a minority position within the company.

It is very rare that these companies can find such a customer; no customer wants to take such a risk. (ii) There would be some interest for getting a majority position, but present management does not want to lose its rights. (iii) The company is divided into parts and management tries to sell these smaller units. However, if they hold rights in their hands, the same situation would occur as in case (i). These approaches, and their problems, show why privatization goes so slowly, with so many contradictions.

MANUFACTURING STRATEGIES IN HUNGARY

What consequences can be drawn from the picture above for manufacturing strategy?

Corporate strategy together with marketing strategy should determine the competitive edges (cost, quality, delivery, flexibility, service) of the company and the basis on which manufacturing can make strategic decisions[1], and since this part of the strategy is missing in almost every Hungarian company, we cannot expect too much from manufacturing strategies either. In a situation where top management makes daily operational decisions without a clear strategic perspective, lower level staff and line managers have no other possibility than to execute these hectic decisions.

Now let us look at the manufacturing strategy of Hungarian companies. We will use the results of two surveys on manufacturing management that were carried out in Hungary, one in 1986 and the other in 1990. The survey itself was initiated by Whybark and Rho [2], within the activities of the Global Manufacturing Research Group (GMRG), and directed in Hungary by Chikán [3,4,5,6]. The surveys contained 78 and 77 companies, respectively, mainly from machine tool and textile industries. Since both the sample size and the type of companies are very similar in the two surveys, comparisons from one time period to the next seem to be meaningful.

According to the previous part of this paper I think there is no manufacturing strategy in Hungarian companies. Therefore, these data will show only the lack of this strategy, or, at best, some promising moves towards the strategy. The logic of description follows the model of strategic decisions as outlined by Skinner [7]. That is:
1. Make or buy
2. Capacity
3. Equipment and production processes
4. Number, size, and location of facilities
    4a. Which product is made in which facility
5. Management systems (for organizational structure, information systems, work force management, etc.)

Make or buy

Under the current new conditions, Hungarian companies are much more cost-sensitive when subcontracting than before. This is reflected in Table 1 (the weights of reasons can vary between 1 and 5). In 1990, they had much different reasons for deciding on subcontracting than in 1986: cost, quality, and speed became more important. This shows a change in company operations, from a focus on production load, the quantitative side of production, to the economic side and even the marketing side, which are now as important as the load was before. That is, the pure quantity orientation, where the only measure of success was the number of goods produced, has started to change.

Table 1. Reasons for subcontracting

Reason	Average weight (1-5)	
	1986	1990
Production load	3.3	2.8
Production difficulty	3.2	3.1
Company policy	2.6	2.6
Lower costs	2.9	3.7
Higher quality	2.5	3.3
Faster production	2.5	2.9

Capacity

Due to the fall of sales, capacity utilization rates of companies in the survey decreased from 75% in 1986 to 59% in 1990. This means that lateness in delivery cannot be the result of capacity problems. This is reflected in Table 2.

Table 2. Reasons for late delivery

Reasons	Average weight (1-5)	
	1986	1990
Insufficient overall capacity	2.3	2.0
Labor shortage	3.0	2.4
Production bottlenecks	3.1	3.1
Materials shortage	4.4	3.9
Quality problems	2.8	2.9
Due date changes	2.5	2.8

Overall capacity is really not a problem. However, production bottlenecks still cause a lot of troubles (its weight did not change). The problems with other resources (material and labor) decreased a lot, and capacity utilization as a problem decreased by 16%. The fact that due date changes are more of a problem indicates a shift in customer impact.

Equipment and production processes

In my opinion, the bottleneck concern of Table 2 is one sign of the missing manufacturing strategy: companies have enough capacity but seem not to have the appropriate equipment. Or if they have the appropriate equipment, then they have problems on the organizational side, in scheduling and control. Wrong equipment is a luxury in every case but even more a luxury when machines are quite old. (In the machine tool industry the net value of machines is only 39% of the total gross value; in textile, the situation is a bit better--59%.) In a situation where financial resources are limited, it is critical to buy machines which eliminate bottlenecks. On the other hand, companies cannot afford to lose even more sales because of organizational problems. The importance of concentrating on production processes is shown in Table 3.

Table 3. Reasons for priority changes on the shop floor

Reasons	Average weight (1-5)	
	1986	1990
Pressure from marketing	3.4	3.9
Pressure from consumers	3.3	3.9
Orders from management	2.9	3.4
Changes in delivery dates	3.7	3.8
Sudden surges in demand	2.8	3.6
Manufacturing problems	3.5	3.5
Material shortages	4.3	3.8
Changes in sales plan	2.8	3.4
Engineering problems	2.7	2.9

Up until the survey of 1986, production was determined by the scarcity of material, manpower, and capacity. The task of production management was only to coordinate these resources. We showed that material problems were decreased, manpower problems were eliminated, and the problem of capacity also decreased because of the falling sales. A customer influence increase was also shown. Table 3 shows much more comprehensively that the importance of the customer dramatically increased while other factors decreased or did not change. Companies are forced to promise shorter delivery times, and need higher flexibility to keep their customers. Today it is not enough to coordinate resources, companies have to concentrate on production processes to react faster with lower costs and higher quality than before.

Number, size, and location of facilities

The location of facilities in Hungary up till the 1960s was determined by factors outside the company. Plant locations were governmental decisions, which were based on a centralization process with the aim to simplify central planning. During these years, companies adjusted their organization to this situation. They developed specialized plants and moved work-in-process among them. However, these plants were located in various parts of the country, making it very difficult to coordinate among them. Furthermore, weak logistics, with a poor infrastructural background, did not really help to overcome distances. However, when companies have both more freedom and real economic forces, they can decentralize and separate processes in order to eliminate this unnecessary coordination. They do it by making limited companies or creating profits centers from the existing plants.

Management systems

Concerning the organizational structure, there has been a shift in production's status. Production went up in the hierarchy, which seems to be a good change from the point of view of manufacturing strategy: the higher production is in the hierarchy, the greater the possibility of a strategic approach. The change is shown by Table 4.

Table 4. Production planning in the organization

Responsibility level for production planning	In % of companies	
	1986	1990
CEO	5.0	5.0
Vice President/Director	25.0	35.9
Department/Division Head	70.0	59.1
Group/Section Manager	0.0	0.0

Work force management has not really improved despite the changes in the labor market. Unemployment requires that employees try to find work, rather than companies look for employees. Productivity changed from approximately 1 million HUF/person in 1986 to 1.4 million HUF/person in the machine tool and from 0.8 to 1 in the textile industry, in nominal terms. Actually, productivity declined in real terms. However, the data can be misleading if we do not take into account that this decline may be the result of sales problems. Companies would be able to produce more, but unable to sell more.

The level of computerization has increased, but that does not mean there are better information systems. Data about internal production processes are still very rare and approximate. Management information systems, including decision support systems, still do generally not exist. Although the developments of information technology have already started, there is small attention paid to training. This fact can slow the promising processes down partly because of a natural resistance and partly because of the lack of knowledge the work force would need in an advanced information technology environment.

CONCLUSIONS

Forming a manufacturing strategy is one of the newest industrial concerns, even in market economies. It got a boost because of the Japanese invasion and success in the U.S. market in the 1980s. It is not surprising that in an economy such as Hungary's, where competitive forces started to work only in the last 10 years, there are still some things to do. This is the case not only for manufacturing strategy but in the field of business strategy in general.

Management changes (towards more professional managers) and the critical situation of large Hungarian companies provide the necessary conditions to create strong manufacturing strategies in the near future. There are some signs that show that this process is starting, and also some facts that make it necessary to go farther in this direction. There are still a lot of problems to solve, however.

In this paper, I have shown that customers and marketing have achieved great recognition (the right kind) within company operations. Economic forces have started to affect production down to its make or buy decisions. Proper decisions on capacity, equipment and processes are very critical today when financial resources and markets are very limited, but expectations of customers are higher and higher. Much more attention should be paid to management systems, especially to work force management (e.g., education) and information systems.

A survey of manufacturing strategies at six companies in the U.S. [8] showed that manufacturing strategy started to be developed when a new CEO or a new president arrived at the company. Other reasons included serious problems in the market, profitability or the focus of the firm. These problems are all met today in Hungary. There are new CEOs at many companies, which are also in a very bad marketing and financial situation, thus conditions are ripe for the development of manufacturing strategy.

REFERENCES

1. T.J. Hill, "Incorporating Manufacturing Perspectives in Corporate Strategy," in: C.A. Voss, *Manufacturing Strategy*, London, Chapman & Hall, 1992.

2.* D.C. Whybark and B.H. Rho, "A Worldwide Survey of Manufacturing Practices," Discussion Paper No. 2, Bloomington, IN, Indiana Center for Global Business, Indiana Business School, May 1988.

3. A. Chikán, "Characterization of Production-Inventory Systems in the Hungarian Industry," *Engineering Costs and Production Economics*, 18 (1990), 285-292. (Also presented at the Fifth International Workshop on Production Economics, Igls, Austria, 1988.)

4. Å. Chikán and K. Demeter, "Production and Inventory Management in the Hungari-

an Industry: An Empirical Study," *AULA Society and Economy*, Quarterly Journal of Budapest University of Economic Sciences, 1990/2, 123-130.

5.   A. Chikán and K. Demeter, "In the Attraction of the Market Economy: Manufacturing Strategies in Hungary," in: C.A. Voss, *Manufacturing Strategy*, London, Chapman & Hall, 1992.

6.*   A. Chikán and D.C. Whybark, "Cross-National Comparison of Production-Inventory Management Practices," in: A. Chikán (ed.). *Inventories: Theories and Applications*, Amsterdam, Elsevier, 1990.

7.   W. Skinner, "Missing the Links in Manufacturing Strategy," in: C.A. Voss, *Manufacturing Strategy*, London, Chapman & Hall, 1992.

8.   A. Marucheck, R. Pannesi, and C. Anderson, "An Exploratory Study of the Manufacturing Strategy Process in Practice," in: C.A. Voss, *Manufacturing Strategy*, London, Chapman & Hall, 1992.

* This article is reproduced in this volume.

ACKNOWLEDGEMENTS

A version of this paper is scheduled for a special issue of the *International Journal of Production Economics* containing the Proceedings of the Seventh International Symposium on Inventories, Budapest, Hungary, August 1992.

Manufacturing Management Practices in Bulgaria Under Conditions of a Centrally Planned Economy

Pavel Dimitrov, University of National and World Economy, Bulgaria
D. Clay Whybark, University of North Carolina-Chapel Hill, USA

ABSTRACT

The machine tool and textile industries are economically important to Bulgaria. This study documents some of the manufacturing management practices in a sample of firms from those industries. They are mostly state owned, so are subject to important changes as Bulgaria moves away from central planning. There are few differences in manufacturing management between the industries and the practices are clearly molded by the central planning of the past. Some differences can be attributed to the size of the firm, but substantial changes in practices will be required of all firms to meet the challenges of the future.

INTRODUCTION

This paper is a contribution to the world-wide study undertaken by the Global Manufacturing Research Group (GMRG) to compare manufacturing planning and control practices in different parts of the world and in different economic environments. The purpose is to identify, document, and present in a unified format a wide range of aspects of production management practices in Bulgaria. The paper reports the results of an investigation of the production planning and control practices in the machine tool and textile industries. It is based on questionnaire data from 1989 from a sample of 15 machine tool and 13 textile enterprises.

The machine tool and the textile industries play important roles in the Bulgarian economy. In 1989 the machine building industry, of which the machine tool industry is a part, had a 17.6% share of the gross output of the whole industry (ranking second after the food industry). It held 19.1% of the total industrial fixed assets, which made it the most capital intensive industrial sector. During the same year, the textile industry produced some 5.1% of the gross industrial output, occupying fifth place. It held 3.9% of the total industrial fixed assets [1].

In 1989, most of the output from these industries was attributed to state owned companies. They produced 98.9% of the machine tool and 99.6% of the textile output in that year. The remaining output was produced by co-operative companies.

The manufacturing management practices revealed in the investigation of these industries are valid for a centrally planned economy. As it will be shown later in the paper, the central planning system has had an enormous impact on the overall strategy of the companies and the production management practices in particular. There are, however, some cultural peculiarities that can be identified by cross-cultural comparisons with other GMRG data and some of these are also reported.

The results of this study complement the findings of studies carried out by the Global Manufacturing Research Group in other countries. These studies can also be utilized in more in-depth analysis and comparisons of the manufacturing management practices among countries, regions, cultures, and economic environments (systems).

As Bulgaria moves away from a centrally planned economy towards a market-oriented economy, the manufacturing management practices will dramatically change. In this respect, the findings of this study can serve as a starting point for future comparative studies of the changes in the manufacturing management practices in the transition towards market economy.

THE DATA BASE

The data collection was based on a pre-described methodology, questionnaire, and format outlined by Whybark and Rho [2]. The questionnaire included sections for company profile, sales forecasting, production planning and scheduling, shop floor control, and purchasing and materials management. There were a total of 96 data items requested of the Bulgarian enterprises. Since the Bulgarian data come from a centrally planned economy, there are some peculiarities that have to be taken into consideration when interpreting the results and when using the data for cross-country comparisons.

It is important to note that the Bulgarian data represent responses from enterprises. According to the Bulgarian national statistical office, an enterprise is a legal entity that maintains a separate bank account and a complete independent set of accounting records, including profit-and-loss and balance-sheet accounts. It enters into contracts, has an independent plan of production, and receives and disposes of income. An enterprise may consist of a number of factories. It can act as an independent company or it can be a part of a company that consists of a number of enterprises.

The reason for gathering data for enterprises, using the Bulgarian definition, is threefold. In the first place, the enterprise is the statistical unit used in the collection of national statistics. This permits the comparison of the sample data with the industry (sectoral) data. Secondly, by 1989 the Bulgarian economy consisted of a few very large companies, each of them consisting of a large number of enterprises. Sometimes these companies were combinations of quite different enterprises in terms of product lines, size, import-export orientation, technological level, and so on. Finally, the enterprise has proved to be a very stable organizational unit that can be easily identified if organizational changes take place. The use of the enterprise data has a shortcoming, however, since no matter how independent an enterprise is, its manufacturing management practices are influenced by the practices of the firm of which it is a part.

The Bulgarian survey included enterprises from the traditional machine tool and the non-fashion textile industries. The sample of the machine tool manufacturers is dominated by enterprises producing milling, turning, drilling, and boring machines; lathes; instruments; dies; and engines. The sample from the textile industry is dominated by enterprises producing products like curtains, sheets, blankets, towels, and industrial cloth. In some cases, however, the product lines of the interviewed enterprises included products outside the traditional machine tool industry or non-fashion textile industry.

The data were gathered in personal interviews and field visits. As a rule, the answers to the questions were given by one person--either a deputy director or an experienced department head, who would consult other specialists on specific questions. Thus, the answers to some of the questions may represent one person's opinion rather than the real situation. All the firms were assured that responses would be held in confidence. This was especially important for the reporting of monetary results. As a result of the promises of confidentiality, complete data for a sample of 15 machine tool and 13 textile enterprises have been obtained. These constitute 6.3% and 9.2%, of the total number of enterprises in the machine tool and textile industry of the country, respectively.

Because of the distortions in prices resulting from the centrally planned economy, and because of the artificially low official exchange rate of the Bulgarian leva to the U.S. dollar, the monetary figures in the study have to be used with great caution, especially for cross-national comparisons. The official exchange rate of the leva to the U.S. dollar, which was used for conversion of the monetary figures in the data, is 1 US$ = 2.73 leva.

The data were collected during the second half of 1990 but covers the 1989 operating year. The central planning system was still active in Bulgaria during 1989 and the production was more or less stable. Thus, the data collected represent the manufacturing management practices applied in the environment of a centrally planned economy.

The Bulgarian data provides possibilities for comparative analysis in the following directions: a) between firms (or groups of firms, according to size or other criteria) within an industry and b) between industries. This paper focuses on the between industry analysis. However, as it will be shown later, the industry differences proved negligible. The methodology applied has been used by several researchers [3,4,5,6].

THE ECONOMIC ENVIRONMENT

The economic environment and the cultural peculiarities of a country are among the most important factors that determine manufacturing management practices, as noted by Chikán and Whybark [7]. Knowledge of the economic environment is of crucial importance for understanding the general information about the companies (e.g., size, domestic or foreign market orientation, and inventory structure), and production management practices.

The Bulgarian data base depicts manufacturing management practices of companies operating in the conditions of a centrally planned economy. The central planning system has been active in Bulgaria for 40 to 45 years and exercises an enormous impact on the production management and control techniques as well as on the overall management of the companies. Since 1963 attempts have been made to reform the system. Every three to five years new specific reform measures have been introduced. These measures attempt to increase the flexibility, autonomy, and responsiveness of the companies to market conditions. Despite nearly constant talk of reform, however, there has been little substantial change in the management and planning system of the economy.

The basic features of the economic environment that have a strong impact on the manufacturing management practices of the companies can be summarized as follows:

1) Although on paper enterprises are legally constituted corporate bodies free from any intervention by the state or having any governmental body specify organizational structures, production systems, and management, the planning and regulating functions of the state remained extremely strong. An important aspect of state intervention was the establishment of a unified, obligatory methodology of planning. This methodology established, among other things, the timing of the planning process, the length of the time horizons, the planning indicators, and the methods for their estimation.

2) The state plan played an extremely important role both in the evaluation of the performance of the managers and in the setting of the wages of the employees. The state plan was the major (in most cases the only) customer of the companies. Therefore, the attention of the companies was directed to fulfilling the state plan rather than meeting the requirements of the market. Moreover, because of the state monopoly in foreign trade, companies producing products for export often did not know their actual foreign customers.

3) The national economy was highly monopolized in all sectors--manufacturing, wholesaling, and retailing. Although the companies were given the right to chose suppliers and customers, the high degree of monopolization meant the right was more nominal than real.

4) The economy operated with severe shortages of materials and supplies, requiring the use of a rationing system. The shortages, coupled with the lack of alternative suppliers, lead to deterioration of buyer-seller relations.

5) Prices, exchange rates, taxes and other monetary factors that have a direct impact on company performance, were set by the government. The distribution of a company's income was also determined by the government. As a result, the companies are almost insensitive to costs.

These features of the economic environment led to the establishment of manufacturing management practices that may seem quite peculiar to western managers and researchers.

GENERAL COMPANY DATA

Table 1 illustrates a comparison between the average industry data and the average data of the companies included in the investigated sample. Like other countries with centrally planned economies, Bulgaria is dominated by large firms. In 1989, an average machine tool company had a total of 624 employees, of whom 483 were factory workers. An average textile company had 828 employees and 717 factory workers.

The general economic indicators show some differences between the two industries. On average, a textile enterprise has a greater number of employees but less investment in machinery and equipment. (The machine tool industry has nearly twice as much capital investment per worker as the textile industry). Correspondingly, the labor productivity (as measured by sales per employee) in the machine tool industry is higher and the capital productivity lower.

Table 1. General industry and sample data

Indicator	Machine tool		Textile	
	Industry*	Sample	Industry*	Sample
Number of enterprises	243**	15	142	13
Average number of employees	624	743	828	917
Average sales***	6110	7821	7109	7940
Equipment***	4926	5951	3907	4066
Sales/employee***	9.79	10.53	8.59	8.66
Sales/capital $	1.24	1.31	1.82	1.95
Capital $/worker	10.20	10.50	5.45	5.47

* Estimated on the basis of the Statistical Yearbook of the Republic of Bulgaria, 1990, pp. 141-164
** Only enterprises from ISIC Sector 382 are included.
*** In thousands of $US

The industries are quite similar in their export orientation. Exports average 31% of total sales in the machine tool industry and 23% in textiles. Most of these exports are to the Soviet Union and other Council for Mutual Economic Assistance (CMEA) countries. Exports to Western countries are very small. Over 85% of the production in both industries is produced to order or to state plan. The remaining 13-14% is produced to stock. A majority of the respondents in both industries indicated that during the last two years the annual growth rate of production in physical terms has been in the range of 0-10%. In several cases, however, the production volume has been declining.

In general, the comparison between the industry and the sample data for the respective sectors indicates that the sample is quite representative, although the companies included in the sample are slightly outperforming the average industry indicators. The sampled firms have larger size, higher labor and capital productivity, and higher capital intensity.

SALES FORECASTING

Figures 1 and 2 depict the major characteristics of sales forecasting practices. The most important subjective factors in sales forecasting seem to be the general company and industry situation and the general economic and political situation. Market research is not extensively practiced. Forecasts are based mostly on past sales and current order backlog. The prevailing forecasting methods used are projections and time series analysis. The share of the responses indicating the use of formal techniques is quite big. However, simple projection is the most often used forecasting technique, being chosen by 47% of the machine tool and 38% of the textile companies. Forecasting, as a rule, is done for individual products both in physical units and monetary terms.

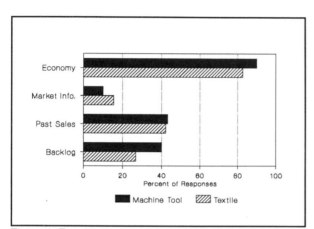

Figure 1. Forecasting inputs

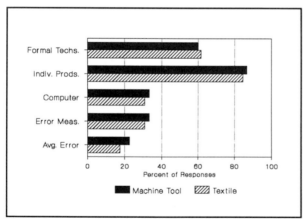

Figure 2. Forecasting techniques

In all cases, forecasts are made for one year with quarterly increments. This reflects the methodology of the central planning system. The forecasts specify ranges rather than exact values. The forecasts are updated an average of twice a year in both industries. About one third of the respondents from both industries claim to use computers in forecasting. The low number of companies that reported their forecast errors and how they were measured suggests that verification of the forecasts is not a regular practice. Forecasts are utilized, in order of importance, for material planning, production planning, budget preparation and, in a few cases, for labor and facilities planning (Figure 3). There are no substantial differences in the forecasting practices between the industries.

It appears that forecasting is an underdeveloped managerial function in Bulgaria. Forecasting is not considered very important when more than 85% of the production is to order or state plan. Moreover, the forecasts must conform to the requirements of the central planning functions rather than to company needs. One of the most visible consequences of the centrally planned economy is that Bulgarian companies lack skills and

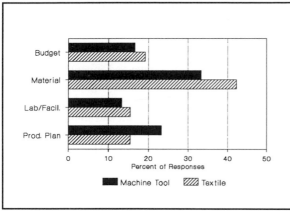

Figure 3. Use of forecasts

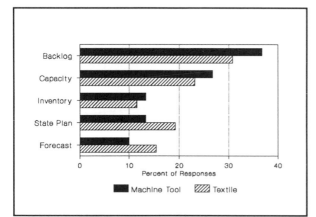

Figure 4. Production planning factors

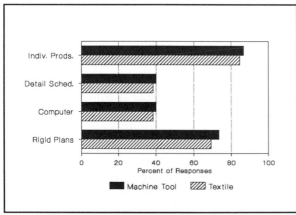

Figure 5. Production planning practices

experience in marketing and market research. These are, of course, key skills for developing useful forecasts.

PRODUCTION PLANNING AND SCHEDULING

Production plans are developed on the basis of the actual order backlog and production capacity. The state plan plays an important role in production planning for the companies producing to state plan (Figure 4). The forecast and inventory level have little importance for production planning.

In all cases, production plans are developed for a year with monthly increments, mostly for individual products, both in physical units and monetary terms. The development of plans in monetary terms explains the great involvement of the administrative/planning department in the production planning activity. Six companies in the machine tool industry and five in textiles prepare more detailed monthly schedules with weekly increments. In both industries, production plans are updated on average three times per year and nearly 40% of the respondents claim to use computers. However the use of computers is moderate and most of the work is done by hand. All of the respondents find the use of computers in production planning useful. Plans are compared with the actual results in all cases. This is not surprising, since this is one of the most important performance indicators of the central planning system. Production plans are quite rigid and considered difficult to change (Figure 5).

Production plans are used in both industries mostly for materials and inventory planning and for subcontracting. Other uses of production plans (e.g., purchasing, scheduling and budget preparation) are less important and somewhat different between the two industries (Figure 6). The domination of the central planning system in the planning function of the companies is obvious. The focus on material and inventory, the length and increments of the plans, and the performance against plan all are derived from the central planning. This also explains the virtual neglect of the market in the planning process.

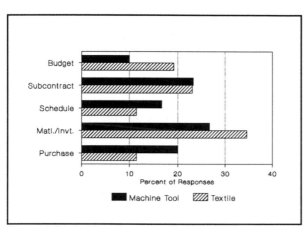

Figure 6. Production plan uses

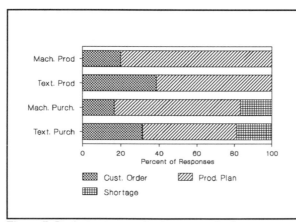

Figure 7. Basis to start production or purchasing

SHOP FLOOR CONTROL

The basis for starting production and purchasing materials is quite similar in the machine tool and the textile industries (Figure 7). Production plans and schedules play a substantial role, followed by customer orders. Shortage lists are also used for starting purchasing in both industries. Smaller firms in both industries are more likely to respond to customer orders. Inventories play almost no role in initiating production or purchasing. Purchase orders are transmitted exclusively in written form.

Scheduling and costing in most cases are based on time standards, although a substantial part of the standards are considered inaccurate (Figure 8). Size also makes a difference, with larger firms tending to have more operations that have time standards. With only a couple of exceptions companies in both industries report that engineering and design changes occur on less than 10% of the orders after production has started.

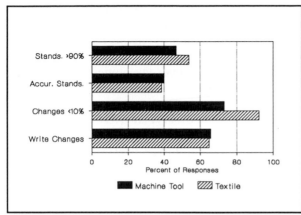

Figure 8. Shop floor practices

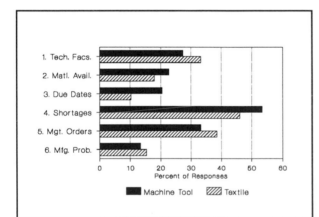

Figure 9. Setting (1-3) and changing (4-6) job priorities

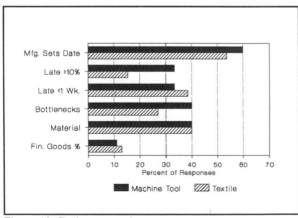

Figure 10. Delivery practices

Technological factors (processing time required, similarity of set-up times, and manufacturing problems) and material availability dominate the establishment and change of shop floor priorities (Figure 9). On average, 65-66% of the changes in priorities are communicated to the production workers by written documents. Marketing and customer concerns play a minor role on the shop floor. Very often production orders are launched according to the availability of materials and supplies rather than in response to changes in the market or customer demands.

The delivery dates to customers are determined mostly by the manufacturers (60% of the machine tool and 54% of the textile companies) and negotiation (Figure 10). Not a single company reported that delivery dates were determined by customers. Only 33% of the machine tool companies and 15% of the textile companies reported that more than 10% of their orders were late. Late orders average less than 3 weeks late, although the average is slightly higher in the machine tool industry than in the textile industry.

The low percentage of the late orders can be explained by the fact that delivery dates are determined by the manufacturers. Under current legislation, an order is considered to be delivered on time if it was shipped in the period (usually a quarter) during which delivery was specified. Even in exceptional cases, the delivery period is still a month. Material shortages and production bottlenecks are the most important reasons for the lateness of orders. Another important factor in the textile industry is the lack of labor capacity, rarely a problem in the machine tool industry. Low finished goods inventories also contribute to lateness.

The most common methods used to meet increases in demand are overtime and building a backlog of orders. In a few cases subcontracting to other companies is used. The reaction to decreases in demand are: undertime (in a few cases reduced work time) and building inventories. Hiring or laying off workers, and leasing capacity are not practiced at all.

The influence of the centrally planned economy is seen again in the shop floor control function. The absence of market or customer influence on manufacturing priorities and on production initiation, and the prominence of technical and material considerations in shop floor control both stem from central planning. The setting of delivery dates by the company and delivery times of a quarter of a year are also vestiges of the central planning system.

PURCHASING AND MATERIALS MANAGEMENT

Bulgarian companies are overstocked. The average inventory turns (sales/inventory) ratio is 1.50 for the machine tool industry and 1.34 for the textile industry, much lower than in western countries. More than two thirds (68% in the machine tool industry and 71% in the textile industry) of the companies' inventory is held as raw materials and purchased parts. Finished goods inventories constitute only 11-13% of the total stock. This inventory structure indicates that Bulgarian companies are acting in a seller's market where the terms of deliveries are dictated by the seller.

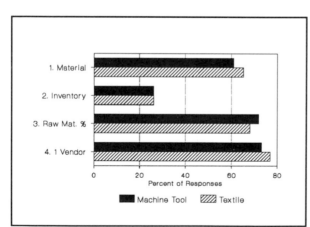

Figure 11. Purchasing involvement (1-2) and practices (3-4)

The situation is aggravated by the supply market's being highly monopolized. Most of the companies in both industries have one vendor per part. Not surprisingly, material shortages and inventory programs are the greatest concern of the purchasing departments in both industries (Figure 11). In a few companies, the purchasing departments are in-

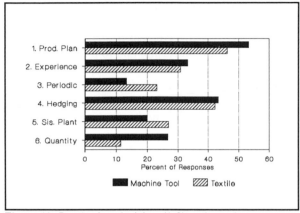

Figure 12. Purchasing decision (1-3) and supply assurance (4-6)

volved in cost reduction programs, as well.

Purchasing is based almost exclusively on the production plan and managerial experience (Figure 12). Some companies indicated that they purchase materials on a periodic basis. In reality, these three methods are applied simultaneously and complement each other. Hedging and large quantity purchasing, augmented by purchasing from sister plants, are the most frequent actions to assure the timely supply of materials and purchased parts. The companies' purchasing strategy is to create high input inventories in order to protect themselves against long supply lead times and delivery interruptions. Production difficulties and cost considerations are the main reasons for subcontracting outside. However, production load and quality are also quite important.

The economic environment in which the companies operate makes the application of modern materials management techniques nearly useless. The companies are not even using the EOQ, let alone concepts like MRP or JIT. The vast majority of the companies responded that they had not even heard of MRP or JIT; the rest naturally considered these concepts inapplicable. The implications of the central planning system are again illustrated by the need to accumulate large levels of raw material inventories and to concentrate on protecting production rather than on serving the market.

ORGANIZATIONAL STRUCTURES AND FUNCTIONS

In this section two organizational aspects of the production management practices are analyzed: the role of the various functions in production management decisions and the level at which these decisions are taken.

Figure 13 depicts the role of the functional groupings in the

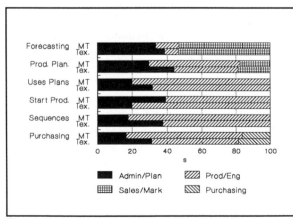

Figure 13. Functional involvement in production management

various production management areas. Two functional groupings are heavily involved in the production management decisions of the companies: production/engineering and administration/planning. The data indicate very low involvement of the sales/marketing functional grouping. Sales/marketing is involved in forecasting and to a minor extent in production planning. The finance/accounting group is not involved in production management decision-making at all. Again, there are not big differences between the two industries. However, it seems that administration/planning has a bigger influence in all areas of production management in the textile industry than in the machine tool industry.

Forecasting is dominated by the sales/marketing department. Administration/planning comes second and production/engineering third. Involved in production planning are administration/planning (greater in textile) and production/engineering (greater in machine tool). Production plans are used by production/engineering and by administration/planning. In the shop floor control activities (starting and sequencing production), production/engineering plays the dominant role. Again, the involvement of the administration/planning department is higher in the textile industry than in the machine tool industry. The purchasing department joins production/engineering and administration/planning in the initiation of purchasing. Unlike in many western companies, the materials management activities of Bulgarian firms are not integrated into a single organizational department.

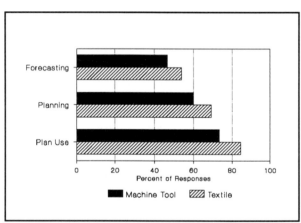

Figure 14. Involvement of department head or lower

The manufacturing management decisions are not taken at the highest levels of company management. This is shown by the involvement of the Department/Division Head or lower management depicted in Figure 14. The use of lower levels is more prevalent in the textile industry than in the machine tool industry. Forecasting is done at a higher level of management than the other managerial functions.

There are some differences in both functional involvement and managerial decision level related to the size of the companies. In smaller companies the production/engineering function plays a more important role and the management decisions are allocated to higher managerial levels. The level of education of the people involved in production management is quite high for all firms in the sample.

SUMMARY AND CONCLUSIONS

No matter how great the differences in products and processes between the two industries, the manufacturing management practices are very similar. The major reasons for the similarities can be found in the unified methodology of the central planning system and the strong influence of the economic environment in which the companies operate. The consequences of these phenomena are:

1) Companies from both industries operate in a highly monopolized seller's market and have adapted their manufacturing management practices to the requirements of that market. Purchasing, material supply, and material shortage considerations play an extremely important role in the decisions in all areas of production management. The purchasing strategy of the companies is to build high input inventories in order to protect themselves against the unreliable supply.

2) Customer considerations play almost no role in any of the areas of production management. The share of the production to stock is almost negligible. Production to order or state plan is by far the most frequent production form. As a consequence, the marketing function is underdeveloped and has little importance. Correspondingly, the manufacturing planning and control system is inflexible and unresponsive to customer needs or demand fluctuations.

3) Despite the high educational level of the persons involved in production management, the managerial methods and techniques applied are outdated and largely based on experience. The knowledge of MRP and JIT is extremely low and the application of computers is limited.

Some differences in manufacturing management practices can be ascribed to the size of the firms. In bigger companies manufacturing planning and control is exercised with greater flexibility, production plans are made in more detail, more formal techniques and computers are used, and the involvement of lower level managers is higher. Bigger companies may be able to remain less flexible to market changes because of their size (greater bargaining power) and their historical connections in the industry.

The manufacturing management practices detailed in this paper describe the starting point from which the introduction of a market economy in the country will depart. During the last several years, these practices have experienced dramatic changes in an attempt to adapt to a completely new economic environment. The planned future studies of the GMRG will give, among other things, deeper insights into this transformation process.

REFERENCES

1. Statistical Yearbook of the Republic of Bulgaria, 1990, pp. 141-164.

2.* D.C. Whybark and B.H. Rho, "A Worldwide Survey of Manufacturing Practices," in D.C. Whybark and G. Vastag (eds.), *Global Manufacturing Research Group, Collected Papers Volume 1*, Chapel Hill, NC, Kenan Institute, University of North Carolina, March 1992.

3.* A.E. Kovacevic, J.C. Lopez, and D.C. Whybark, "Manufacturing Practices in Chile," Discussion Paper No. 40, Bloomington, IN, Indiana Center for Global Business, Indiana Business School, April 1990.

4.* B.H. Rho and D.C. Whybark, "Manufacturing Practices in Korea," Discussion Paper No. 43, Bloomington, IN, Indiana Center for Global Business, Indiana Business School, April 1990.

5.* B.H. Rho and D.C. Whybark, "Comparing Manufacturing and Control Practices in Europe and Korea," *International Journal of Production Research*, 28, No. 12 (Fall 1990), 2393-2404.

6.* B.H. Rho and D.C. Whybark, "Comparing Manufacturing Practices in the People's Republic of China and South Korea," in D.C. Whybark and G. Vastag (eds.), *Global Manufacturing Research Group, Collected Papers, Volume 1*, Chapel Hill, NC, Kenan Institute, University of North Carolina, March 1992.

7.* A. Chikán and D.C. Whybark, "Cross-National Comparison of Production-Inventory Management Practices," *Engineering Costs and Production Economics*, 19, 1-3 (May 1990), 149-156.

* This article is reproduced in this volume.

Manufacturing Practices in the Soviet Union

Alexander Ardishvili, IMEMO, Academy of Sciences, Russia
Arthur V. Hill, University of Minnesota, USA

ABSTRACT

The purpose of this paper is to identify and compare the production planning and control practices in the small machine tool and non-fashion textile industries in the Soviet Union. The survey results provide valuable background information for researchers and practitioners who are seeking ways of helping the Soviet economy move towards a more effective system. The findings on the emerging linkages between production and the market are particularly interesting and provide insights into Soviet manufacturing practices that will be of use to prospective international joint venture partners. In addition, insights from the Soviet experience also help scholars develop a more general theory of manufacturing excellence that can be of use in any manufacturing context. The study is particularly timely as the Soviet command-control systems disintegrate and many world leaders are beginning to promise technical assistance to help restructure the Soviet economy.

INTRODUCTION

This paper presents an analysis of the manufacturing planning and control practices in the small machine tool and non-fashion textile companies in the Soviet Union. Forty-nine companies were surveyed in March and April 1991. The survey is a part of a world-wide project being conducted in North America, South Korea, Japan, the People's Republic of China, Australia, Chile, western and eastern Europe by the Global Manufacturing Research Group. General information about the project can be found in a papers by Whybark [1] and Whybark and Rho [2].

This paper concentrates exclusively on the data from the Soviet Union. Subsequent papers will compare Soviet manufacturing practices to those in other countries.

The survey results are useful for four main audiences. First, the results reveal problems in Soviet manufacturing practices in general and in the two industries in particular. This information is of value to researchers, educators, and policy makers who are seeking ways to provide technical and educational assistance in converting the Soviet economy to a more effective system.

Second, the results may be useful to individual companies in the Soviet Union by providing them with valuable information for comparison of their practices with those of other companies within the country and particular industry.

Third, the survey is connected with the issue of international joint ventures. Presently, when considering joint venture opportunities in the Soviet Union, foreign companies tend to analyze the general political and economic situation in the country, legislative climate, financial projections for the project, and level of technical skills of the prospective partner. Very little attention is paid to the peculiarities of manufacturing practices in the Soviet Union. At the same time, the experience of joint ventures in other countries indicates that the long-term success of the project depends heavily on compatibility of manufacturing philosophies and practices of both partners and on their ability to find common ground for cooperation. The necessity of such understanding is particularly important in the case of the Soviet Union, where manufacturing practices are in many cases strikingly different from western practices. These differences are conditioned not only by cultural background, but also by the sociopolitical legacy of the last 70 years. The authors of the survey had difficulty in explaining some basic concepts and assumptions to the respondents; this fact in itself is evidence of the above assertion.

Fourth, studying Soviet manufacturing can help academics who study manufacturing to better understand the essence of what makes for a good manufacturing operation no matter where it is found. Insights from the Soviet experience lead us to develop a more general theory of manufacturing excellence.

This study is particularly timely as the Soviet Union dissolves and many world leaders are promising technical assistance to help restructure the Soviet economy.

We begin with some background on Soviet production management. The survey methodology and results are then presented. The results sections include company profile, sales forecasting, production planning and scheduling, shop floor control, and purchasing and material scheduling. We will then identify the most important problems and discuss possible solutions to these problems.

BACKGROUND ON SOVIET PRODUCTION MANAGEMENT

The pre-reform Soviet economy was characterized by centralization of planning and decision-making which resulted in a passive role for enterprise management. Central distribution of scarce inputs and outputs made the task of negotiating with central planning authorities in order to secure needed inputs and minimize output quotas one of the most important activities of an enterprise manager [3].

A significant characteristic of the Soviet economy (and of any socialist economy) is a "soft" budget constraint. A budget constraint is "soft" when the strict relationship between expenditure and earnings is relaxed and excess expenditures of an enterprise are covered by some other organization (e.g., the state in a socialist economy). Hungarian economist Kornai [4] argues that the existence of soft budget constraint is associated with the paternalistic role of the state towards economic organizations. This leads to the absence of financial incentives for quality improvement, technological innovations, timely deliveries, or minimization of inventories.

Another important characteristic of the Soviet economy is the shortage of producer and consumer goods. These shortages are caused by the practices of input maximization, output minimization, and pricing followed by Soviet enterprises [5]. The shortages result in hoarding materials and excessive inventories.

The economic reform which started in 1987 has given the enterprises more freedom in pricing and production planning, increased financial responsibility, and allowed for establishment of direct contacts between enterprises. However, the halfway character of the reforms has not eliminated many of the major drawbacks of the existing system and some new problems have appeared including: disrupted centralized channels of distribution of supplies, severe shortages of nearly everything, absence of legal mechanisms for enforcing direct contracts between enterprises, and continued dependence of state-owned companies on central planning and distribution authorities.

Our goal is to understand to what extent the above general conditions affect the production management practices. This understanding may enable us to determine what changes are needed to facilitate introduction of a new production management system that is more compatible with a market-driven economy.

SURVEY METHODOLOGY

Our survey instrument was a Russian translation of the Global Manufacturing Research Group (GMRG) survey which has been used in several regions of the world. The survey includes 65 questions, many of which have several parts. We added explanations for many questions in order to make the survey understandable to Soviet managers.

The GMRG survey instrument was translated from English into Russian by the first author and was carefully checked by our colleagues at Moscow State University. The data gathering was organized by the Center for Opinion Research (COR) at the Moscow State University. The population chosen included machine tool companies and non-fashion textile companies (manufacturers of curtains, sheets, blankets, towels, etc.) from all major regions of the Soviet Union. The sample of companies was randomly selected from the COR database which consists of more than 160,000 records.

The COR sent out Russian-language mail questionnaires to 400 companies. However, the extremely low response rate led us to abandon this approach. Teams of interviewers were then sent to interview key managers in the selected enterprises. This method was much more successful. The final sample included 18 machine tool companies and 31 textile companies. About 80% of the people interviewed had the title of either "general director" or "director". The general procedure involved the director calling in specialists to help answer the survey questions.

This is the first time that Soviet manufacturing data of this type has ever been collected and published in the west. This effort was achieved only through tremendous effort on the part of our colleagues at COR who had to break through many barriers to gather this data and to get it out of the country.

The survey instrument included company profile questions and questions on the four major areas of production planning and control: 1) forecasting, 2) planning and scheduling, 3) shop floor control, and 4) purchasing/material management. The results and analysis for each section follows.

SURVEY RESULTS

This section will present the survey results with some interpretation of the results. Section 5 will analyze the problems in more depth.

Company Profiles

Average sales, average number of employees, and average number of factory workers for the Soviet machine tool companies in our sample are substantially less than those of the textile companies. (See Table 1 and Figure 1.) The number of employees varied significantly (from a minimum 50 employees in both industries to 10,930 in textiles and 3,638 in machine tools). Sales also varied widely (from less than $1 million sales to $329 million).

Table 1. Company profiles

	Overall	Machine tool	Textile
Number of respondents	49	18	31
Average number of employees	1245	920	1401
Average number of production workers	981	576	1176
Average sales in 1990 ($millions)	44	15	59
Average exports in 1990 ($millions)	0	0	0
Average investment ($ millions)	4	3	6
Average days of training/person/year	3	4	3
Average capacity utilization	75%	68%	78%
Percent make-to-stock	15%	31%	7%
Percent make-to-order	54%	42%	61%
Percent make-to-plan	35%	42%	32%

Export sales for both industries indicated that the majority of companies surveyed were oriented exclusively towards national markets. Only four companies in the total sample had any exports last year (three of these were in the machine tool industry). Not a single respondent reported annual export sales of over $1 million.

Training of employees is apparently not a high priority for either industry. Twenty-seven percent of the machine tool respondents and 58% of the textile respondents reported that

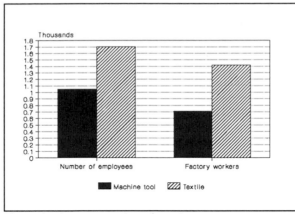

Figure 1. Average number of employees

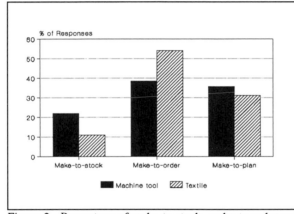

Figure 2. Percentage of make-to-stock, make-to-order and make-to-plan production

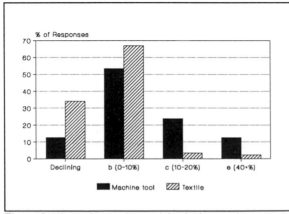

Figure 3. Annual growth rate (physical output)

they did not have any employee training programs at all. The average number of days spent on training per employee per year was between four and five days for both machine tool and textile industries.

Most companies relate to their "customers" as a make-to-order or make-to-state plan supplier. (See Figure 2.) The percent of make-to-stock production was higher for machine tools than for textiles (23% compared to 12%). Many respondents treated orders from other companies and make-to-state-plan orders as identical.

Both industries were experiencing very slow sales growth. Ninety percent of respondents indicated that sales were either declining or growing at a rate between 0 and 10% annually. The situation in textiles was particularly bad. Only one textile company reported growth in the 10 percent or more per year category whereas 34% indicated a decline in sales. (See Figure 3.)

Sales Forecasting

The two most important factors in forecasting sales were customer information/market research (57% overall) and past sales (71%). Fifty-three percent of companies surveyed do not use any formal techniques for forecasting. Forty-two percent use simple projections. Both percentages were about the same for the two industries.

Forecasts were used for production planning by 53% of the respondents and for sales planning by 45%. Textiles used forecasts more in production planning, whereas machine tools used forecasts more for sales planning.

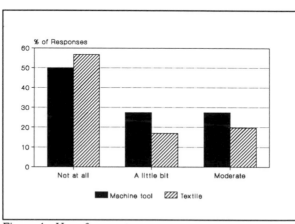

Figure 4. Use of computers

Production Planning and Scheduling

The two most important factors in developing production plans for both industries were the actual customer orders and production capacity. There was a difference between the industries with respect to the functional group which prepares production plans. In textiles this was mainly the Administrative/Planning department (87% of responses). In machine tools, the Production/Engineering group was involved more often (53%). The role of Sales/Marketing departments was insignificant in both cases. This is very different from companies in market-driven economies.

Fifty-four percent of the companies surveyed did not use computers for planning (50% in machine tools and 57% in textiles). (See Figure 4.) This was due to the lack of availability and cost of installation of computers more than reluctance to use them. Ninety-five percent of the respondents believe that computers would be helpful.

Production plans were used more for managerial activities (such as budget preparation and facilities planning) than for operational activities (such as material/inventory planning or scheduling). Plans were revised on average only 2.6 times a year with little differences between the two industries. The modal production plan for both industries had a 12-month horizon. It was divided into increments of 1.5 months on average.

Shop Floor Control

The two industries were substantially different with regard to the functional group that initiates the production. In textiles, it was mainly the duty of the Administrative/Planning department (76% of companies) whereas in machine tools it is primarily Production/Engineering (61%).

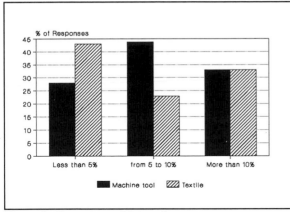

Figure 5. Percentage of late orders

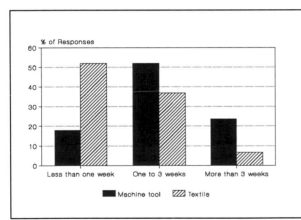

Figure 6. Average lateness

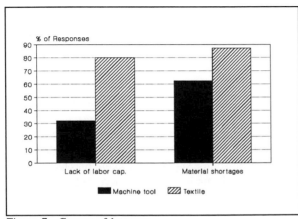

Figure 7. Causes of lateness

The most important factors in establishing shop floor priorities were customer orders (48% overall) and production plans (52% of all companies). In machine tools, customer orders played a much more significant role (72% of companies) than production plans (28%). In textiles, the situation was reversed (32% and 68% respectively).

Purchase orders were transmitted to suppliers mostly in written form (about 90% of the responses). The percentage of orders transmitted orally in the machine tool industry was higher (23%) than that in textiles (5%). In the west, Electronic Data Interchange (EDI) is a very popular method for communicating purchase orders and other information. However, none of our respondents have used computers for this purpose.

Fifty-four percent of respondents indicated that 5-20% of orders were late. The difference between the two industries is minor. Most late orders were late by one to three weeks. (See Figures 5 and 6.) The two most important causes of lateness were the lack of labor capacity and material shortages. (See Figure 7.)

When capacity increases were necessary, both industries hired additional workers or used overtime. There was no significant difference between the two industries on this point.

Answers (and the frequent absence of answers) to the questions about methods used in time of decreased demand indicate that such a problem is rare in the Soviet economy at this point.

The delivery date for customer orders was determined mostly by negotiations in both industries. This is quite different from the predominant practice in the west where companies typically quote a standard lead time. The biggest difference seems to be that Soviet companies have very unreliable supply lead times and therefore cannot promise fixed lead times to their customers. This problem is accentuated by the fact that supply lead times are long and therefore less predictable. Negotiation with both suppliers and customers, therefore, is a central part of the way that Soviet companies must do business in the current environment.

Purchasing and Materials Management

About 57% of the inventory was in the raw materials and purchased parts (about 37% in machine tools and 68% in textiles). Finished goods inventories were about 17% without significant differences between industries. Purchasing was based mainly on production plans (76% of responses overall).

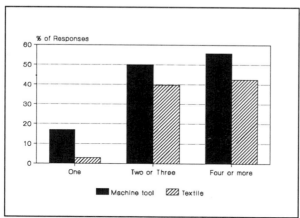

Figure 8. Number of vendors per part

The two most important actions taken to assure the timely supply of materials and purchased parts were long-term contracts and multiple sourcing (77% of overall responses for both variables). But there was a significant difference between the two industries in that, long-term contracts had first priority for textiles (82% of responses) whereas multiple sourcing had first priority for machine tools (85%). The textile industry also used large quantity purchases (41%).

Averages for both industries indicate that the majority of companies had four or more vendors per purchased part. (See figure 8.) It is very interesting to see that this command-control economy tends not to have a sole-source supplier for many purchased parts. Evidently the uncertainty of supply gives manufacturers high incentive to have multiple vendors.

The level of exposure to current concepts and methods for operations management was generally very low. About 80% of all respondents indicated that they have never heard of the JIT (Just-In-Time) concept. In the textile industry this percentage was even higher (90%). The remainder of the respondents claimed to know something about the concept; but even these respondents did not see any possibility of implementing JIT at this point.

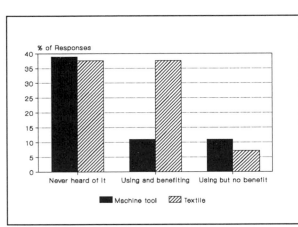

Figure 9. Exposure to MRP

Forty percent of respondents have never heard of MRP (Materials Requirements Planning) with almost no difference between the two industries. Twenty-seven percent claim to be using it and benefiting from it. However, we speculate that those who claim to understand MRP and/or use MRP are probably not using MRP as we know it in the United States and Europe.

Fifty-eight percent of the respondents claim to use the EOQ (Economic Order Quantity) concept. More textile respondents (67%) claimed to use EOQ than machine tool respondents (44%). These findings are summarized in Figures 9 and 10.

Inventory held for protection against unexpected demand fluctuation was 2.9 weeks of demand on average. The safety stock was larger in machine tools (4.2 weeks) and smaller in textiles (2 weeks).

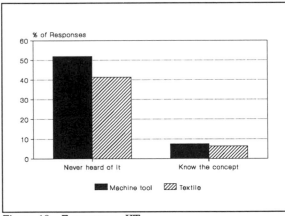

Figure 10. Exposure to JIT

Inventory turnover (annual sales/inventory investment) was much higher in textiles than in machine tools (22 compared to 5). These turnover ratios are not bad by world standards. However, our direct experience with Soviet factories suggests that inventories are generally planned to be as large as possible. We reconcile these two facts with the observation that it is difficult to have large raw materials inventories because supplies are often very difficult to find and finished goods inventories are often kept very low because customers are desperate to take delivery. Typical Soviet manufacturing companies are large vertically integrated firms that value work-in process inventories at the materials cost. As a result, the turnover ratio tends to be higher that it would with western accounting practices.

ANALYSIS OF PROBLEMS

In this section we analyze the problems in the four production planning and control areas (forecasting, planning and scheduling, shop floor control, purchasing/material management) by looking at three periods: pre-reform, reform (1987 to present), and post-reform (the desired condition). Analysis of the reform period is based directly on the survey results presented in the last section.

Forecasting

Pre-reform period: Forecasting was almost non-existent on the enterprise level.

Reform period: Some enterprises make forecasts. However, top executives play a limited role in forecasting, forecasts have limited use in planning and production management, and formal computer-based forecasting techniques are not used.

Post-reform period: Forecasting will be based on market research and computer-based formal techniques. Forecasting will have a significant role in developing production plans and inventory levels.

Planning

Pre-reform period: Enterprise management was passive. Plans were developed based on central plans. Enterprise-level forecasts and market research were not used in planning. The finance and marketing people played a weak role in planning. No department integrated the overall plans.

Reform period: Enterprises have some degree of freedom of management as they change from a reactive to a proactive role, but are still largely dependent on state orders. The need for forecasts is recognized but formal techniques are still not applied. The soft budget constraint continues to exist and marketing and sales continue to play a limited role in the company.

Post-reform period: Planning will be based on forecasts, market research, financial projections, and facilities planning. Production, marketing, and finance will have integrated plans. Companies will be driven by an integrated set of plans at all levels in the planning process. Finance and marketing will play a much more important role than before. Top-level managers will play an important role in developing and monitoring plans.

Shop Floor Control

Pre-reform period: This period was characterized by a "crisis management" approach with extensive use of overtime and crisis mobilization of all resources to meet plan requirements at the end of each planning period. Managers needed to negotiate with higher level organizations and to find informal ways to solve the problems. Managers also needed to get cooperation from the labor force by using authoritarian methods.

Reform period: "Crisis management" still prevails, but managers are driven more by monetary incentives. Negotiating skills play a more important role as the emphasis changes from negotiations with central agencies to direct negotiations with suppliers.

Post-reform period: A pro-active role will become the norm as enterprises learn to prevent trouble rather than "manage by crisis." Use of formal computer-based techniques for manufacturing control will become widespread. Labor management will have more flexibility in combining monetary incentives with persuasion.

Purchasing and Materials Management

Pre-reform period: Materials were not purchased, but rather received through a centralized distribution system. The role of management was to secure a bigger share of allocated resources through negotiations with planning authorities and to get additional resources through informal bargaining with other enterprises. General shortages led to hoarding and excessive inventories.

Reform period: The survey data indicate the presence of typical features of a shortage economy. Companies are hoarding materials because they want to protect themselves against material supply uncertainties and to have something for barter exchange with other companies. In order to get a rough measure of hoarding, we devised a ratio (R) which is equal to the total inventory (in rubles) divided by the finished goods inventory. As can be seen from Table 2, most of the Soviet respondents had a high value of R.

Table 2. Percentage of respondents by raw materials/finished goods ratio (R)

Country/Area	R<2	2<R<5	R>5	Totals
Soviet Union	14.0%	33.3%	52.6%	100%
Hungary	6.5%	32.1%	61.5%	100%
Western Europe	54.0%	26.0%	20.0%	100%

This was also true for a GMRG study done in Hungary by Chikán and Demeter [6]. In contrast, most of the respondents in the western European GMRG surveys had a low ratio. The Soviet data had an average R of 9.4.

Post-reform period: The role of management will require new skills as managers spend less time searching for new sources of materials and bargaining with central authorities and more time managing systems that use formal techniques for managing materials replenishment.

Other Problems

In addition to the observations presented above, the survey enabled us to identify the following problem areas for manufacturing firms in the Soviet Union:

Delivery Problems: The main obstacles for timely delivery are unreliable transportation and production difficulties at the supplying companies. Compared to GMRG data from western companies, the percentage of late deliveries in our sample of Soviet companies was very high. Existing planning systems do not have any mechanisms for correcting these problems.

Negotiations were used as the primary method of determining delivery dates. Long unreliable lead times create a need for bargaining, persuasion, bartering, incentives, and bribes for dealing with both enterprise and ministry management officials. This task is complicated by the absence of economic disincentives for delays.

The Roles of Managers: The involvement of senior management in inventory control was generally high. Inventory control was viewed as the primary means to control overall production. This was particularly true for raw materials inventory management. One of the most important tasks for a senior manager was to overcome delays for raw materials. Soviet managers had little concern for forecasting and planning demand because the state was usually the source of the demand and the state will usually tolerate long lead times. Soviet manufacturing managers rank inventory control/purchasing as first in importance, production planning and control as second, and forecasting as the least important.

The Manufacturing/Market Interface: One of the most important observations from the survey is that there is a weak linkage between manufacturing practices and the market. Sales and marketing people were rarely involved in important decisions and are only rarely mentioned as the users of production plans. Customer plans played little role in forecasting and planning. Make-to-order production constituted a relatively small percentage of overall output. Only 70% of companies claimed to revise their plans according to the fluctuations in demand.

The machine tool industry was more market-driven than the textile industry. To illustrate this point, we have chosen several questions related to the manufacturing/market interface (Table 3). Responses on all of these questions indicate that neither industry was particularly market-driven and that the machine tool industry is more market-driven than the textile industry.

Table 3. Sample survey questions related to the manufacturing/market interface

	Overall	Machine tool	Textile
What group produces the sales forecasts?			
Administrative/Planning	37	31	40
Production/Engineering	39	54	32
Sales/Marketing	29	31	28
Who orders to start production?			
Administrative/Planning	50	15	70
Production/Engineering	47	62	35
Sales/Marketing	19	31	13
What is the basis for the order?			
Customer order	49	62	36
Production plan	51	31	64
What group orders the purchase of material?			
Administrative/Planning	38	15	50
Production/Engineering	57	54	54
Sales/Marketing	14	23	8
What is the basis for purchasing?			
Customer Order	50	53	29
Production plan	71	53	76
Production schedule	12	8	14
Shortage list	6	7	5
Inventory position	3	23	4
What group establishes the sequence for releasing jobs?			
Administrative/Planning	33	46	25
Production/Engineering	75	54	85
What is the basis for setting priorities?			
Processing time	28	23	23
Similarity of set-up	49	54	45
Marketing Preference	37	46	27

(continued)

Table 3, cont.

	Overall	Machine tool	Textile
What is the most frequent method for purchasing raw materials?			
Periodic basis	14	23	5
Fixed amounts	14	15	14
Based on experience	19	31	9
For specific orders	8	0	9
According to a plan	78	54	91
Distributed by the state	11	0	14

CONCLUSIONS

As we mentioned in our introduction, the results of this study should be of use to a wide variety of audiences including 1) Educators, 2) Soviet Factory Managers, 3) Managers of Joint Ventures in the west, and 4) Operations Management Researchers seeking to develop a more general theory of operations management. We will organize our conclusions for each audience in the following sections.

Educators

Providers of education for Soviet managers have a great challenge ahead. Our study found that Soviet factories need a massive restructuring before the "post-reform" ideal is achieved. Restructuring is needed for both production planning and control systems and the way that Soviet managers think about production planning and control.

Clearly the most urgent requirement for manufacturing management education in the Soviet Union relates market-driven systems. Current systems in the Soviet Union are ineffective at assessing customer needs. This indicates that one of the goals of management development should be education in the use of market research (and marketing in general), forecasting techniques, order entry, master scheduling, logistics, and basic inventory management concepts.

The role and methodology of planning at the enterprise level will have to change dramatically. Master scheduling should not be based on central plans, but rather on the assessment of market demand, open orders, competition, and financial projections. Management must be able to link marketing, finance, and production plans in order to develop integrated plans throughout the company. The survey showed that no such integration exists at this point. Neither marketing nor finance currently play a significant role in the production planning process, managers do not integrate plans, and plans are seldom revised.

Substantial changes will be required in purchasing systems. Managers have to learn to develop stable supplier networks and to rely on a smaller number of suppliers than before. Attitudes and practices concerning setting delivery dates, dealing with lateness, and purchasing should be changed. Our survey reveals that the current methods are completely conditioned by the distortions of a shortage-plagued command control economy. Just-in-time principles need to be taught now so that they can be implemented as soon as they become feasible in the Soviet economy.

Production control practices also need serious consideration. At present, detailed plans for months, weeks, and days are not existent in many enterprises. Formal methods of control are not used and non-production departments are only minimally involved in this process.

A new approach to inventories is needed. In the past, capital tied up in inventory was not of much concern to managers. This can be explained by the soft budget constraint. In the new environment, managers must have a better appreciation of the cost of inventories and the cost of capital. They must know how to make economic tradeoffs related to inventories, lot sizes, transportation options, etc. and know how to use financial information intelligently.

Soviet Factory Managers

It is clear that the Soviet companies in our study are in the embryonic stage of the development of market-driven mechanisms for production planning and control. Soviet manufacturing companies (enterprises) must begin to develop and implement mechanisms for linking production with the market in order to speed up the transition toward a market-driven economy.

These changes in the Soviet Union should be supported by more extensive use of computers and formal techniques in forecasting and planning and by greater exposure to up-to-date methods of production and inventory management concepts and techniques such as Just-in-Time, MRP, and basic independent inventory management concepts such as EOQ, reorder points, etc.

The mere introduction of more sophisticated techniques by itself cannot lead to desired change. The entire concept of production planning and control needs to be reoriented from a command-control to a market-driven philosophy.

Our study found that education and training were very low priorities in the Soviet companies that we surveyed. We assert that education absolutely must become a top priority as the Soviet manufacturing planning and control systems go through a major "perestroika" in the next few years.

Managers of Joint Ventures in the West

Our study results suggest that joint venture partners from the west will have much difficulty in working with Soviet manufacturing companies as long as the Soviet "order - entry" systems continue to be so unresponsive to schedule changes and other customer - driven requests.

Operations Management Theory

If our "theories" of operations management are "good," they should be as useful (or more useful) in a Soviet context as they are in a European or American context. It has been said that "there is nothing more practical as a good theory." We do not have much in the way of practical theory to offer to the Soviets. This suggests to us that we need to have a more sophisticated descriptive contingency theory that allows us to classify Soviet manufacturing companies and then help them develop appropriate planning and control systems. We also need a better normative theory that describes how a manufacturing company should move through an appropriate "life cycle" of production and planning control systems.

Hoarding and hedging lead to large order quantities, "lumpy" demand, highly uncertain demand, and ultimately to severe shortages. Severe shortages then encourage more hoarding and hedging. This vicious cycle is very much like the well-known "lead time syndrome" discussed by Plossl [7]. The lead time syndrome is the cyclic process of customers increasing the order backlog in response to suppliers increasing lead times and suppliers increasing lead times in response to customers increasing the order backlog. The solution to the hoarding/hedging problem and to the lead time syndrome is the same. Customers need to communicate projected demands to their suppliers and should only buy what they need. Suppliers should not increase lead times in response to increased demand.

The centrally planned economy of the Soviet Union has failed to provide for the people of the Soviet Union. The large Soviet planning agencies with large computers and sophisticated linear programming optimization could not manage the complex distribution network efficiently. Maybe the lesson to be learned here is that big (and centrally planned) is not necessarily beautiful. Small entrepreneurial "focused factories" (Skinner, [8]) closely connected to the market can out-perform large "rationalized" networks of factories.

Our conclusions about the lack of a good manufacturing/marketing interface in the Soviet Union leads us to note that the malaise of manufacturing in the United States and western Europe is also partially attributable to poor links between production and the market. Solid links between manufacturing and marketing seems to be a key to long-term success of all manufacturing systems everywhere in the world.

General Conclusions

We believe that "perestroika" (restructuring) of manufacturing systems in Soviet block countries and in Soviet Republics will play a key role in the progress (or the collapse) of these economies. If these manufacturing systems cannot provide competitive goods for both national and international markets, the "revolutions" during the 1985-1991 era will only result in more revolutions, suffering, and pain for the 300 million people who live in the area.

REFERENCES

1. D.C. Whybark, "An Analysis of Global Data on the Impact of the Market on Manufacturing Practices," Discussion paper No. 38, Bloomington, IN, Indiana Center for Global Business, Indiana Business School, 1989.

2.* D.C. Whybark and B.H. Rho, "A Worldwide Survey of Manufacturing Practices," Discussion Paper No. 2, Bloomington, IN, Indiana Center for Global Business, Indiana Business School, May 1988.

3. C. Vlachoutsicos and P. Lawrence, "What We Don't Know About Soviet Management," *Harvard Business Review*, November-December 1990.

4. J. Kornai, *Vision and Reality. Market and State*, New York, Harvester, 1990.

5. M. Bernstam, *Anatomy of the Soviet Reform*, Hoover Institution Reprint Series, No. 116, Stanford, 1988.

6. A. Chikán and K. Demeter, "Production and Inventory Management in the Hungarian Industry—An Empirical Study," *AULA: Society and Economy* (Quarterly Journal of Budapest University of Economic Sciences), 12, 2, 1990, 123130.

7. G.W. Plossl, *Production and Inventory Control - Principles and Techniques*, Englewood Cliffs, Prentice-Hall, 1985.

8. W. Skinner, *Manufacturing in the Corporate Strategy*, New York, John Wiley & Sons, 1978.

* This article is reproduced in this volume.

ACKNOWLEDGEMENTS

A version of this paper is scheduled for publication in the *International Journal of Operations and Production Management* and has appeared as Working Paper 91-10, Carlson School of Management, University of Minnesota, Sept. 1991.

SECTION THREE

BETWEEN REGION COMPARISONS

In this section, all the papers present comparisons of practices between two or more regions of the world. The first four papers are bilateral comparisons, while the remaining report multilateral results. Those presenting bilateral comparisons primarily document manufacturing practices that differ significantly from one region to the other (varying from the contrast between China and the western countries, to that between North America and western Europe). The other papers relate the differences to specific aspects of the manufacturing mission or differences in the cultural or economic climate in the regions. In most cases, the research demonstrates that the differences between regions far outweighs the differences between industries.

Boo-Ho Rho and D. Clay Whybark
Comparing Manufacturing Practices in the People's Republic of China and South Korea

One of the early papers in the work of the group was this comparison of the practices in the People's Republic of China to those in South Korea. The data were compared such that differences between the countries was the key focus, but pointing out where differences between the industries in each of the countries might stand out. The key differences highlight the impact of central planning in China in contrast to the market based economy in Korea. As the amount of trade between these two countries continues to increase, the desirability of creating joint ventures will grow. Once the political difficulties are overcome, accommodating the differences in manufacturing practices will be important to the success of such ventures.

Gyula Vastag and D. Clay Whybark
Reducing the Gap in Manufacturing Practices Between North America and the People's Republic of China

Several papers have indicated enormous differences in manufacturing practices between the North American and Chinese manufacturing firms. This paper addresses those differences and comments on the requirements for reducing those differences in order to bring Chinese manufacturing practices up to world standards. The data from the Global Manufacturing Research Group was supplemented by interviews of manufacturing firms in several parts of China to draw conclusions and make recommendations. A profile of the differences in performance is provided and priorities for management attention in China are given.

Boo-Ho Rho and D. Clay Whybark
Comparing Manufacturing Planning and Control Practices in Europe and Korea

Another early GMRG paper is this one presenting a comparison of South Korea with western Europe. In the western European data, there are several countries included, some with very small overall representation in the sample. Nevertheless, there are large differences between the European countries, taken as a group, and Korea. They are great enough to overcome within region variance due to differences between European countries. Some of the differences highlight the export and market oriented focus of the Korean firms. The conclusion provides insights into specific areas where potential partners might want to look for problems.

Gyula Vastag and D. Clay Whybark
Comparing Manufacturing Practices in North America and Western Europe: Are There Any Surprises?

This paper, comparing western Europe and North America, is one of the more recent bilateral comparisons. The postponing of this analysis can easily be explained by the feeling that the two areas were too similar overall to have many differences in manufacturing practices. After all, both regions are highly developed, heavily industrialized and are very close to each other culturally. This paper uses a conservative approach for making comparisons. The conservative nature of the approach means that those differences that were found could be very important to companies trying to arrange partnerships, especially now that there is a rush to establish a presence in Europe in anticipation of the Common Market. Doing business in western Europe is not quite the same as in North America, but is still very close.

Robert B. Handfield and Barbara Withers
A Comparison of Logistics Management in Hungary, China, Korea and Japan

A broad definition of logistics is used in this article to explore the performance of firms in Hungary, China, Korea and Japan. The political and economic conditions in those countries are used to set a backdrop for the comparison of some of the data in the survey. In particular, the article looks at the performance of the firms in these regions and the underlying materials management activities that define them. This provides the basis for recommendations for training to improve the logistics performance of firms interested in international collaboration.

Scott T. Young, K. Kern Kwong, Cheng Li and Wing Fok
Global Manufacturing Practices and Strategies: A Study of Two Industries

This paper looks at the relationship between manufacturing strategy and practices in China, Korea, Japan, western Europe and North America. The research is conducted for both the textile and machine tool industries in these areas. After an evaluation of the differences in practices in the two industries in each of these regions, the relationship to manufacturing strategy is developed. The implications for manufacturing practices of the different manufacturing strategies are then presented.

Attila Chikán and D. Clay Whybark
Cross-National Comparison of Production-Inventory Management Practices

Cross-national comparisons of several regions of the world are used to draw conclusions in this paper. Included are western (Europe), eastern (Korea and China) and combined (Hungary) countries. This mixture provides insights into the differences that may arise due to economic system (centrally planned versus free market) and culture (East versus West). Indeed the paper does report some differences that can be argued and are due to each of these distinctions and Hungary provides some examples that seem to combine aspects of both. Hungary is a country experiencing substantial change and has experience with western culture and centrally planned economies. These are both clearly reflected in this comparison of manufacturing practices.

Comparing Manufacturing Practices in the People's Republic of China and South Korea

Boo-Ho Rho, Sogang University, Korea
D. Clay Whybark, University of North Carolina-Chapel Hill, USA

ABSTRACT

This paper presents a comparison of the manufacturing planning and control practices between South Korea and China in the non-fashion textile and small machine tool industries. It uses a subject of data collected from countries around the world. Specifically, this paper presents the key differences in practice between South Korean firms and Chinese firms in the two industries. The major differences stem from the residual state planning activities in the Chinese firms, particularly in the textile industry. This means there is substantially less influence of market factors on production activities in China than Korea. It also means high level administration/planning people in China are involved in production planning and forecasting activities to a much greater extent than in Korea. The lessons of these findings are as important for an emerging China as they are for Koreans who look forward to improved relations with the Chinese and possible joint ventures in the future.

INTRODUCTION

This paper presents an analysis of two sets of data from a large data base of manufacturing planning practices gathered from companies around the world. Two industries were surveyed, the non-fashion textile industry and the small machine tool industry. Both of these industries are found throughout the world in both advanced and developing nations. Data has been collected from North America, Western and Eastern Europe, South Korea, and the People's Republic of China; and plans for gathering additional data in South America and Japan have been made. This paper, however, only uses the data from China and South Korea. Detailed information on the data gathering methodology for the total study, including details on the questionnaire, format, and data availability, can be found in Rho and Whybark [1].

A number of factors combined to generate interest in gathering the data which was used in this study. Among these factors was an interest in manufacturing productivity and the key role of manufacturing planning and control practices in enhancing productivity. This interest led to the gathering and analysis of data in China and South Korea. These within-country analyses have shown important relationships between manufacturing planning and control practices and performance in China [2,3], and differences among firms in Korea [4]. Also of interest were comparisons between countries for firms contemplating joint ventures and other business relationships with firms represented in the countries surveyed.

This paper reports on a cross-geographical comparison. The importance of gathering empirical data from the field is very clear when the objective is to compare practices between different parts of the world. It is even more important when part of the interest concerns doing business with firms in these areas. Although theoretical models were used for developing the questions to be asked of managers, it is actual practices that are the interest of this paper. In order to gain a preliminary understanding of the differences between China's and Korea's planning and control practices, questions for which there appeared to be large and/or interesting differences were selected for analysis. Statistical tests of the differences between the responses were performed and are reported in the Appendix.

The choice of China and South Korea for this comparison was based upon two factors: first, the proximity of Korea and China and, secondly, Korea's emergence as an industrial power. Also, both countries are increasingly involved in joint ventures with European, Japanese and American firms. The Korean interest in the study was founded on their interest in their country's manufacturing productivity considerations. The Chinese also were interested in doing analyses of their own data, but were more highly motivated to learn if Korea's approaches to manufacturing planning and control may have played a role in its development. The study does highlight some areas in which the Chinese could profitably invest in order to improve some of their practices.

DATA GATHERING

The South Korean data gathering activity was supported by Sogang University and the Korea Productivity Center. The effort was one that involved both a mail survey and personal interviews. The responses to the mail survey were disappointingly small, and the decision was made to send teams of interviewers into the field to gather the data from the companies directly. The data, therefore, is from a reasonable portion of the population of firms producing small machine tools and non-fashion textiles. In the textile industry, integrated firms from spinning and weaving through sewing are included. Small tool manufacturers include firms that produce small drill presses, milling machines and other forming equipment, some of which is computer-controlled. The sample is representative of the Korean firms doing business in these industries, with more than 30 textile companies and nearly 90 machine tool companies included.

Sampling in China was supported by the Shanghai Institute of Mechanical Engineering (SIME), and the Shanghai Enterprise Management Association. A pilot of the survey form was tested in both industries and then was mailed out, again the response was disappointing. Interviewers were sent out to firms in each industry (from Shanghai, nearby provinces, and some northern and southern Chinese locations) to get the final sample. The textile firms include some with various portions of their output specified by state plans. There are integrated manufacturers, as well as some fairly small finishing firms. The small machine tool manufacturers range from specialized product producers to fairly large, complete line manufacturers. Again, some have fairly high levels of state planning. The sample is representative with about 50 firms in each industry.

The questionnaire included sections for general company information, forecasting, production planning, scheduling and chop floor procedures, materials management and

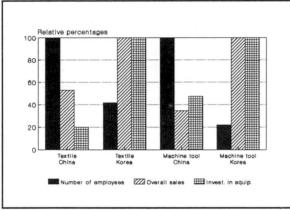

Figure 1. General company data comparisons

purchasing. This paper describes the findings for each of these categories.

GENERAL COMPARISONS

Overall sales are lower in the Chinese companies, but they have a much higher employment level than the Korean companies. There are about twice as many Chinese employed in the textile industry and four times as many in machine tools. A similar phenomena is found with respect to the investment in production equipment, where the Chinese have a generally lower level of investment. Thus the Chinese have a substantially lower rate of output per employee and a lower investment per person than the Korean companies. These comparisons are shown in Figure 1.

In addition to the make-to-order and make-to-stock categories that describe Korean production, Chinese firms have another category: make-to-state-plan. This category signifies that, for the products involved, the state dictates the quantity to be produced and provides the resources for the production. While Korean firms mostly produce to order (about 70% in the textile industry and 80% in the machine tool industry), Chinese firms produce basically either to state-plan or to order (in the textile industry, 66% to state-plan and 26% to order; in the machine tool industry, 24% to state-plan and 55% to order).

These are not exclusive categories in either industry or country. Most firms have a mixture of make-to-stock and make-to-order products. In China many firms will have some portion of production in each of these categories, as well as in make-to-state-plan. The textile industry has a greater portion of state plan production than the machine tool industry in China.

FORECASTING PROCEDURES

The forecasts are prepared by personnel in the same function and at the same level in both industries in Korea and in the machine tool industry in China. They rely on low levels of management (about 85% use department heads or lower) in the marketing function (about 80% done by marketing) to produce the forecasts. The Chinese textile firms use higher level executives outside the marketing area for forecasting as seen in Figure 2.

There are substantial differences between Chinese and Korean firms in how the forecasts

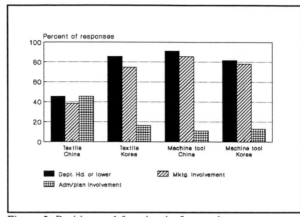

Figure 2. Position and function in forecasting

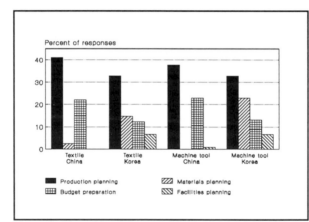

Figure 3. Use of forecast

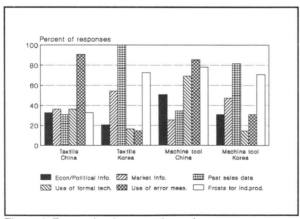

Figure 4. Forecasting inputs and practices

are used. Few Chinese firms use forecasts for material, manpower or facilities planning while about 30% of the Korean firms do. Chinese firms make more use of forecasts in budget preparation (22% compared to 13%) and in new product planning (10% compared to 5%). Slight differences occur between Korean and Chinese firms in their use of forecasts for sales planning (about 25% for China and 20% for Korea). The greatest use of forecasts, as seen in Figure 3, is for production planning in both countries (about 33% in Korea and 40% in China).

Figure 4 summarizes the results concerning the information used to prepare the forecasts. The industry differences are small compared to the country differences. For example, Korean firms use more subjective market information (averaging about 54% compared to 31% for China) but less information on the current status and trends in the economic and political situation (about 29% compared to 43%) than the Chinese firms. In terms of objective factors, Korean firms use past sales data almost exclusively while Chinese firms use a combination of past sales, order backlog and industry statistics. Computers are seldom used for forecasting in either country (10% or less), but there is greater use of formal forecasting techniques in China. Forecast error is measured more frequently in China than in Korea (about 90% compared to 26%).

There is a difference in the way the products are aggregated for forecasting by the Chinese textile industry. About 68% of these firms forecast on a product group basis while about 75% of the Korean firms and the Chinese machine tool firms forecast on an individual product basis.

Perhaps as interesting as these differences are some of the similarities. For example, the forecasting period averages about a year, and the increment is about a quarter in both countries. Forecast revisions are performed between 3 and 4 times per year in Korea and from 2 to 3 times per year in China. About half of the firms in both Korea and China provide a range for their forecasts, as opposed to a single point. Also, about 60% of all firms forecast in physical units rather than monetary value.

PRODUCTION PLANNING AND SCHEDULING

The primary factors used in developing Korean production plans are the forecast and actual orders (about 45% each), while in China they are production capacity (37%), state-plan (22%) and actual orders (27%). The forecast plays a very minimal role in production planning in China (less than 10%) while the state-plan is one of the major factors (29% in the textile industry and 14% in the machine tool industry).

In looking at the units used for production planning, again the Chinese textile firms stand out. They use product groupings as opposed to the more detailed individual product units used in the other cases. Figure 5 shows this and that there is slightly more computer usage for production planning in China.

There are some differences between Korea and China in the function of the people who prepare and use the production plans (see Figures 6 and 7). In Korea, production personnel are the primary preparers and users of the production plans. In China, however, it is different between the two industries. In the textile industry, it is predominantly administration/planning personnel, while, like Korean firms, it is production personnel in the Chinese machine tool industry. In fact, production personnel play very minimal roles (8%) in preparing the plans in the Chinese textile industry. Even in the machine tool industry in

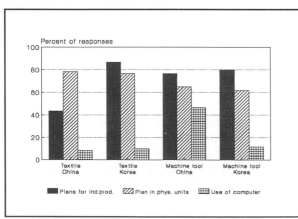

Figure 5. Production planning practices

China administration/planning personnel play a relatively more important role in preparing and using the plans than in Korea.

Figures 6 and 7 show that Korean firms involve lower level people more frequently in the preparation and the use of the plan in both industries. Many more plan preparers and users are at or below the group/section level of management in the Korean firms than in the Chinese firms. Department/divisional head level of management or higher involvement is more pronounced in China, especially in the textile industry where about 50% are even vice president/director or above.

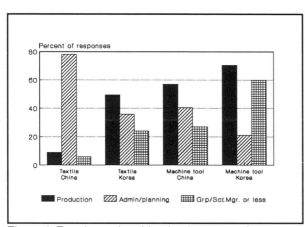

Figure 6. Function and position in plan preparation

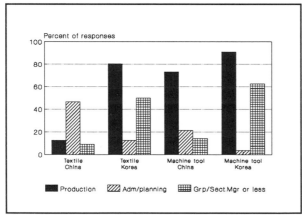

Figure 7. Function and position in plan use

The modal production plan is for one year in both countries. The plan is divided into increments of about two months with most values one month and one quarter. The plan is revised monthly in Korea, but only every four months in China. Some 80% to 95% of the firms in both countries revise plans when there is demand fluctuation. The production plan is used primarily for operational activities such as scheduling or material planning rather than planning activities like budgeting, facilities investment or staffing in both Korea and China.

SHOP FLOOR CONTROL

There are important differences between China and Korea, and some differences between industries in China in managing shop floor activities. In most Korean and Chinese machine tool firms, it is production people who initiate production activities and establish job release sequences. The Chinese textile industry departs from this pattern by having administration/planning people perform these functions as shown in Figure 8.

In Korea, marketing and customer concerns dominate in establishing and changing shop

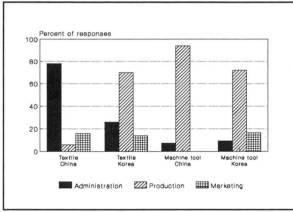

Figure 8. Involvement in shop-floor activities

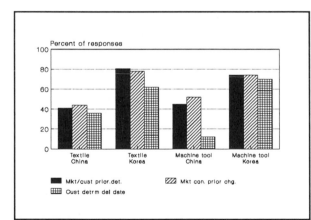

Figure 9. Market influences on the shop floor

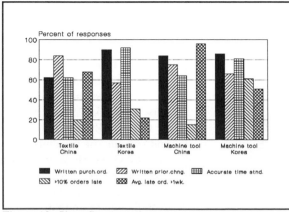

Figure 10. Shop floor practices

floor priorities, and in determining delivery dates (see Figure 9). In China, shop floor priorities and customer delivery dates are also influenced by manufacturing factors like the processing time required and the availability of material.

Figure 10 summarizes several shop floor practices. Data is transmitted to suppliers in written form (no computer transmission is used) in both countries, but there is more written communication of priority changes to the shop floor in China than in Korea. Although Koreans think more highly of their time standards than the Chinese, the Chinese firms had fewer orders late than the Korean firms (much fewer in the machine tool industry). The Chinese, however, had many more orders over one week late.

The causes of lateness covered a wide spectrum in both countries, with some differences in the magnitude of the responses as seen in Figure 11. Capacity problems appear to be the primary causes of lateness across countries and industries, although material shortages hurt the textile industry in China. Also, in China, transportation problems caused delays while quality problems affected Korean firms.

When capacity changes are necessary, there are a variety

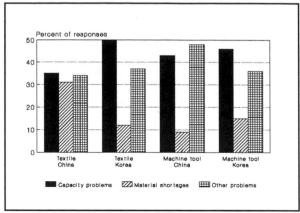

Figure 11. Causes of lateness

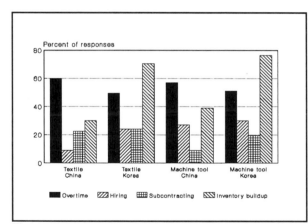

Figure 12. Means of capacity change

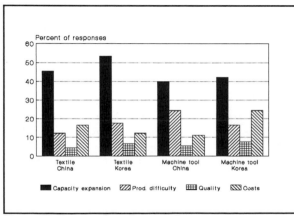

Figure 13. Reasons for subcontracting

of responses in both countries Overtime, subcontracting/leasing, and hiring are all used for capacity increases. During times of reduced demand, inventory build-up is a common response, but in China the work time per day is sometimes reduced or undertime is incurred (see Figure 12). In the Chinese firms, the work time reduction is often used for training and study. Subcontracting is used mainly for capacity expansion in Korea. In China it is used for both capacity expansion and capability (faster or more specialized production) in China, as shown in Figure 13.

PURCHASING AND MATERIALS MANAGEMENT

About 50% of the inventory holdings of both Korean and Chinese firms is at the raw materials/purchased parts end of the continuum, with the exception of the Chinese textile industry where it is less than 40% (see Figure 14). The purchasing of these raw materials is largely based on customer orders in Korea, but mainly on the production plan in China. State distribution is also cited as a procurement method in China (18% in the textile and 7% in the machine tool industries). Single sourcing has not been widely accepted as a supply guarantee mechanism, even though more than 50% of

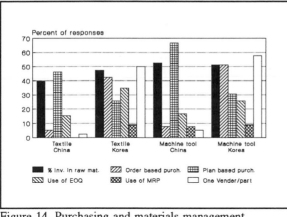

Figure 14. Purchasing and materials management

the Korean responses indicated only a single vendor per part. About 60% of the Chinese firms have four or more vendors per part.

There appears to be slightly more use of the EOQ in Korea than in China, but almost no use of MRP or JIT in either country. In fact, 40% or more of the firms in both countries have never heard of either MRP or JIT.

SUMMARY AND CONCLUSIONS

There are differences between the Chinese and the Korean approaches to manufacturing planning and control although the Chinese machine tool industry was often closer to the Korean firms than the Chinese textile industry. Probably the clearest explanation of this observation is the degree of state planning in the Chinese textile firms (66%) as opposed to the machine tool firms (24%). The effect of state planning can be seen in many responses. The use of high level administration/planning function executives for the forecasting and production planning activities (while the production people have roles more restricted to the operations areas) is one illustration.

The lack of detail in the forecasts and production plans for the Chinese textile firms provides another example of the state planning influence. Similarly, the use of forecasts for facilities and materials planning as opposed to production planning reflect the state-plan-based organization. Production plans themselves are revised much less frequently in China, reflecting the rigidity of the state planning process. Even the forecast error measurement and use of formal forecasting techniques (greater in China) are related to a state-plan requirement. The fact that the machine tool industry is a little more dynamic than the textile industry corroborates this theory of state-plan effect.

Although there are clearly differences in the infrastructure between Korea and China (the transportation problem in China, for example, reflects the road and rail situation there), there are lessons to be learned by the Chinese from these results. One clear lesson is that the Korean firms are closer to the customers. The degree of customer and market influence on the shop floor and planning responsiveness is substantially higher in Korea. Likewise, the labor productivity, in terms of sales output, is substantially greater in Korea.

The Koreans, on the other hand, should take note of the use of modern techniques in

China. As state planning is reduced, the Chinese may make more effective use of these techniques to modernize their manufacturing organizations. The rigidity of Chinese plans may prove an advantage over the flexibility of the Korean plans when one considers the use of levelized frozen plans for JIT applications. Finally, as both countries seek ways to normalize relations and engage in joint ventures, the differences in practice discussed here are important to recognize in attempting any mutual projects. It is more important to acknowledge the differences and that the parties find agreeable compromises than to argue for any particular approach. This study's contribution is the bringing of the facts of actual practice to light. After all, the dispelling of ignorance is the first step in improving practice.

REFERENCES

1.*   D.C. Whybark and B.H. Rho, "A Worldwide Survey of Manufacturing Practices," Discussion Paper No. 2, Bloomington, IN, Indiana Center for Global Business, Indiana Business School, May 1988.

2.   J.G. Chen, "Operations management in Chinese textile enterprises," Master's thesis, Department of Systems Engineering, Shanghai Institute of Mechanical Engineering, 1987.

3.   X.J. Yan, "Operations management in Chinese machine tool manufacturing firms," master's thesis, Department of Systems Engineering, Shanghai Institute of Mechanical Engineering, 1987

4.   B.H. Rho and D.C. Whybark, "Production planning and control in Korea," Pan-Pacific Conference IV, Taipei, Taiwan, 1987.

* This article is reproduced in this volume.

ACKNOWLEDGEMENTS

The authors would like to thank the Korea Productivity Center, the Sogang Institute for Economics and Business (Korea), the Shanghai Institute of Mechanical Engineering (China), IMEDE (now IMD, the international Institute for Management Development, Switzerland), and the Business School at Indiana University (USA) for supporting the project.

APPENDIX
(Values are percentages unless otherwise indicated)

	Korea		China		Significance	
Item	Textile	Machine	Textile	Machine	Level	Test
General data						
Sales ($)	15.3M	14.4M	8.2M	5.1M		
Employment	484	468	1159	2155		
Products equipment ($)	8.1M	10.0M	1.7M	4.7M		
Sales/person ($)	37.6K	29.6K	6.2K	2.5K	<.001	*t*
Equipment/person ($)	12.K	11.1K	1.1K	2.1K	<.001	*t*
Production to order	73	82	26	55	<.001	*t*
Production to state plan	0	0	66	24	<.001	*t*
Forecasting						
Marketing involvement	75	80	37†	86	<0.001	K-S
By lower management	85	83	46†	91	<0.001	K-S
Use:						
Budget	13	13	22	23	<0.001	K-S
Material	15	23	2	0		
Manpower	4	3	0	0		
Facilities	7	7	0	1		
New products	7	3	9	14		
Sales plan	22	18	26	25		
Production plan	33	33	41	38		
Marketing data	55	49	36	26	<.005	K-S
Econ/political data	21	32	34	52	<.100	K-S
Past sales	100	83	33	36	<.001	K-S
Computer use	11	15	14	7	No	K-S
Formal techniques	17	14	37	69	<0.001	K-S
Error measure	15	31	92	86	<0.001	K-S
Individual products	74	71	32†	78	<0.001	K-S
Revision/year	3.7	3.4	3.6	4.2	No	*t*
Units of production	58	58	67	64	No	K-S
Production plan						
Use:						
Forecast	51	42	2	9		
Actual order	40	47	22	30	<.001	K-S
Production capacity	2	2	40	33		

(continued)

Appendix, cont.

Item	Korea		China		Significance	
	Textile	Machine	Textile	Machine	Level	Test
Production plan use (cont):						
State plan	0	0	29	14		
Individual products	84	79	42†	76	<0.001	K-S
Computer use	10	12	9	30	No	K-S
Preparation:						
Administration	36	21	77†	40		
Production	48	70	8†	57	<.001	K-S
Lower man. prepares	24	59	5	27	<.001	K-S
Use						
Administration	11	3	48†	23		
Production	81	93	13†	74	<.001	K-S
Lower management use	51	64	8	14	<0.001	K-S
Plan increment (months)	1.3	1.7	1.6	2.2	<.020	*t*
Revisions/year	9x	10x	3.1x	2.4x	<.001	*t*
Plan use operational	58	64			No	K-S
Shop floor						
Initiate production	64	73	6†	93	<0.001	K-S
by Administration	24	10	73†	7		
Production sequence	92	95	12†	63	<0.001	K-S
by Administration	25	15	74†	27		
Market/customer determines priority	82	75	42	40	<0.001	K-S
Market changes priority	80	63	44	47	<.020	K-S
Customer determines due date	60	70	36	14	<0.001	K-S
Write purchase orders	90	85	63	83	No	K-S
Write shop orders	65	58	84	75	No	*t*
Good standards	91	81	63	64	<0.050	K-S
>10% Late orders	30	61	20	15	<0.001	K-S
>1 week late	21	51	67	95	<0.001	K-S
Due to Material	13	15	31	9	<0.001	K-S
Due to Quality	31	28	6	10		
Due to capacity	50	47	35	43		
Due to transportation	0	0	26	24		

(continued)

Appendix, cont.

Item	Korea		China		Significance	
	Textile	Machine	Textile	Machine	Level	Test
Capacity increase:						
Hire	24	29	9	26		
Overtime	49	50	59	57		
Undertime						
Subcontracting capacity	53	42	45	39	<.100	K-S
Capability	18	16	24	37		
Materials management						
Raw material	48	51	39†	52	<0.010	*t*
Purchase to order	43	51	5	7	<.001	K-S
to plan	25	31	46	66		
Single vendor	50	57	2	5	<0.001	K-S
EOQ use	35	26	15	17	No	K-S
Number know JIT	40	51	66	49	<.100	K-S
Number know MRP	41	51	62	53	No	K-S

* The level of significance is for the test between Korea and China (industry data aggregated) except where noted. The *t* test (*t*) and Kolmogorov-Smirnov (K-S) tests were used for numerical and nominal comparisons, respectively. The indication 'No' is used when the value exceeds a significance level of 0.10.
† Level of significance is for a test between the Chinese Textile firms and all other cases.

Reducing the Gap in Manufacturing Practices Between North America and the People's Republic of China

Gyula Vastag, Budapest University of Economic Sciences, Hungary
D. Clay Whybark, University of North Carolina-Chapel Hill, USA

ABSTRACT

This paper provides a comparison of manufacturing practices between firms in the People's Republic of China and North America. The background was a project to help the Chinese identify areas where improvements in manufacturing practice were needed in order to increase exports of products into the world market. A profile of Chinese manufacturing practices is provided and recommendations for change are made in the areas of market planning, market response, manufacturing activities and asset management.

INTRODUCTION

Considerable attention is being given to China's role in the world economy. The views of observers range from overly optimistic (e.g., China as an East Asian development model) to more cautious ones. The more cautious observers (e.g., Fischer [1]) emphasize that without substantial changes in the Chinese economic and political system, economic growth cannot be sustained as the era of traditional, export-led growth based solely on cheap labor comes to an end.

Over the next decade, progress in developing China's economy will depend in part on its capacity to produce, export, and market higher-quality manufactured goods in regions outside Asia. The United States currently is--and the emerging North and South American free-trade area will increasingly become--an important market for Chinese-manufactured products. The value of China's exports to the United States has steadily increased from a little more than $3 billion in 1984 to more than $10 billion in 1990. The question is whether or not China will be able to maintain or increase its levels of exports in an environment that is changing dramatically. The objective of this paper is to answer part of this question from the point of view of Chinese companies. What manufacturing practices should be changed in Chinese companies for them to compete in the North American market? We answer this question by comparing the Chinese companies to their North American counterparts. If Chinese companies want to compete in this market or want to create joint ventures, they should adapt or develop manufacturing practices that are similar to, or better than, those of the North American companies.

This paper presents the results of a survey of manufacturing practices in the People's Republic of China and North America. The design of the research was done jointly by the researchers from the Kenan Institute at the University of North Carolina, USA, and the National Research Center for Science and Technology Development (NRCSTD) in Beijing, China. The authors wish to thank Dr. Gao Xiaosu and Mr. Wu Minglu, both of NRCSTD, and Mr. Sun Yongjian of the State Science and Technology Commission, for their collaboration in the design of this research. This joint collaboration proved valuable in defining and assigning the various tasks in the project, increasing mutual understanding of the project scope and objectives, and making best use of the comparative strengths of each of the researchers.

In addition to presenting the results, the data and methodology used to determine manufacturing practices that were different between North American and Chinese textile and machine tool firms in general are described. The presentation will first provide a discussion of the data base used for the study and the procedure used to determine a general difference between the two regions. Finally, the results are discussed in terms of the broad managerial implications and some recommendations for the enterprise.

DATA

The North American data used for this study were gathered in four areas of the United States and in Canada. The Chinese data come from companies mostly located in or near Shanghai and in north and south China. The survey was carried out in the late 1980s in state-owned enterprises.

The distribution of the sample by region and industry for the North American data is provided in Table 1. Canada and the Western United States are under-represented in the sample, relative to the population distribution of the firms. The remaining areas, however, are representative of the population distribution. In the judgement of the research group, this bias was deemed immaterial to the generality of the manufacturing practices represented in the North American sample. The judgement was quite different for the sample of Chinese firms, however. First, the data all came from the Shanghai area and there are possible differences in practice between Shanghai and other areas of China. Secondly, the distribution of ownership of firms has been changed significantly since the late eighties as a result of the reforms in China.

Table 1. Distribution of the sampled companies by region and industry group

Region	Machine tool	Textile	Total
People's Republic of China	44	56	100
North America	45	50	95
Total	89	106	195

To evaluate the impact of the potential bias in the Chinese data, the research group analyzed comprehensive case studies written in 1992 on five Chinese companies. Also, the same survey that was used earlier was carried out at two of the five companies, the Beijing Printing and Dying Plant and the Shanghai Surgical Instruments Factory, and the new data were compared with the earlier survey data.

In general, we conclude that there are no significant differences in manufacturing practices because of ownership or time. This indicates that our conclusions, based on the earlier survey, can serve for making broad managerial recommendations.

The questionnaire used for gathering the data on manufacturing practices contains 65 questions that are mainly concerned with manufacturing planning and control. There are 95 original variables, and others were derived for the study. Some of the derived variables describe relationships between the original variables (e.g., sales per employee or inventory turnover). Others result from individually analyzing each possible response to a multiple-response question (e.g., how is capacity increased?). The variables are associated with the five sections of the questionnaire (General Data, Sales Forecasting, Production Planning, Shop Floor Control, and Materials Management) and fall into three measurement categories (ratio-scaled, ordinal, and nominal). Overall, 119 variables (43 ratio-scaled, 43 ordinal, and 34 nominal) were analyzed for this paper. Table 2 shows the distribution of variables by measurements scales and questionnaire sections.

Table 2. Distribution of measurement scales in the questionnaire sections

Questionnaire section	Ratio	Ordinal	Nominal	Total
Company profile	16			16
Sales forecasting	4	3	8	15
Production planning	10	8	9	27
Shop floor control	6	30	11	47
Materials management	6	2	6	14
Total	42	43	34	119

THE ANALYTICAL FRAMEWORK FOR THE COMPARISONS

The framework presented here groups the data into basic units for statistical comparisons, accounts for the different scales for the variables, and uses different statistical approaches depending on the assumptions required of the data and the type of variable.

We start our discussion of the analysis by describing the grouping of the data for analysis and introduce the concept of a pure regional effect. We next describe the types of variables from the questionnaire that must be accommodated by the analysis. The remainder of this section describes the statistical approaches used for each type of variable considered. Although our examples are concerned with comparing the two regions, the approach is quite general for bilateral comparisons.

The four cells shown in Table 3 represent a natural way to group the data since it was gathered in the two industries and in both regions. In Table 3, 'ch' and 'na' represent the two regions. The symbols 'chm' and 'cht' denote the machine tool and textile industry groups respectively from China. Similarly, the machine tool and textile groups from

North America are denoted by 'nam' and 'nat' respectively. This is not the only way to group the data, however. Based on the responses to the questionnaire, other detailed subsamples could be developed by size of firm, by type of ownership, by export orientation, or by portion of in-house production.

Table 3. Grouping the data

Region	Machine tool (m)	Textile (t)
China (ch)	chm	cht
North America (na)	nam	nat

The groupings shown in Table 3, however, permit us to make five different types of comparisons:
   (a) manufacturing practices between different regions, namely 'ch' and 'na'
   (b) practices between the industries, namely machine-tool and textile
   (c) regions within the industry (e.g., 'chm' or 'cht' compared with 'nam' and 'nat')
   (d) industries within the country (e.g., 'chm' with 'cht')
   (e) cross-industry-region comparisons (e.g., 'chm' compared with 'nat')

Each of these comparisons has a certain rationale and, at the same time, focuses on a different issue. The analysis in (a) emphasizes the regional differences and eliminates the industrial ones by combining the two types of industries (this analysis is stronger if it has been determined that the industries can be pooled for the statistical analysis). The analysis for (b) combines different regions in order to focus on the industrial characteristics (again, stronger if the regions can be pooled). The types of comparisons in (c), (d), and (e) are the most elemental. Industrial differences are not considered in (c), and regional ones are not considered in (d). In (e) the regional and industrial differences are mixed.

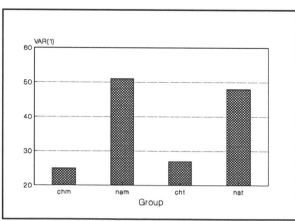

Figure 1. The concept of pure effect

The concept of pure effect

In this paper we use the groupings of Table 3 to make regional comparisons using the (c) type of analysis. This does not require any prior determination of whether or not the data can be pooled. To make the comparisons, we use the concept of a pure effect. The pure effect analyzed in this paper is illustrated in Figure 1 as a hypothetical example.

In this example the groups represent the four cells of Table 3. The responses shown for VAR(1) in the figure are hypothetical. The lengths of the bars represent the averages of the variable for each group. Suppose that the differences between the regions in different industries are significant (compare 'chm' to 'nam' and 'cht' to 'nat'). Since the differences are in the same direction in both industries, we define this as a pure regional effect. Obviously, a pure industrial effect can be defined by direct analog. In the example in Figure 1, it appears that there is no statistically significant difference in responses to VAR(1) between the industries within the same region (compare 'chm' to 'cht' and 'nam' to 'nat'). This is not a condition for defining a pure effect. As long as the regional differences are significant and in the same direction, we have a pure effect regardless of what the direction or significance is for the industrial values.

Using hypothetical examples, Table 4 gives further illustrations of pure regional and industrial effects, and their interaction. Using the pure effect to define differences is very conservative since it may overlook some interesting interactions between industry and region. On the other hand, since the effect is "pure," the description and interpretation of the differences are straightforward. As a method of identifying regional differences that warrant discussion, it works quite well. The use of the subgroups like those of Table 3 and the concept of pure effect are not limited to specific types of variables, measures of central tendency (location), or statistical tests.

This approach is an extension of the approaches used by others to analyze differences between regions. The approach is quite useful for other purposes as well. For example, a firm considering a joint venture might be interested in some of the other combinations, particularly those that involve regional differences for their industry--regardless of the other effects. The use of selected subgroups allows such comparisons and the exploration of other combinations of interest. In this analysis we focus on the pure regional effects, leaving discussion of pure industrial and/or interaction effects for further work.

Structure of the analysis

The different scales require different methods for determining statistical differences between the responses. Since we have indicated that we will only describe the differences that result from a pure effect, we need to determine when both regional comparisons are significant. In the next paragraphs we will discuss the methods used for making these comparisons for each type of measurement scale.

A common method of testing for statistically significant differences between the mean responses of two such groups of *ratio-scaled data* is to use the Student's t-test. For the t-test to be valid when comparing two samples, the response populations must be normally distributed have approximately equal variances. A preliminary analysis of the ratio-scaled variables in the data base revealed that these assumptions are not met (obvious non-normality and different variance levels) in many cases.

To overcome the lack of normality and the heteroscedasticity, we turn to nonparametric

Table 4. pure and "mixed" regional and industrial effects

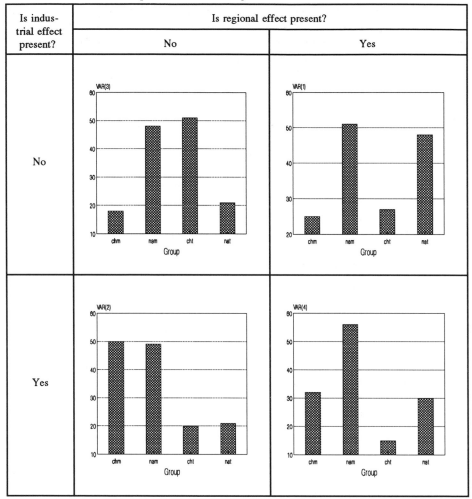

tests. To test for statistically significant differences in the location parameter (the nonparametric equivalent of mean or median) of a ratio-scaled variable, we use the Kruskal-Wallis one-way analysis of variance of ranks. This is the nonparametric counterpart of the t-test (see Daniel [2] or Siegel [3] for details).

Discovering that responses from different groups are significantly different is only the first part of the question. The second is determining whether or not they are different in the same direction--the other condition necessary for a pure regional effect. This is rather easily determined for ratio variables since the mean or median values can be consulted directly. Another way of determining the direction of differences is to use a box plot to visualize them.

The use of box plots opens up another test of significance. Wilkinson [4,5] describes how McGill, Tukey and Larsen use the confidence intervals around the median, shown in the box plots, to determine if there is a significant difference between them. They indicate that when the confidence intervals around two medians do not overlap, you can be confident at about the 95% level that the two population medians are different. Although the median test is not as powerful as the Kruskal-Wallis test, it can be used with the box plot to visualize significant differences.

Ordinal-scaled variables are the most numerous in the questionnaire. Because the responses are ordered (the variable is measured on an ordinal scale), a meaningful, cumulative percentage can be developed. A test for determining significant differences between these cumulative distributions is the Kolmogorov-Smirnov test. The test is based on the maximum difference between the cumulative distributions of the two groups.

The question of interest in analyzing *nominal-scaled data* is whether or not there is a significant difference between the response patterns of the groups. A natural way to perform this analysis is with the Kolmogorov-Smirnov test. The difficulty is with the order of the categories of response. The maximum difference in the cumulative distributions can change with any change in the order of the categories. Since there is no fixed order for the categories, a conservative approach would be to test all permutations of the orders of the categories and see if the minimum of the maximum differences was significant. There are substantial computational difficulties with carrying out this idea, however.

Fortunately, we have two other analytical alternatives for comparing nominal variables. The first of them is the chi-square test of homogeneity, which tests the hypothesis that the distribution of the responses to the question is the same in both groups. The second is the binomial test, which can be used since the responses for each category are either yes or no. The binomial test compares the percentages of each group's responses in a single category.

Sample size is a concern with this test, as it is with the tests for the other types of variables. For example, the rule of thumb for the chi-square test is that no more than 20% of the cells should have fewer than five expected responses. One way to attempt to overcome this difficulty is to combine several categories into one to create cells of sufficient sample size. Although some data is lost in doing this, in most cases the differences in the response patterns can still be tested. If it is not the case, we can focus on one of the categories of the variable and use the binomial test. Stacked bar charts can be used to illustrate the differences in the nominal and ordinal variables.

In the analysis of nominal data, we usually have to combine categories. That means only some parts of the data are explicitly evaluated. Care must be taken in framing the question to be answered in order to guide the choice of what is included in the analysis and in the interpretation of the results.

RESULTS

In the analysis a very stringent test was used. For both industries, the required significance level to declare a pure regional effect was 5%. Even so, the differences between Chinese and North American firms were statistically significant for 41 variables out of the total of 119. Moreover, the actual significance levels were less than 3% in all cases. This is not surprising since our null hypothesis was that these countries are very much different. To put this in perspective, only 7 of the 119 variables are significantly different, at the 5% level, between North America and Western Europe [6].

Although the Chinese sample of this survey was not statistically representative of the whole country and it did not reflect the distribution of the different types of ownership well, we believe that the overall results are correct. We expect that new joint ventures are probably much closer to the North American results than the state-owned enterprises in our sample, but it will not significantly change the overall picture: China's manufacturing companies have a long way to go to catch up with their North American counterparts.

The first section of the questionnaire contains general data on the companies. Table 5 shows the means of the significant differences in this section. The Chinese companies are much larger and have much lower sales productivity than their North American counterparts. These differences can be attributed to the central planning structure (focusing on large state-owned enterprises and full employment) and to the underdevelopment of the Chinese economy.

The differences in sales forecasting are shown in Table 6, which gives the means for ratio-scaled variables and the medians for the ordinal-scaled variables.

In the case of the nominal-scaled variables, the differences in the distribution are briefly summarized. Although Chinese managers use more sophisticated forecast-error measurement techniques, the forecasts are not as dynamic nor are they generated by personnel as high in the organization as those in North America.

In China, the percentage of companies where the position of the sales forecaster was vice president or higher is 9% in the machine-tool industry and 28% among the textile companies compared to 62% and 77%, respectively, in North America. Most of the differences can be explained by the plan-driven behavior of the Chinese companies compared to the market-oriented approach of the North American ones. If the survival of the company depends mostly on market forces, then the demand forecasts are modified more frequently, the sales forecaster is higher in the hierarchy, and the companies focus more on micro-issues that are directly related to their survival. In China, the primary goal of the companies is still to meet plan targets--often expressed in physical units--and not to satisfy market demand.

Table 5. Differences in general data variables

Variable	chm	nam	cht	nat
Employees	2,155	240	1,159	505
Factory workers	1,251	113	946	320
Sales per employee (in million US$)	0.003	0.117	0.008	0.474
Sales per factory worker (in million US$)	0.005	0.273	0.011	0.550

Table 6. Differences in sales forecasting

Variable	chm	nam	cht	nat
Number of forecast modifications per year	2.32	6.64	3.03	6.67
Position of the sales forecaster	Dept. Head	Vice President	Dept. Head	Vice President
Subjective factors in the forecast	In China the emphasis is more on macroeconomic and political factors, while in North America it is on micro (company, industry) issues.			
Forecast for individual products or product groups, in monetary or physical units	In China the forecast usually is made for physical units and individual products, while in North America it is in monetary terms and product groups.			
Two most important uses of the forecast	In China the main purpose of the forecast is for inventory and production planning, while budget is included in North America.			
Measure of the forecast error	In China the forecast error is measured as a percentage error, while in North America usually it is not measured.			

Table 7 shows the differences in production planning. As with forecasting, the Chinese production plans are more rigid than those made by the North American firms--they are revised only about three times a year. More companies in China use frozen production plans (75% in the machine-tool industry and 66% in the textile industry) than companies in North America (24% and 26%, respectively).

Most of the Chinese companies do not use computers at all in production planning (69% and 89%, respectively), while in North America the proportion of companies that do not use computers at all is 9% and 34%, respectively. In China, the companies do not use computers mostly because they are not available; therefore, about 80% of the respondents think that they would be useful. In North America the situation is different: if the companies do not use computers, it is because they tried and were not satisfied with the results--only 20% of the machine-tool companies and 36% of the textile companies think that they would be useful. The Chinese companies are more internally oriented, and the plans are made for physical units.

Table 7. Differences in production planning

Variable	chm	nam	cht	nat
Number of production plan revisions per year	2.5	10.2	3.2	30.0
Education of the production planner	Advanced degree	College	College	College
Use of frozen production plan	Yes	No	Yes	No
Use of computers in production planning	Not at All	Moderate	Not at All	A Little
Helpful to use computers (if not used)	Yes	No	Yes	No
Two factors for production plan	In China, production-related inside factors (e.g., inventory management) dominate production planning; while in North America, production is more controlled by such outside factors as demand.			
Plan for products or product groups, in monetary or physical units	Generally there is more emphasis on physical units and products in China than in North America.			
Production planner's functional grouping	In North America, the production planners usually are from the Production Department or from Administration. In China they may come from other departments as well.			

It is interesting that in China most of the production planners have advanced degrees (57% in the machine-tool industry and 46% in the textile industry), while in North America it is very rare (13% and 10%, respectively).

Table 8 summarizes the differences on part of the shop floor-control section of the questionnaire (the area of greatest detail in manufacturing activities). This section has the greatest number of variables that have a pure regional effect. Generally, the Chinese companies are much more formal and driven more by the production plan and less by the customer's needs than their North American counterparts.

Table 9 lists the differences between the perceived importance of the factors that change priorities on the shop floor. In North America 89% of the machine-tool companies and 80% of the textile companies feel a high pressure from the customers to change priorities. In China, this number is 30% and 13%, respectively. About 56% of the companies in China feel a high pressure to change the production priorities because of material shortages. The comparable proportion in North America is about 36%. The proportion of companies feeling high pressure from changes in the sales plan is relatively low in China: 25% in the machine-tool industry and 23% in the textile compared to the North American 11% and 6%, respectively.

Table 10 shows the difference in the importance of the means of changing the production capacity of the company. In North America 44% of the machine-tool companies and 34% of the textile companies felt that laying off workers to decrease production capacity was of high importance. In China, it is of very low importance (only 2% of the companies

Table 8. Differences in shop floor control

Variable	chm	nam	cht	nat
Percentage of priority changes communicated by document	77.0	44.4	84.1	38.8
Percentage of operations with time standards	90+%	60-90%	60-90%	60-90%
Percentage of operations scheduled using time standards	90+%	60-90%	60-90%	20-60%
Percentage of orders delivered late	Less than 5%	10-20%	Less than 5%	5-10%
Function of people starting production	In North America, production and administration are the main departments in both industries. In China the industry determines the functional background: in machine-tool it is production, in textile it is administration.			
Basis for starting production on an order	The Chinese companies are driven by the production plan, while North American firms are usually driven by the actual customer order.			
Function of group that purchases	In North America in both industries, it is almost evenly distributed between production and administration. In the Chinese machine-tool industry it is production; in the textile industry it is sales and other.			
Basis of the purchase order	In North America the purchase order is initiated by the customer order, while in China it is based on the production plan.			
Basis of sequencing jobs	In North America the role of the customer order due date is more important than in China where there are other considerations.			
Determination of delivery date	In China the delivery date is negotiable as opposed to being set by customers or company.			
Factors influencing delivery time	In China the production load and the importance of the customer are the main factors; while in North America the product complexity, material availability, and the production load are important.			

Table 9. Importance of factors to change priorities on the shop floor

Variable	chm	nam	cht	nat
Pressure from customers	Medium	High	Medium	High
Material shortages	High	Medium	High	Medium
Changes in sales plan	Medium	Low	Medium	Low

Table 10. Importance of factors to change capacity of the company

Variable	chm	nam	cht	nat
Lay off the extra workers to decrease capacity	Not useful	Moderate	Not useful	Moderate

in the machine-tool industry and 5% of the textile companies thought that it is highly useful).

Table 11 shows the differences in the materials management practices. The Chinese companies' inventory value is lower than their North American counterparts; and they make more use of multiple vendors than the American companies. The North American practices may be a consequence of following the current modern manufacturing concept of single sourcing.

This analysis of the manufacturing practices in the People's Republic of China and North America discloses big differences between the two regions that are common to both industries (non-fashion textile and small machine tool). Although there are probably variations in intensity of these differences among different forms of ownership and geographical location in China, the fact remains that wide differences in practice are quite common. The strength of this conclusion is underscored by the conservative nature of the test that was used in identifying those differences that were identified.

Table 11. Differences in materials management

Variable	chm	nam	cht	nat
Total inventory (million US$)	3.919	9.685	0.963	7.731
Number of vendors per purchased part	2.6	2.0	2.6	1.9
Existence of materials manager position	It is very rare that the Chinese companies do not have this position, while it is not common in North America.			
Exposure to MRP	It is not known in China, but is in North America.			
Two actions to assure timely supply	In North America, companies use long-term contracts and multiple sourcing; while in China hedging and single sourcing are the most typical methods of assuring timely supply.			
Two reasons for subcontracting	Higher-quality and faster production are the reasons in China versus cost considerations and production problems in North America.			
Exposure to JIT	It is known in North America, but not known in China.			

GENERAL CONCLUSIONS AND RECOMMENDATIONS

Many of the variables that define the differences in manufacturing practices between China and North America may not be material to the objective of entering the U.S. market. In order to focus on those practices that matter to performance, a subjective test (based on experience with manufacturing companies around the world) was used to extract the key variables from the study. The test was applied to each of the 41 variables that passed the pure regional effect criteria used to define manufacturing practice differences.

Two very clear conclusions emerge from the analysis. The first is that the practices between North America and China are different on a number of important dimensions. The second is that not all the categories for which performance might be affected are adverse to China. Still, practice and policy changes are required for the Chinese firms to be competitive with those in North America or to enter into productive joint ventures with North American firms.

The first priority in affecting the ability of the Chinese firms to enter the U.S. market is to develop a market awareness and response capability. This will not be an easy task after more than 40 years as a centrally planned economy. Better mechanisms to bring the market requirements into the enterprise for planning and responding purposes will be required. In addition there are a number of manufacturing activities that will need to be modified if smooth joint ventures are to be achieved. Finally, a whole different view of the productive role of assets will need to be created in order to enter U.S. markets and/or collaborate with North American firms.

Some tentative recommendations can be made on the basis of the findings in this study. They fall naturally into the four categories indicated in Table 12. Each will be covered in the discussion in the order presented in Table 12.

Improving market planning

The preliminary recommendations in this area fall into two categories. The first is in the area of systems for forecasting and planning for demand, the second is in setting the delivery dates and sequences of production orders. The forecasting and planning systems of Chinese firms need to be made more flexible and market oriented. This means changing the incentives and practices that currently preclude market and customer information (by nature uncertain) from being incorporated into the forecasts and plans. It also means that customer information and desires must be incorporated into the setting of due dates and determining the sequences for production. Much of this will require major policy changes and education in the practices required.

Awareness of markets, market forces and customer needs must be heightened to meet today's standards of global competitiveness. To do this requires policy changes that facilitate more information flow between customers and enterprises and allow managers more opportunity to get to know their customers. In addition to the policy changes, the managers will need help in learning how to listen to customers and incorporate that information into their plans to meet the customers requirements.

Responding to the market

The Chinese firms show a great deal of rigidity in responding to the market changes that inevitably occur. The forecasts and plans are infrequently revised, probably a carryover from the centrally planned economy system. Clearly, plans need to be made more responsive to the market and must be revised on a more frequent basis if the Chinese firms are

Table 12. The key manufacturing practice variables

Category	Variables	Less desirable	Profile			More desirable
Market planning	Forecaster position	low	*			high
	Forecast factors	political	*			industry
	Planning factors	internal	*			market
	Production planning	frozen	*			flexible
	Job sequence setting	other factors	*			customer order
	Delivery date promise	firm sets		*		customer sets
Market response	Revise forecasts	seldom	*			often
	Revise plans	seldom	*			often
	Freeze production plans	yes		*		no
	Start manufacturing by	plan	*			customer order
	Start purchasing by	plan		*		customer order
	Priorities change by	other	*			customer
	Change delivery time by	other			*	customer
Manufacturing activities	Know modern techniques	no	*			yes
	Have material manager	no			*	yes
	Assure supply by	multi-vendor			*	one-vendor
	Lease outside capacity	yes			*	no
	Subcontract because of	cost/problems			*	quality/speed
Asset management	Sales/person	low	*			high
	Sales/equipment $	low	*			high
	Excess capacity	hoard			*	lease

to meet today's global competition. Also, forecasts that meet these needs will be required. Changes will be needed that allow the incorporation of information from the market into the firm and transmit its responses back to the market.

It is not only necessary to change practices that relate to the revision of the forecasts and plans, but the daily priorities by which the shop works must reflect current needs. A clear prerequisite to setting current priorities is to change policies that block that feedback from the marketplace and prevent it from reaching the factories. This means opening up channels between customers and the companies and helping the firms know what to do with the information that comes in. This, again, is something that many firms are not accustomed to doing and will require substantial shifts in attitudes and practices.

Changing manufacturing activities

In the area of manufacturing activities, there are two aspects to the recommendations. The first has to do with the techniques used by Chinese managers and the second with the practices they apply on the shop floor. Policy changes will be required to change the effectiveness with which manufacturing techniques are used in China. The first of these is a need to increase awareness among Chinese managers of the current techniques used by world class manufacturers. This is necessary if only for understanding the requirements of global customers or for communicating with joint venture partners.

Training in the new techniques is only one part of the change needed, however. Policies that require the use of techniques that are not well understood and may not be helpful must be reversed. For example, the government policy of using ISO 9000 standards or the requirement to use specific management techniques like linear programming or systems analysis is dysfunctional for many firms. It simply diverts energy from more basic needs.

The position of materials manager seems to be widely used in China, but only on the supply side of the market. The material manager impact on the supply side is seen in the subcontracting and leasing capability of the firms. Restricting the material manager to the supply side could be a major underutilization of a potential resource in the sales market.

In order to carry out the new manufacturing management mandate, the plans for manufacturing activities must be communicated to management (and customers) as well as to internal manufacturing management. This means that practices and systems will need to be changed. The persons that prepare the plans must be those most suited to the needs of the enterprise, not necessarily those most politically well positioned. Similarly, a broad range of persons must be able to incorporate market forces into the activities on the factory floor. This means policy changes with regard to who can communicate with people outside the work unit and increased cross-department communication inside the company.

Asset management

This area is the simplest to articulate, but requires the most difficult policy and practice changes. Policies that reward the hoarding of human and equipment assets, as opposed to using them productively, must be reversed. This also implies recognizing that enterprise responsibility for social welfare may require a different form of accounting in order to be able to cost products competitively for the global markets. These are very significant policy changes indeed.

SUMMARY

In all of these tentative recommendations, there is both a need for policy changes and for mechanisms for changing the practices. The practices may be harder to change. In order to change practices, changes are required in education, incentives, and behavior, as well as in policy. There is ample evidence, in this study and many others, that simply requiring the enterprise to make the changes may not achieve the results desired.

REFERENCES

1. W.A. Fischer, "China's Potential for Export-Led Growth," Chapel Hill, NC, International Private Enterprise Development Research Center, Kenan Institute, University of North Carolina, August 1990.

2. W.W. Daniel, *Applied Nonparametric Statistics*, 2nd ed., Boston, PWS-Kent Publishing Company, 1990.

3. S. Siegel, *Nonparametric Statistics for the Behavioral Sciences*, New York, McGraw-Hill, 1956.

4. L. Wilkinson, *SYSTAT: The System for Statistics*, Evanston, IL, SYSTAT, Inc., 1990.

5. L. Wilkinson, *SYGRAPH: The System for Graphics*, Evanston, IL, SYSTAT, Inc., 1990.

6.[*] G. Vastag and D.C. Whybark, "Comparing Manufacturing Practices in North America and Western Europe: Are There Any Surprises?" in: R.H. Hollier, R.J. Doaden and S.J. New (eds.), *International Operations: Crossing Borders in Manufacturing and Service*, Amsterdam, Elsevier, 1992.

[*] This article is reproduced in this volume.

ACKNOWLEDGEMENTS

The authors would like to acknowledge Citibank for supporting Gyula Vastag as a Citibank International Fellow at the Kenan Institute at the University of North Carolina during preparation of this paper. Also the International Private Enterprise Development Research Center of the Kenan Institute and the National Research Center for Science and Technology Development in Beijing, China, supported some of the data analysis and field work. A version of this paper appeared as an International Private Enterprise Development Research Center working paper, November 1992.

Comparing Manufacturing Planning and Control Practices in Europe and Korea

Boo-Ho Rho, Sogang University, Korea
D. Clay Whybark, University of North Carolina-Chapel Hill, USA

ABSTRACT

This paper presents a comparison of the manufacturing planning and control practices between firms in South Korea and western Europe in the non-fashion textile and small machine tool industries. The key differences in practice were found in the details of forecasting and production planning, the market influence on shop-floor activities, the organizational level used in planning and control, and the level of sophistication in terms of computers and techniques used. The implications of these findings are important for firms with plans to enter into joint ventures or other cooperative arrangements. The data used in this study are from a large data base of actual manufacturing practices collected from countries around the world.

INTRODUCTION

This paper presents an analysis of two sets of data from a large data base of manufacturing planning practices gathered from companies around the world. Two industries were surveyed, the non-fashion textile industry and the small machine tool industry. Both these industries are found throughout the world in both advanced and developing nations. Data were collected from western and eastern Europe, Japan, South Korea, North America, and the People's Republic of China; and plans for gathering additional data in South America, Central America, Russia and Australia have been made. This paper, however, only uses the data from western Europe and South Korea. Detailed information on the data-gathering methodology for the total study, with details on the questionnaire, format, and data availability, can be found in Whybark and Rho [1].

A number of factors combined to generate interest in gathering the data which was used in this study. Among these factors were an interest in manufacturing productivity and the key role of manufacturing planning and control practices in enhancing productivity. This interest led to the gathering and analysis of data in China and South Korea. These within-country analyses have shown important relationships between manufacturing planning and control practices and performance in China [2,3] and differences among firms in Korea [4]. Another factor was an interest in comparisons between countries by firms contemplating joint ventures and other business relationships with firms located in the countries represented in the survey.

This paper presents the results of cross-geographical comparisons. The importance of gathering empirical data from the field is very clear when the objective is to compare practices between different parts of the world. It is even more important when part of the interest concerns doing business with firms in these areas. Although theoretical models were used for developing the questions to be asked of managers, it is their actual practices that are of interest in this paper. In order to better understand the similarities and differences in the regions' manufacturing planning and control practices, a relatively straightforward analysis of the questions for which there were large or interesting differences was performed for this paper. A summary of the statistical results is given in the Appendix.

The choice of western Europe and South Korea for this comparison was based on two factors. The first was the emergence of Korea as an industrial power, and the second was the increasing interest by western European firms in engaging in joint ventures, sourcing, and other business arrangements in South Korea. The Korean interest in the study was founded on their interest in their country's productivity considerations. The western Europeans also were interested in doing internal comparisons, but were more highly motivated to learn what different approaches to manufacturing planning and control they may have to face with joint-venture partners. The study does highlight some areas for which understanding the differences will be important and which may require substantial patience. There are also areas of potential for making technological improvements in some of the Korean practices.

DATA GATHERING

The South Korean data-gathering activity was supported by Sogang University and the Korea Productivity Center. The effort involved both a mail survey and personal interviews. The response to the mail survey was disappointingly low, and the decision was made to send teams of interviewers into the field to gather the data directly from the companies. The data, therefore, is from a substantial portion of the population of firms producing small machine tools and non-fashion textiles. In the textile industry, some integrated firms (from spinning and weaving through sewing) are included as well as some smaller less integrated companies. Small tool manufacturers include firms that produce small drill presses, mills, other forming equipment and some computer-controlled equipment. The sample is representative of the Korean firms doing business in these industries, with 89 machine tool and 33 textile companies included.

The European sampling was supported by IMD (the international business school in Lausanne, Switzerland). The European data were much more difficult to gather, and comprise a considerably smaller percentage of the companies involved in the industries. The survey involved a mail questionnaire (in the appropriate language), again with a disappointing response, and a telephone solicitation and follow-up through the trade associations that represent the industries. On the basis of these follow-up phone requests, the final sample was secured. The textile firms include some integrated manufacturers, as well as some fairly small finishing firms; the small machine tool manufacturers range from very small specialized product designers and producers to fairly large, complete-line manufacturers. Again, the sample is representative, with 34 machine tool firms and 24 textile firms.

The questionnaire included sections for general data on the firm itself, forecasting procedures, production planning and scheduling procedures, shop-floor procedures, and purchasing and materials management. The findings in each of these categories are presented in the following section.

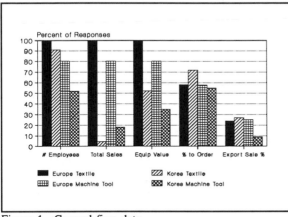

Figure 1. General firm data

GENERAL COMPARISONS

The companies in both industries in Korea have slightly fewer employees than those in Europe, but there is a wide variation in the numbers. Both overall sales and investment in production equipment were substantially less in the Korean firms (see Figure 1). The result is that the Korean firms have substantially smaller investments in equipment per employee and lower sales output per employee.

Some firms in both industries produce strictly to order and others strictly to stock, with many producing a mixture of both. The majority of the production is to order, with Korea having slightly more than Europe in both industries (see Figure 1). Capacity utilization was, on average, about the same (77%-86%) in all cases. The average textile firm exports about 25% of their total sales. The Korean machine tool firms export an average of 14% of sales, much less than their European counterparts whose exports are an average of 32% of sales. This is one of the few instances in the survey where the difference between industries (although only in machine tools) was more pronounced than between countries.

FORECASTING

The preparation of forecasts is about the same in both industries, but does vary between Europe and Korea. Although the sales/marketing function produces the forecasts about 70% of the time in both Korea and Europe, there is a difference in the level at which the forecast is made. In Europe more executive attention is paid to forecasting than in Korea. About 40% of the Korean firms produce forecasts at the lowest managerial level (group or section), as opposed to about 20% in Europe. At the other extreme, only about 3% of Korean forecasts are made at the president or CEO level, compared with about 30% in Europe.

Another difference exists in the use of forecast data between European and Korean firms. The European firms use forecasts to prepare budgets much more frequently than Korean firms (about 30% as compared with 13% respectively). The major use of forecasts in

Korean firms is production planning, which is the second most frequent use, after budget preparation, for European firms.

Perhaps as interesting as these differences are the similarities. For example, the forecasting horizon averages about 1 year, and the increment mostly ranges from 1-3 months in both Korea and Europe. Forecast revisions are performed quarterly. About half of the firms in all cases provide a range for their forecasts, as opposed to a single point.

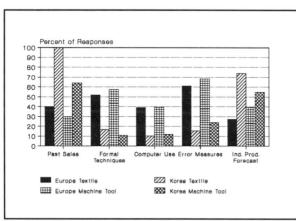

Figure 2. Forecasting data

Figure 2 summarizes several aspects of forecasting that differ between Korea and Europe (the industry differences are small). For example, Korean firms consider past sales data as a much more important objective data input to their forecasting than European firms. European firms use a combination of past sales, order backlog, and industry statistics. There is a much higher use of computers and formal techniques in Europe than in Korea, as would be expected. Many more European than Korean firms measure forecast errors. More than 70% of the Korean firms forecast on an individual product basis, while less than half the European firms do so. The European firms forecast product groups of lines instead. A higher portion of Korean than European firms forecast in physical units instead of monetary terms.

PRODUCTION PLANNING AND SCHEDULING

The input factors for production planning are primarily order backlog and forecast with production capacity also an important factor in Europe. The forecast is a more important factor in Korean firms than it is in Europe (as expected from the forecasting results). As is shown in Figure 3, the plan itself is developed for individual products more frequently in Korea than in Europe, as was also the case with the forecasts. There is a greater use of computers for production planning in Europe than in Korea.

In both industries, production personnel are the primary preparers and users of the production plans. The Korean firms in general, however, tend to involve lower level people more frequently in the preparation and use of the plan. As seen in Figure 3, with the exception of the preparation of the textile plan, nearly twice as many plan preparers and users are section managers or below in the Korean firms as in the European firms.

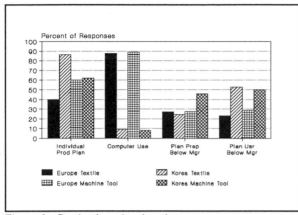

Figure 3. Production planning data

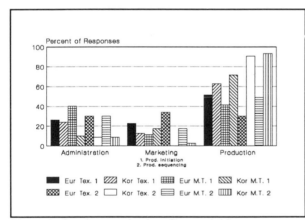

Figure 4. Involvement in shop-floor activities

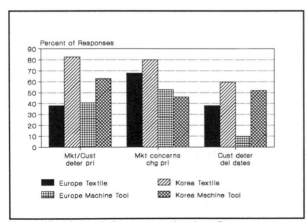

Figure 5. Market influences on the shop floor

Korean production plans cover about 12 months, while in Europe, they average only about 9 months. The increment averages about 1 month, but is slightly higher in Korea than Europe. There are some similarities in planning, however, that are important. The plans are revised monthly, apparently accommodating demand fluctuations, since over 80% of the responses indicate that plans are revised when there is a demand fluctuation. The production plans are used primarily for operational activities like scheduling or material planning rather than for planning activities like budgeting, facilities investments, or staffing in both Korea and in Europe.

SHOP FLOOR ACTIVITIES

There are important differences between Europe and Korea in managing shop floor activities. In initiating production and establishing job sequences, functional involvement is predominantly production in Korea while in Europe, marketing and administration play a substantial role (see Figure 4). Figure 5, however, shows other shop floor activities for which production has less influence in Korean firms than in European firms. In Korea, marketing and customer influences dominate in establishing priorities, determining delivery dates and in changing priorities. In Europe, factors like processing time required and material availability have a more important influence on these items. Thus the initiation of production activities in Korea is dominated by production but market oriented factors dominate in due date and priority considerations while the forces are more functionally balanced in Europe.

The transmittal of data to suppliers is mostly in written form across industries and countries. The European firms, however, have more oral or computer transmission of data than Korean firms. Similarly, there is more written communication to the shop floor in Korea than in Europe. Figure 6 summarizes these and other shop-floor comparisons. The Koreans seem to think more highly of their time standards, but, in both Korea and Europe, many companies confessed to having difficulty in using time standards for management purposes.

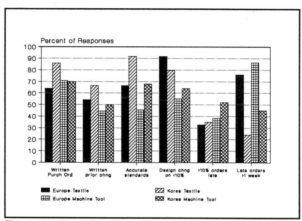

Figure 6. Shop floor data

Although the pattern of engineering or design changes on orders in the shop appear to be different (there are fewer changes in Europe than Korea for textiles, while the reverse is true in machine tools) it is not significant. In terms of lateness, however, the European companies tend to have fewer orders late but their late orders are substantially later (in terms of time) than those of the Korean firms.

When asked to specify the cause of lateness, the companies' answers covered a wide spectrum, with capacity, material availability and quality a problem for everyone. Labor capacity in Europe and machine capacity and quality in Korea were especially important. The means for making capacity changes covered a wide response set. Overtime is used frequently for capacity increases, while inventory is built during times of reduced demand. Apart from overtime, the Korean firms tend to hire people, while in Europe the backorders are allowed to increase. When subcontracting is done in the textile industry, the tendency is for it to be used for capacity expansion in Korea but to be used for strategic considerations such as cost and quality in Europe. In the machine tool industry, both the Korean and European firms seem to subcontract for capacity and cost purposes.

PURCHASING AND MATERIALS MANAGEMENT

The inventory position of Korean firms is slanted to raw materials/purchased parts end of the continuum (see Figure 7). About 50% of total inventory for each industry falls in this category compared to less than 40% in Europe. The purchasing of this raw materials is based on customer orders or the production plan in a majority of cases with Korea placing slightly more emphasis on customer orders. Single sourcing has not been widely accepted as a supply-guarantee mechanism, even though more than 50% of the Korean firms indicated only a single vendor per part.

In general, there is more use of the economic order quantity (EOQ), material requirements planning (MRP) and just-in-time (JIT) among European firms than Korean. In fact very few firms in Korea use either MRP or JIT. Up to 50% of the European machine tool firms use EOQ and/or MRP. Interestingly, in both industries, about 40% of the European firms are benefitting from MRP while about 30% of the Korean firms are having trouble introducing the practice.

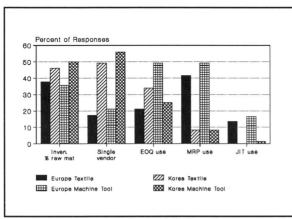

Figure 7. Purchasing and materials data

SUMMARY AND CONCLUSIONS

There are some important differences in practice disclosed in this study. Surprisingly, the differences are greater, in general, between South Korea and Europe than between the textile and machine tool industries. There were issues where there were industry differences, but not many. The great disparity between products, processes and markets between these industries can explain these differences.

The most important differences between Korea and Europe occur in the involvement of marketing, the detail, and level of management used for forecasting and planning; the influence of marketing and customers on shop-floor activities; and the level of computer usage and technique application by the firms. The implications of these findings are quite clear. In planning joint ventures, western European firms must take into account the differences in organizational level and sophistication South Korean firms use in their activities. In the day-to-day management of the activities the differences in responsiveness to customer requests may cause real problems unless addressed. These fundamental distinctions must be faced in order to minimize the difficulty in integrating the practices in western Europe with those in South Korea. Additionally, there is room in the Korean firms for increasing their knowledge and use of computers, and of some of the newer techniques that are available for forecasting, planning and scheduling.

The primary conclusion arising from this study is, not surprisingly, that caution must be exercised in any European-South Korean joint venture. Secondly, it is important to understand in detail some of the practices of the joint-venture partner in order to integrate systems and approaches. Finally, there is an opportunity for training and technology transfer in order to increase the sophistication of joint-venture partners.

REFERENCES

1.* D.C. Whybark and B.H. Rho, "A Worldwide Survey of Manufacturing Practices," Discussion Paper No. 2, Bloomington, IN, Indiana Center for Global Business, Indiana Business School, May 1988.

2. J.G. Chen, "Operations management in Chinese textile enterprises," Master's thesis, Department of Systems Engineering, Shanghai Institute of Mechanical Engineering, 1987.

3. X.J. Yan, "Operations management in Chinese machine tool manufacturing firms," Master's thesis, Department of Systems Engineering, Shanghai Institute of Mechanical Engineering, 1987.

4. B.H. Rho and D.C. Whybark, "Production planning and control in Korea," Pan-Pacific Conference IV, Taipei, Taiwan, 1987.

* This article is reproduced in this volume.

ACKNOWLEDGEMENTS

The authors would like to thank the Korea Productivity Center, the Sogang Institute for Economics and Business, the Shanghai Institute of Mechanical Engineering, IMEDE (the international business school in Lausanne, Switzerland), and the Business School at Indiana University for supporting the project. A version of this paper has been published in the *International Journal of Production Research*, 28, No. 12 (Fall 1990), 2393-2404.

APPENDIX
(Values are percentages unless otherwise indicated)

Item	Korea		Europe		Significance*	
	Textile	Machine	Textile	Machine	Level	Test
General data						
Sales ($)	15.3M	14.4M	402.1M	53.0M		
Employment	484	468	531	595		
Products equipment ($)	8.1M	10.0M	14.9M	19.2M		
Sales/person ($)	37.6K	29.6K	890.1K	142.9K	<0.03	*t*
Equipment/person ($)	12.1K	11.1K	234.4K	33.8K	<0.02	*t*
Production to order	73	82	59	72	<0.005	*t*
Export ratio†	28	14	24	32	No,<0.0005	*t*
Forecasting						
Marketing involvement	75	80	61	69	No	K-S
By lower management	40	56	9	25	<0.001	K-S
Use						
Budget	13	13	27	30		
Material	15	23	18	10		
Manpower	4	3	3	14		
Facilities	7	7	3	1	<0.05	K-S
New products	7	3	3	3		
Sales plan	22	18	17	17		
Production plan	33	33	27	21		
Forecast period (months)	12	12	12.7	17.2	<0.0005	*t*
Forecasting increment†	2.2	2.0	1.5	2.9	<0.02,<0.05	*t*
Revision/year	3.7	3.4	3.6	4.2	No	*t*
Range	56	55	41	45	No	K-S
Past sales	100	83	42	33	<0.001	K-S
Computer use	11	15	38	42	<0.001	K-S
Formal techniques	17	14	52	62	<0.001	K-S
Error measure	15	30	62	73	<0.001	K-S
Individual products	74	71	27	43	<0.001	K-S
Physical units	58	58	37	47	<0.001	K-S
Order backlog	40	47	27	27		
Forecast	51	42	18	19	<0.001	K-S
Production capacity	2	2	25	26		
Individual products	84	79	40	61	<0.001	K-S

(continued)

Appendix, continued

Item	Korea		Europe		Significance*	
	Textile	Machine	Textile	Machine	Level	Test
Administration	36	21	33	29		
Production	48	70	30	49	No	K-S
Marketing	16	10	26	17		
Lower management prepares†	24	59	27	29	No,<0.025	K-S
Use						
Administration	11	3	33	16		
Production	81	93	48	54	<0.10	K-S
Marketing	7	4	11	16		
Lower management use	51	64	23	28	<0.001	K-S
Plan period (months)	12	12	9.5	8.4	<0.001	*t*
Plan increment (months)	1.3	1.7	1.1	0.8	<0.001	*t*
Revisions/year	9	10	11	11	No	*t*
Plan use operational	58	64	76	69	No	K-S
Shop floor						
Initial production	64	73	46	42	<0.05	K-S
Production sequence	92	95	31	50	<0.001	K-S
Market/customer determines priority	82	75	48	47	<0.001	K-S
Market changes priority	80	63	68	59	No	K-S
Customer determined due date	60	70	38	18	<0.001	K-S
Write purchase orders	90	85	62	69	<0.05	K-S
Write shop orders	58	65	52	35	<0.005	*t*
Good standards	91	81	65	42	<0.001	K-S
<10% Engineering changes†	79	76	91	53	No,No	K-S
>10% Late orders	30	61	29	35	<0.10	K-S
>1 week late	21	51	75	85	<0.001	K-S
Due to						
Machine capacity	28	24	4	6		
Labor capacity	11	15	23	18		
Material	13	15	21	19	<0.005	K-S
Quality	32	29	19	15		
Due-date changes	6	8	8	5		
Capacity increase						
Hire	24	29	7	17		
Overtime	49	50	53	43	<0.1	K-S

(continued)

Appendix, continued

Item	Korea		Europe		Significance*	
	Textile	Machine	Textile	Machine	Level	Test
Capacity decrease						
Layoff	8	4	9	21		
Undertime	15	17	30	21	<0.001	K-S
Reduce time	0	0	17	26		
Inventory	69	75	35	24		
Subcontracting capacity†	71	58	40	47	No,No	K-S
Strategic	29	42	60	53		
Materials management						
Raw material	48	51	39	36	<0.01	*t*
Purchase to order	43	51	20	42	NO	K-S
Purchase to plan	25	31	35	33		
Single vendor	50	57	18	22	<0.001	K-S
EOQ use†	35	26	22	50	No,<0.10	K-S
JIT use	0	1	13	21	No	K-S
MRP use	9	9	42	50	<0.001	K-S

* The level of significance is for the test between Korea and Europe (industry data aggregated) except where noted. The *t* test (*t*) and Kolmogorov-Smirnov (K-S) tests were used for numerical and nominal comparisons, respectively. The indication 'No' is used when the value exceeds a significance level of 0.10.
† Level of significance is for the test between industries: first value for textile, second for machine tool.

Comparing Manufacturing Practices in North America and Western Europe: Are There Any Surprises?

Gyula Vastag, Budapest University of Economic Sciences, Hungary
D. Clay Whybark, University of North Carolina-Chapel Hill, USA

ABSTRACT

This paper focuses on the regional differences in the manufacturing practices between North America and western Europe using the data from the Global Manufacturing Research Group. The general belief has been that there would be few differences between these two regions relative to their comparison with countries like Korea or the People's Republic of China so no analysis of these "siblings" has yet been done. This research tests the broad hypothesis that there are few significant differences in the manufacturing practices of western Europe and North America. The hypothesis holds for the stringent test used in this paper to disclose the differences. In general, we can say that no surprises were uncovered, but knowledge of the few significant differences might be of importance to some firms.

INTRODUCTION

The Global Manufacturing Practices Project was begun in 1986. It involves gathering and analyzing data on manufacturing practices in companies around the world. The questions come from companies in two industries: non-fashion textiles and small machine tools, principally from the materials planning and control functions [1]. These two industries are found virtually everywhere in the world. Information obtained from surveys in North America, Australia, Bulgaria, Chile, the People's Republic of China, Finland, Hungary, Japan, Mexico, Korea, the former USSR and western Europe is now contained in the project's data base.

A continuing theme in the analysis of the data base is the comparison of practices between various regions, although many other questions have been raised. This paper compares practices in North America with those in western Europe. Other comparisons have been made, usually involving regions expected to be quite different [2,3]. The "lateness" of this work can easily be explained by the expected similarity of the two regions (both are highly developed, industrialized, and are very close to each other culturally). This gave rise to the general feeling that analyzing the two "siblings" should take a low priority, simply because there would be few differences.

The research reported in this paper uses an analytical framework specifically developed for bilateral comparisons using the project's data base. A conservative approach, called a "pure" regional effect, was defined to assure a stringent test before declaring and describing any difference. This approach means that any differences found in manufacturing practices between western Europe and North America would be statistically significant and consistent between industries.

THE DATA

In this paper, we compared two big, industrialized and developed parts of the world: North America and western Europe. The North American data was gathered in four regions of the United States and in Canada. Table 1 shows the geographical distribution of the companies. Except for Canada and the Western U.S. region, which are under-represented, the sample distribution is similar to the population distribution of firms.

Table 1. Distribution of the North American companies by region and industry.

Region	Machine tool	Textile	Total
Northeast US	12	19	31
Southeast US	5	17	22
Midwest US	26	6	32
Western US	2	5	7
Canada	-	3	3
Total	45	50	95

Table 2 provides the distribution of the western European companies by country. Although the western European sample comes from 10 countries, it is not representative of all of western Europe. The countries in the sample are all from the most developed northern part of western Europe.

Table 2. Distribution of the western European companies by country and industry

Country	Machine tool	Textile	Total
Austria	-	2	2
Belgium	1	-	1
France	13	5	18
Germany	4	3	7
Holland	2	-	2
Ireland	-	1	1
Scotland	1	-	1
Sweden	2	5	7
Switzerland	5	2	7
United Kingdom	6	6	12
Total	34	24	58

Table 3 summarizes the data by region and industry group.

Table 3. Distribution of the sampled companies by region and industry group

Region	Machine tool	Textile	Total
North America	45	50	95
Western Europe	34	24	58
Total	79	74	153

The questionnaire (see [1]) contains 65 questions distributed among the five sections shown in Table 4. There are 95 original variables and additional variables were derived from them for the study. Some of the derived variables are relationships between original variables (for example, sales per employee or inventory turnover). Others result from individually analyzing each possible response of a multi response question (e.g., how do you increase capacity?, overtime, hire, etc.). Some of the variables are measured on ratio scales, others on ordinal scales and some on nominal scales. The distribution of the measurement scales for the variables in the different sections of the questionnaire is shown in Table 4.

Table 4. Distribution of measurement scales for the variables

Questionnaire Section	Ratio	Ordinal	Nominal	Total
Company profile	16	-	-	16
Sales forecasting	4	3	8	15
Production planning	10	8	9	27
Shop floor control	6	30	11	47
Purchasing	6	2	6	14
Total	42	43	34	119

The different scales require different methods for determining statistical differences between the responses. In the next section we outline the analytical procedure and give a brief overview of the techniques used. Most statistical computations were performed with the SYSTAT and SYGRAPH software [4,5].

STRUCTURE OF THE ANALYSIS

An underlying general assumption in the GMRG project is that the differences in manufacturing practices between the companies can largely be explained by the following three basic factors:
  (1) Regional/cultural: which should be seen by comparing companies from different geographical areas.
  (2) Industrial: which should be found between the machine tool and textile companies.
  (3) Political/economical: which should be shown by comparing companies operating in different types of economies (market or centrally planned).

In this paper we focus on the regional/cultural factors since: (i) our primary objective is to show the regional differences between North America and western Europe, (ii) the industrial differences can be treated separately using the same analytical framework, (iii) both North America and western Europe are market economies and in this respect the third differentiating factor simply does not exist. A brief description of the framework used to analyze the regional differences given next. A detailed explanation is provided in [6].

The four groups shown in Table 3 permit us to make five different types of comparisons. From among them we use the two most directly focused on regional differences, namely comparing regions within the industry (e.g., compare the North American machine tool or textile companies with the western European ones). This approach does not require any prior determination of whether the industry data can be pooled.

We use the concept of a "pure effect" to make the comparisons between the two regions (North America and western Europe). There is a pure regional effect present for a variable if there are statistically significant differences in *both* industries and these differences are in the *same direction*. For example, a pure regional effect exists if the mean value of a variable is higher in North America than in western Europe for both industries or, for another variable, if the proportion of responses for a certain practice is higher for one of the regions in both industries. In this paper we only describe those variables for which there is a pure regional effect, recognizing that we will possibly leave many other interesting differences for future analysis.

For testing the statistical differences between the responses of the two regions, we used procedures appropriate to the measurement scale of the variable. For variables measured on a ratio scale it is common to use the Student's t-distribution to determine significant differences between mean responses. For the test to be valid, however, there are normality and homoscedasticity assumptions to be met. A preliminary analysis of the data indicated that these assumptions were not often met due to skewness and/or different variances in the two regions' data.

Since the t-test assumptions do not broadly hold, we used the Kruskal-Wallis one way analysis of variance of ranks to test the differences in the location parameter (mean or median) of ratio scale variables. The Kruskal-Wallis test is the nonparametric counterpart of the t-test [7]. It tests differences in the distributions of responses of two groups by comparing the ranks of the individual responses.

For those ratio scale variables where significant regional differences were found, we prepared a box plot. Box plots show the medians, quartiles, extreme points, other information on the distributions, and whether the differences between distributions are in the same direction. We also applied the median test. The median test is not as powerful as the Kruskal-Wallis test but can be visualized with the box plot [4].

For analyzing the differences in the variables measured on an ordinal scale we used the Kolmogorov-Smirnov test. The Kolmogorov-Smirnov test evaluates differences that may exist between distributions. The approach is to calculate the maximum difference between the cumulative distributions of the responses for each region. The larger the differences the more probable that the distributions are significantly different [7]. For variables that were found to be significantly different, a stacked bar chart was used to visualize the distributions and determine if the directions were the same for both industries.

In the case of the nominal scale variables we used two methods depending on the expected number of responses per cell. When the required minimums were met (less than 20% of the cells have fewer than five expected responses), we used the chi-square test of homogeneity to determine if the response patterns were different. In some cases, this required combining responses for some choices into an "other" category. When the requirements weren't met, we tested the differences in the proportions of responses for an individual response category using the binomial test [7] and the North American sample as the base for the comparison. For both tests, we used stacked bar charts to illustrate the variables for which significant differences were found.

The binomial test requires us to focus on a single response from a group of responses (essentially lumping the remaining responses into an "other" category). To guide the choice as to which response(s) to evaluate, we started with the modal response and evaluated as many categories as possible without going below the minimum required expected number of responses.

The structure of the remainder of the paper follows the sections of the questionnaire: general data (company profile and some derived variables), sales forecasting, production planning and scheduling, shop floor control, and purchasing and materials management. We provide summary statistics for all variables using the means for ratio scale variables, and modal responses for ordinal and nominal variables. The results of the appropriate statistical tests are provided as well. Variables that have a pure regional effect have been illustrated with box plots or stacked bar charts.

GENERAL DATA

The general data variables come from questions belonging to the company profile section of the questionnaire plus several derived ratio scale variables. The derived variables are: total sales (domestic sales+export sales), sales per factory worker (total sales/factory workers), sales per employee (total sales/total employees), export ratio (export sales/total sales), factory worker to employee ratio (factory workers/total employees) and inventory turnover (total sales/inventory investment). The financial data are given in millions of U.S. dollars, the make-to-stock, make-to-order and utilization rates in percentages. The abbreviation 'nam' denotes the group of the North American machine tool companies, 'nat' is for the textile companies from the same region while 'wem' and 'wet' are for their western European counterparts.

Table 5. Mean values of the ratio scale general data variables[*]

Variable	'nam'	'wem'	'nat'	'wet'
Employees	240	596	505	531
Factory workers	113	299	320	401
Domestic sales[†] (million US$)	30.24	29.76	17.30	34.66
Export sales (million US$)	2.81	24.10	2.12	9.55
Total sales (million US$)	33.05	53.86	19.42	44.21
Sales per factory worker	0.273	0.429	0.550	0.118
Sales per employee	0.117	0.143	0.474	0.100
Export ratio	0.143	0.367	0.118	0.271
Factory worker to employee ratio	0.523	0.517	0.798	0.908
Inventory turnover	4.151	6.289	6.872	4.069
Make-to-order percentage	74.53	71.88	75.74	59.22
Make-to-stock percentage	25.36	28.12	24.16	42.09
Capacity utilization (%)	71.40	76.93	76.43	81.96
Equipment investment (million US$)	6.106	19.22	4.056	14.91
Days of training	4.793	2.130	9.091	3.455

[*] Only 15 variables are listed here. The sixteenth is the make-to-state plan percentage, which does not apply to either of the regions.
[†] The means for the domestic and export sales were computed using the responses from those companies which did not have missing values in either their domestic or export sales. When neither are missing the total sales volume is the sum of these values.

Table 6. Significance levels for the ratio scale general data variables using the Kruskal-Wallis test

Variable	Machine tool	Textile
Employees	0.065	0.318
Factory workers	0.108	0.350
Domestic sales	0.675	0.206
Export sales	0.013	0.021
Total sales	0.915	0.105
Sales per factory worker	0.068	0.484
Sales per employee	0.023	0.806
Export ratio	0.001	0.023
Factory worker to employee ratio	0.954	0.044
Inventory turnover	0.060	0.042
Make-to-order %	0.767	0.033
Make-to-stock %	0.712	0.028
Capacity utilization %	0.051	0.006
Equipment investment	0.630	0.436
Days of training	0.001	0.456

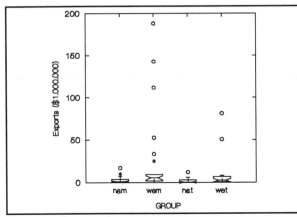

Figure 1. Export sales

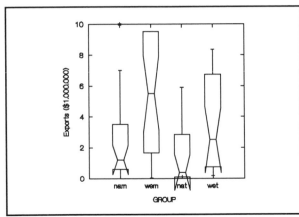

Figure 2. Export sales (expanded scale)

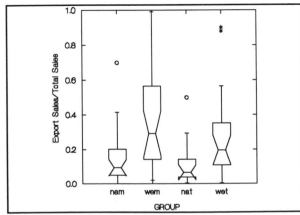

Figure 3. Export sales/total sales

Table 5 shows the mean values of the general data variables. Only 15 of the 16 variables indicated in Table 4 are shown here. The 16th variable is "Make-to-state plan," one that doesn't apply in either of these regions. Table 6 shows the significance levels of the variables in the regional comparisons. Figures 1-5 show the box plots for those variables which have a possible pure regional effect.

We use Figures 1 and 2 to illustrate box plots. An important measure for each box is the interquartile range, the distance from the upper quartile to the lower quartile (i.e., the distance between the top and bottom of the box). In Figure 1, the circles at the top of the plot denote values which are beyond three times the interquartile range above the box. The asterisks show those observations which are more than 1.5 times the interquartile range above the box. The vertical line above and below the box is terminated at the highest and lowest data points within 1.5 times the interquartile range.

The range of zero to 200 million U.S. dollars, which incorporates all of the observations, hides important details. Figure 2 shows the same variable with the range reduced. This figure shows the details more clearly. The top of each box is the upper quartile cut off. The bottom is the lower quartile cut off. The horizontal line in each box marks the median. The boxes are notched at the median and return to full width at the lower and upper 95% confidence interval levels. In some of the cases where we reduced the range in order to get a better picture of the boxes, some of the points may lie outside the figure and thus not be shown.

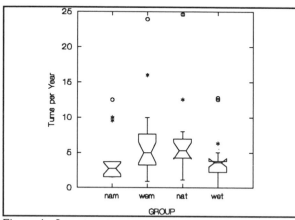

Figure 4. Inventory turnover

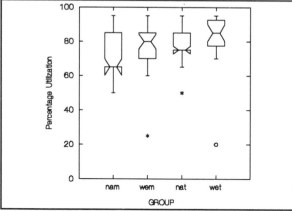

Figure 5. Capacity utilization

There are four variables from Tables 5 and 6 that might have pure regional effects: export sales, export ratio, inventory turnover, and capacity utilization. For export sales, the Kruskal-Wallis test of ranks indicates the two regions are significantly different in each industry. Since the confidence intervals of the two regions for the textile industry overlap in the box plot of Figure 2, the median test indicates that the difference in the medians is not significant at the 5% level. However, we will consider this variable as having a pure regional effect and that the western European companies' export sales are greater, on the average, than the North American ones.

The analysis is more conclusive for the export ratio. The data in Tables 5 and 6 indicate that a higher portion of total sales is exported by western European firms than those in North America. The box plot in Figure 3 corroborates this finding. So, we conclude that western European export activity is greater than North American, even though one of the tests for one of the industries did not achieve a 95% level of confidence.

For inventory turnover, the directions of the differences were different. This is seen clearly in Figure 4. Thus the regional effect is not "pure." The final variable for which there might be a pure effect is capacity utilization. The direction is the same for both industries. The utilization is higher in western European firms than in North American ones. The Kruskal-Wallis test for the machine tool industry did not quite achieve a 95% level of confidence. The box plot in Figure 5 shows the effect clearly, however, and indicates that the median test is significant at the 0.05 level for both industries.

The box plots are helpful in spotting instances where the assumptions necessary for the t-test may not be met. In Figure 5, for example, we can see the skew in the utilization, particularly for the North American data. The use of the t-test under such conditions could lead to highly suspect conclusions. In fact, the t-test results, unlike the Kruskal-Wallis test, indicate no significant differences between regions. The significance levels for the comparisons in the machine tool and textile industries 0.102 and 0.141 respectively for the t-test as opposed to 0.051 and 0.006 for the Kruskal-Wallis test.

We end up with three variables that are significantly different in this section: export sales, export ratio and capacity utilization. In all cases the levels of activity are greater in western European firms than those in North America.

SALES FORECASTING

Table 7 shows the mean values of the ratio scale variables from the sales forecasting section of the questionnaire.

Table 7. Mean values of the ratio scale sales forecasting variables

Variable	'nam'	'wem'	'nat'	'wet'
Length of the forecast interval (months)	13.73	17.18	10.61	12.68
Length of the forecast increment (days)	75.38	86.16	42.51	45.81
Number of forecast modifications per year	6.64	5.08	6.67	4.00
Forecast error (%)	16.94	16.87	15.61	18.40

The length of the forecast interval is the period of time covered by the forecast, measured in months. The increment is the length of the periods into which the forecast is divided, measured in days. The number of forecast modifications per year is provided along with the average forecast error percentage.

The Kruskal-Wallis test results shown in Table 8 indicate very few significant differences, and no pure regional effects. The similarity of the regions' forecasting practices can be further underscored by closer examination of the data.

We illustrate this with the box plot of the length of the forecast interval in Figure 6. In three of the four groups, there is no "box." This signifies that the interquartile range is zero and that at least 50% of the points lie on the median. For all four boxes, the median is the same. For these data, the interpretation is clear. The modal forecast interval is one year. Using the mean data of Table 7 alone does not disclose the high frequency of the one year planning horizon.

Table 8. Significance levels for the ratio scale sales
forecasting variables using the Kruskal-Wallis test

Variable	Machine	Textile
Length of the forecast interval (months)	0.080	0.239
Length of the forecast increment (days)	0.700	0.831
Number of forecast modifications per year	0.385	0.0500
Forecast error (%)	0.232	0.349

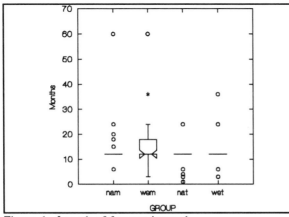

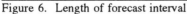

Figure 6. Length of forecast interval

Table 9 shows the modal responses for the ordinal scale sales forecasting variables. The organizational position (e.g., president, section manager, etc.) of the person that makes the forecast is shown, as well as the weight of subjective factors in the forecast itself and whether computers are used to produce it. Table 10 contains the significance levels for the regional comparisons of these variables using the Kolmogorov-Smirnov test. Few significant differences were found and there are no pure regional effects.

Table 9. Modal responses of the ordinal scale sales forecasting variables

Variable	'nam'	'wem'	'nat'	'wet'
Position of forecaster	Vice pres.	Dept. head	Vice pres.	Pres.
Weight of subjective factors in forecast	40-60%	40-60%	20-40%	40-60%
Use of computers for forecasting	no	no	no	no

Table 10. Significance levels for the ordinal scale sales forecasting variables
using the Kolmogorov-Smirnov test

Variable	Machine tool	Textile
Position of sales forecaster	0.267	0.087
Weight of subjective factors in forecast	0.959	0.731
Use of computers for sales forecasting	0.776	0.053

Table 11. Modal responses of the nominal scale sales forecasting variables

Variable	'nam'	'wem'	'nat'	'wet'
Functional grouping of sales f'caster	Sales, Marketing	Sales, Marketing	Sales, Marketing	Sales, Marketing
Subjective factors in f'cast	Company info and market research	Company info and market research	Company info and market research	Company info and market research
Objective factors in f'cast	Past sales and current order backlog	Past sales and current order backlog	Past sales and current order backlog	Past sales and current order backlog
Formal techniques used	None	None	None	None
F'cast for individual products or product groups, in monetary or physical units	Prod. groups, monetary terms	Prod. groups, monetary terms	Prod. groups, monetary terms	Prod. groups, monetary terms
Specification of range or single value	Range, single value (tie)	Single value	Range	Single value
Two most important uses of f'cast	Bud. and prod. planning	Bud. and inventory or bud. and facil. plan. (tie)	Inventory and prod. planning	Bud. and inv. plan. or facil. and man. plan. (tie)
Measure of f'cast error	None	None	None	None

Table 12. Significance levels for the nominal scale sales forecasting variables
using the chi-square test

Variable	Machine tool	Textile
Functional grouping of sales f'caster (sales)	0.236	0.272
Subjective factors in f'cast (company info and market research)	0.067	0.358
Objective factors in f'cast (past sales and current order backlog)	0.035	0.927
Formal techniques used (none)	0.751	0.313
F'cast for individual products or product groups, in monetary or physical units (product groups and monetary terms)	0.564	0.541
Specification of range or single value (range)	0.693	0.432
Two most important uses of f'cast (budget and inventory planning or budget and prod. Planning)	0.338* (0.726)	0.348* (0.091)
Measure of the f'cast error (none)	0.180	0.786

* More than one fifth of the cells have expected frequencies less than five so these chi-square significance values are suspect. The significance levels obtained by the binomial test are in parenthesis.

The modal responses for the nominal scale sales forecasting variables are shown in Table 11 and the significance levels for the regional comparisons are provided in Table 12. The nominal scale variables include information on the department of the forecaster, what factors (subjective and objective) are used in the forecast and whether formal techniques are used. Also shown are the units of the forecast (products or monetary, groups or individual products), whether the forecast is given as a range or just a single value, what the forecast is used for and what method is used to measure forecast error. As with the ordinal scale variables, few significant differences were found and there were no pure regional effects.

No pure regional effects were found in this section of the questionnaire.

PRODUCTION PLANNING AND SCHEDULING

The ratio scale production planning variables include the length and increment of the production plan, the number of revisions per year and how much of a demand fluctuation is required to trigger a revision of the plan. Variables for the length and increment of three more detailed plans (schedules) are also included. Table 13 shows the mean values for all the ratio scale variables in this section.

Table 13. Mean values of the ratio scale production planning variables

Variable	'nam'	'wem'	'nat'	'wet'
Length of production plan (months)	11.757	9.040	7.674	9.533
Length of production plan increment (days)	29.579	26.958	26.690	31.867
Number of production plan revisions per year	10.171	11.286	29.643	11.789
Demand fluct. limit (%) to change prod. plan	19.545	14.375	12.727	11.125
Length of detailed plan No. 1 (weeks)	47.333	12.944	19.200	6.875
Length of detailed plan No. 1 increment (days)	11.889	18.077	31.080	6.625
Length of detailed plan No. 2 (days)	173.364	97.33	88.474	14.677
Length of detailed plan No. 2 increment (days)	16.000	8.429	7.222	2.500
Length of detailed plan No. 3 (days)	102.500	81.833	38.875	7.833
Length of detailed plan No. 3 increment (hours)	140.000	6.000	24.483	32.000

Table 14 shows the significance levels for the regional comparisons of these variables, using the Kruskal-Wallis test. There are a few significant differences within an industry, but no pure regional effects were found.

Table 15 shows the modal responses for the ordinal scale variables. The variables describe the organizational positions of the plan preparer and user, and the education level of the preparer. Information on the use of techniques like revising when there is a demand fluctuation, comparing plan to actual, freezing the plan or using computers in preparing the plan is also given. One variable contains information on whether the use

Table 14. Significance levels of the ratio scale production planning variables
using the Kruskal-Wallis test

Variable	Machine tool	Textile
Length of production plan (months)	0.021	0.287
Length of production plan increment (days)	0.540	0.377
Number of production plan revisions per year	0.539	0.048
Demand fluct. limit (%) to change the production plan	0.161	0.340
Length of detailed plan No. 1 (months)	0.002	0.450
Length of detailed plan No. 1 increment (weeks)	0.359	0.087
Length of detailed plan No. 2 (days)	0.724	0.203
Length of detailed plan No. 2 increment (days)	0.184	0.186
Length of detailed plan No. 3 (days)	0.948	0.203
Length of detailed plan No. 3 increment (hours)	0.016	0.511

Table 15. Modal responses for the ordinal scale production planning variables

Variable	'nam'	'wem'	'nat'	'wet'
Position of the production planner	Dept. head	Dept. head	Vice president	Dept. head
Education of the production planner	College	High school	College	High school
Position of the production plan user	Dept. head	Dept. head	Vice president	Dept. head
Revision upon demand fluctuation	Yes	Yes	Yes	Yes
Comparison of plan to actual results	Yes	Yes	Yes	Yes
Use of frozen production plan	No	No	No	No
Use of computers in production planning	Extensive	Extensive	Not at all or moderate (tie)	Moderate
Helpful to use computers (if not used)	No	No	No	No

Table 16. Significance levels for the ordinal scale production planning variables
using the Kolmogorov-Smirnov test

Variable	Machine tool	Textile
Position of the production planner	0.815	0.345
Education of the production planner	0.000	0.000
Position of the production plan user	0.909	0.313
Revision upon demand fluctuation	0.273	0.981
Comparison of plan to actual results	0.295	0.804
Use of frozen production plan	0.618	0.003
Use of computers in production planning	0.413	0.499
Helpful to use computers (if not used)	0.514	0.129

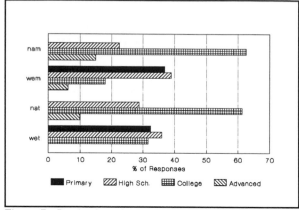

Figure 7. Production planners' education

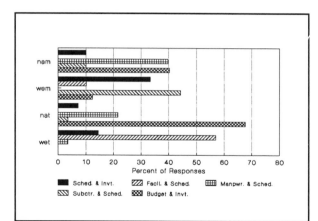

Figure 8. Two production plan uses

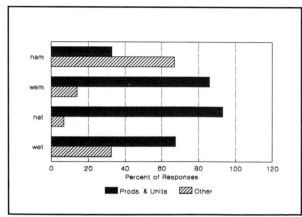

Figure 9. Production plan No. 3

of computers would help for those firms where it is not currently used. The Kolmogorov-Smirnov test results in Table 16 show that there is a potential pure regional effect regarding the education level of the production planners.

The bar chart in Figure 7 shows that the North American firms have more production planners with college training and fewer with only a primary school education than the western European companies. This result is likely a workplace reflection of the differences between the European and the American formal education systems. Another difference is the European apprentice system which places some young people in a program of practical skills training and work experience as preparation for a job in a manufacturing company. People from this program would be classed as primary or, at best, high school graduates, which could greatly understate their manufacturing skills.

The nominal scale production planning variables contain information on the factors used in production planning, the units of the production plan (products or monetary, groups or individual products), the uses of the plan, the functional and educational field of the production planner, and the functional field of the plan user. The units of three more detailed plans is also given. The modal responses for each of these variables is given in Table 17. The chi-square and binomial tests of the nominal variables are provided in Table 18. Both tests indicate that there are regional differences in what the firms consider the two most important uses of the production plan.

The bar chart in Figure 8 was prepared to verify that this is a pure regional effect. In the North American firms in both industries, operations scheduling is combined with inventory or manpower planning while facility planning is combined with subcontracting or operations scheduling in western Europe. Strengthening the pure regional effect is the observation that the combination of subcontracting and facility planning is not mentioned by North American firms while being quite important for western European ones. The opposite is seen for the combination of manpower planning and operations scheduling.

Table 17. Modal responses of the nominal scale production planning variables

Variable	'nam'	'wem'	'nat'	'wet'
Two factors for prod. plan	Customers' plan and forecast	Prod. capacity, forecast	Customers' plan and forecast	Prod. capacity, forecast
Plan for products or product groups, in monetary or physical units	Products, physical units	Products, physical units	Products or product groups (tie), physical units	Product groups, physical units
Prod. planner's functional grouping	Production	Production	Administration	Administration or production (tie)
Prod. planner's educational field	Business	Engineer	Business	Business
Two uses of the prod. plan	Manpower plan, op. sched. or op. sched. invt. plan (tie)	Facil. plan., op. sched.	Op. sched., inventory plan	Subcontracting, facil. plan.
Prod. plan user's function	Prod.	Prod.	Prod.	Prod.
Detailed plan No. 1 for products or product groups, in monetary or physical units	Products, physical units	Products, physical units	Products, physical units	Products or product groups (tie), physical units
Detailed plan No. 2 for products or product groups, in monetary or physical units	Products, physical units	Products, physical units	Products, physical units	Products, physical units
Detailed plan No. 3 for products or product groups, in monetary or physical units	Products, physical units	Products, physical units	Products, physical units	Products, physical units

Table 18 shows a significant difference in the variable describing the factors used in producing the production plan for both industries using the chi-square test. The binomial test was run and no pure regional effect was found. In contrast, the binomial test for the units used in detail plan No. 3, indicates a potential pure regional effect, while the chi-square test did not. Figure 9 shows that there is no pure regional effect, however. The figure shows the percentage of firms from each group that make their detailed plans in physical units of individual products (the modal choice of all groups) as opposed to some other combination. We can see there is no pure regional effect since the percentages increase between regions in one industry and decrease in the other.

Table 18. Significance levels for the nominal scale production planning variables
using the chi- square test

Variable	Machine tool	Textile
Two factors for prod. plan	0.012*	0.024*
	(0.250)	(0.049)
Plan for individual products or product groups, in monetary or physical units	0.453	0.812
Prod. planner's function	0.556	0.042
Prod. planner's field of education	0.283	0.311*
		(0.204)
Two uses of the prod. plan	0.018*	0.000*
	(0.000)	(0.000)
Prod. plan user's function	0.096	0.642*
		(0.209)
Detailed plan No. 1 for individual products or product groups, in monetary or physical units	0.589*	0.813*
	(0.093)	(0.266)
Detailed plan No. 2 for individual products or product groups, in monetary or physical units	0.302	0.675*
		(0.293)
Detailed plan No. 3 for individual products or product groups, in monetary or physical units	0.036*	0.133*
	(0.006)	(0.055)

* More than one fifth of the cells have expected frequencies less than five so these chi-square significance values are suspect. The significance levels obtained by the binomial test are in parenthesis.

In this section we end up with two variables that have a pure regional effect: the education level of the person who prepares the production plan and the use to which that plan is put. The formal educational levels are higher among North American firms. The plan is used for operations scheduling combined with inventory or manpower planning in North American firms and facility planning combined with subcontracting or operations scheduling in western European ones.

SHOP FLOOR CONTROL

Table 19 shows the mean values for the ratio scale shop floor control variables. They include typical production lead times, whether priority change information is communicated to the production personnel orally or in written form, delivery times promised to customers (either fixed by the industry, or typical minimum and maximum times), and the change in physical volume over the last two years.

Table 19. Mean values of the ratio scale shop floor control variables

Variable	'nam'	'wem'	'nat'	'wet'
Production leadtime (days)	167	163	26	43
Oral priority change (%)	49.395	49.333	55.521	47.609
Written priority change (%)	44.442	31.200	38.750	52.391
Fixed delivery time (days)	146	76	25	33
Min. delivery time (days)	100	60	19	15
Max. delivery time (days)	289	233	55	66

Table 20 contains the results of the Kruskal-Wallis tests on the ratio scale variables. Few significant differences were found and there are no pure regional differences between North American and western European firms for the ratio scale variables.

Table 20. Significance levels for the ratio scale shop floor control variables using the Kruskal-Wallis test

Variable	Machine tool	Textile
Production leadtime (days)	0.371	0.085
Oral priority change (%)	0.786	0.398
Written priority change (%)	0.149	0.144
Fixed delivery time (days)	0.060	0.237
Min. delivery time (days)	0.051	0.526
Max. delivery time (days)	0.144	0.878

There were four ordinal scale shop floor control variables covering time standards: their existence, their use in scheduling and costing, and their accuracy. In addition, variables in this section cover the percent of fabricated parts produced in house, whether the blueprint travels with the work in the shop, how many in-process orders are subject to engineering changes, what percent of the orders are delivered late, and the lateness of the late orders. The modal responses are shown in Table 21.

The Kolmogorov-Smirnov test results presented in Table 22 indicate that there are two variables which might have pure regional effects, time standard based costs and time standard accuracy. Figure 10 shows the percentages of the products with costs based on time standards for each of the groups. A pure regional effect is shown. The general conclusion is clear, the western European firms have a higher percentage of product costs based on time standards. Even in the detail, the regional effect is the same in each industry for every category except the 20%-60% one.

Table 21. Modal responses for the ordinal scale shop floor control variables

Variable	'nam'	'wem'	'nat'	'wet'
Fabricated parts produced in house (%)	0-20% or 40-60% (tie)	40-60%	80+%	80+%
Operations with time standards	90+%	60-90%	60-90%, 90+% (tie)	90+%
Operations scheduled using time standards	90+%	60-90%	90+%, 60-90% (tie)	90+%
Costs based on time standards	90+%	90+%	90+%	90+%
Accuracy of time standards	Some are accurate, some not	Some are, some not	Some are, some not	Close to actual
Engineering changes on work-in-process	Less than 10%	Less than 10%	Less than 10%	Less than 10%
Orders delivered late	10-20%	5-10%	5-10%	5-10%
Average lateness	1-3 weeks	1-3 weeks	1-3 weeks	1-3 weeks
Growth rate in the physical volume of production	0-10%	10-20%	0-10%	0-10%

Table 22. Significance levels for the ordinal scale shop floor control variables using the Kolmogorov-Smirnov test

Variable	Machine tool	Textile
Fabricated parts produced in house	0.993	0.885
Time standards for operations	0.155	0.003
Operations scheduled using time standards	0.557	0.006
Costs based on time standards	0.034	0.068
Accuracy of time standards	0.016	0.023
Engineering changes on work-in-process	0.282	0.996
Orders delivered late	0.005	0.235
Average lateness	0.226	0.918
Growth rate in the physical volume of production	0.391	0.051

A pure regional effect is found for the accuracy of time standards, as seen in Figure 11. The accuracy of the time standards is higher for western European firms and the regional effect for every category of response is the same for both industries.

Two of the questions in this section had more complicated responses than any others in

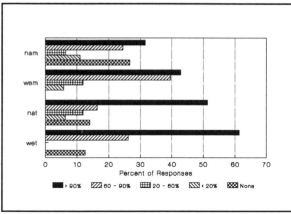

Figure 10. Costs based on time standards

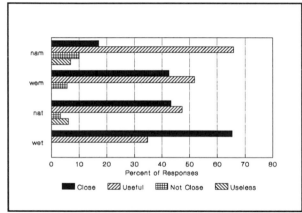

Figure 11. Accuracy of time standards

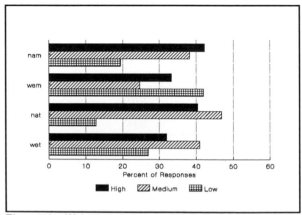

Figure 12. Weight of delivery date changes

the questionnaire. One of these had to do with the factors that changed priorities once an order was being produced, the other with means for changing capacity. The responses were a weighting of the importance for several alternatives. For these questions, the distribution of weights for each of the possible responses was analyzed separately for each alternative.

Table 23 gives the modal weights for the variables that can change priority on an order once production has started. The weights (high, medium or low) indicate heavy, moderate or little influence on priorities. Table 24 shows the results of the Kolmogorov-Smirnov test for those variables. Only one variable, changes in delivery date, is a possibility for a pure regional effect. Figure 12 indicates that there is one. The delivery date changes have a heavier weight in changing the priorities in North American firms than in western European ones.

The weights for each alternative means to increase or decrease capacity (high, moderate, not useful) indicate whether it is highly, moderately or not useful to the company. The modal weights are shown in Table 25. Table 26 reports the results of the Kolmogorov-Smirnov tests. They indicate one variable may have a pure regional effect: laying off extra

Table 23. Modal weights for the responses for the priority change
shop floor control variables[*]

Variable	'nam'	'wem'	'nat'	'wet'
Pressure from marketing	Medium, low (tie)	Low	High	High
Pressure from customers	High	High	High	High
Orders from management	High	Low	High	High
Changes in delivery dates	High	Low	Medium	Medium
Sudden surges in demand	Medium	Low	Medium	High, low (tie)
Manufacturing problems	Medium	Low	Medium	Medium
Material shortages	High	High, low (tie)	High	Low
Changes in sales plan	Low	Low	Low	Low
Engineering problems	High, medium (tie)	Low	Low	Low

* A tenth variable, other, was not included because there were not enough responses.

Table 24. Significance levels for the priority change shop floor
control variables using the Kolmogorov-Smirnov test

Variable	Machine tool	Textile
Pressure from marketing	0.430	0.897
Pressure from customers	0.218	0.992
Orders from management	0.049	0.418
Changes in delivery dates	0.067	0.050
Sudden surges in demand	0.021	0.173
Manufacturing problems	0.514	0.071
Material shortages	0.975	0.030
Changes in sales plan	0.665	0.071
Engineering problems	0.013	0.167

workers. Figure 13 confirms that there is a pure regional effect for this variable. The modal weight for this variable in the North American firms is high, while it is not useful in the western European firms.

Six nominal variables in this section cover the functional field of the person and the basis for starting production, purchasing material, and sequencing production jobs. Others involve the transmittal of purchase orders to suppliers (oral, written, or other), the determination of the delivery date promised to the customer (customer, negotiated, or the firm), the factors that affect the delivery date promise, and the two most frequent causes of lateness. The modal responses for these variables is presented in Table 27.

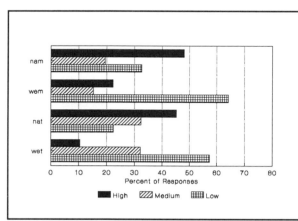

Figure 13. Weight of layoffs

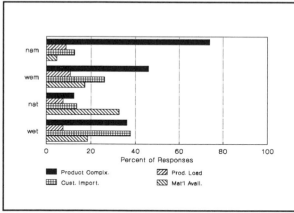

Figure 14. Delivery time influences

Table 28 provides the results of the chi-square and binomial tests for the shop floor control nominal variables. The results of the binomial test indicate that there may be a pure regional effect for the factors that affect the delivery date promise. Figure 14 shows that there is no clear regional pattern, hence no pure regional effect, however.

For this section we end up with four significant variables. Two relate to time standards, their use for costing products and their accuracy (both higher in western European firms). Delivery date changes have a higher weight in changing production priority in North American firms as does laying off workers to decrease capacity.

Table 25. Modal weights for the responses for the alternatives
to change the capacity shop floor control variables[*]

Variable	'nam'	'wem'	'nat'	'wet'
Hire additional workers	Moderate	Not useful	High	Not useful
Use overtime	High	High	High	High
Subcontract production	High	High	Not useful	Moderate
Backorder the production	Not useful	Not useful	Moderate	Moderate
Lease temporary capacity	Not useful	Not useful	Not useful	Not useful
Lay off extra workers	High	Not useful	High	Not useful
Use undertime (idleness)	Not useful	Not useful	Moderate, not useful (tie)	Not useful
Reduce the worktime	Not useful	Not useful	Moderate	Moderate, not useful (tie)
Build inventory	Moderate	Moderate	Moderate	Moderate
Lease capacity to others	Not useful	Not useful	Not useful	Not useful

*   An eleventh variable, other, was not included because there were not enough responses.

Table 26. Significance levels for the alternatives to change the capacity
shop floor control variables using the Kolmogorov-Smirnov test

Variable	Machine tool	Textile
Hire additional workers	0.208	0.000
Use overtime	0.443	0.713
Subcontract production	0.902	0.022
Backorder the production	0.653	0.397
Lease temporary capacity	0.669	0.004
Lay off extra workers	0.031	0.001
Use undertime (idleness)	0.902	0.323
Reduce the worktime	0.284	0.022
Build inventory	0.986	0.810
Lease capacity to others	0.715	0.034

Table 27. Modal responses for the nominal scale shop floor control variables

Variable	'nam'	'wem'	'nat'	'wet'
Function of people Starting Production	Production	Administration	Administration	Production
Basis for starting production	Customer order	Customer order	Customer order	Customer order
Function of group that purchases	Production	Administration	Administration	Administration
Basis of the purchase order	Customer order	Prod. plan	Customer order	Customer order
Purchase order transmittance to supplier	Written	Written	Written	Written
Blueprint travels with product	Yes	Yes	No	It depends on the product
Function that sequences jobs to production	Production	Production	Admin.	Admin., Prod., Sales (tie)
Basis of sequence	Customer order due date	Customer order due date	Customer order due date	Customer order due date
Determination of delivery date	Company	Negotiated	Customer	Negotiated
Factors influencing delivery time	Product complexity	Product complexity	Load, available materials (tie)	Product complexity
Two causes of lateness	Material shortages, lack of labor capacity	Transportation, quality problems	Material shortages, production bottlenecks	Material shortages, quality problems

Table 28. Significance levels for the nominal scale shop floor control variables using the chi-square test

Variable	Machine tool	Textile
Function of people starting production	0.620	0.056
Basis for starting	0.254	0.536
Function of group that purchases	0.838	0.060
Basis of the purchase order	0.032	0.956
Purchase order transmittance to supplier	0.030	0.412
Blueprint travels with product	0.269* (0.116)	0.002

(continued)

Table 28, continued

Variable	'nam'-'wem'	'nat'-'wet'
Function that sequences jobs	0.420	0.001
Basis of sequence	0.179	0.070
Determination of delivery date	0.012	0.118
Factors influencing delivery time	0.026	0.226* (0.064)
Two causes of lateness	0.120* (0.610)	0.024* (0.078)

* More than one fifth of the cells have expected frequencies less than five so these chi-square significance values are suspect. The significance levels obtained by the binomial test are in parenthesis.

PURCHASING AND MATERIALS MANAGEMENT

The purchasing and materials management ratio scale variables cover inventory amounts (raw material, work-in-process, finished goods and total), the number of vendors per purchased part and the level of finished goods safety stock. The mean responses for these variables are given in Table 29. The Kruskal-Wallis test results for them are presented in Table 30. No variable indicates a potential pure regional effect.

Table 29. Mean values of the ratio scale materials management variables

Variable	'nam'	'wem'	'nat'	'wet'
Raw material inventory (%)	35.528	36.897	37.140	39.048
Work-in-process inventory (%)	44.250	43.448	29.721	21.524
Finished goods inventory (%)	19.556	19.690	31.674	39.429
Total inventory (mill.US$)	9.685	10.567	7.731	232.766
Number of vendors per purchased part	1.977	1.844	1.940	1.955
Finished goods safety stock (weeks)	7.706	5.500	7.882	6.053

Table 30. Significance levels for the ratio scale materials management variables using the Kruskal-Wallis test

Variable	Machine tool	Textile
Raw material inventory (%)	0.721	0.847
Work-in-process inventory (%)	0.879	0.069

(continued)

Table 30, continued

Variable	Machine tool	Textile
Finished goods inventory (%)	0.921	0.101
Total inventory (million US$)	0.176	0.004
Number of vendors per purchased part	0.260	0.926
Finished goods safety stock (weeks)	0.238	0.120

Only two ordinal scale variables are found in this section. One indicates whether the firm uses the EOQ, the other whether there is a person performing the function of materials manager. The modal responses for these variables is shown in Table 31. The Kolmogorov-Smirnov test results are given in Table 32. Although both comparisons for the textile industry are significant, no pure regional effect is indicated.

Table 31. Modal responses for the ordinal scaled materials management variables

Variable	'nam'	'wem'	'nat'	'wet'
Usage of EOQ	No	No	No	No
Materials manager position	Yes	No	No	No

Table 32. Significance levels for the ordinal scale materials management variables using the Kolmogorov-Smirnov test

Variable	Machine tool	Textile
Usage of EOQ	0.868	0.006
Materials manager position	0.249	0.006

The nominal variables in this section include information on the activities in which purchasing is involved, the company's exposure to MRP and JIT, the basis for purchasing decisions, the actions taken by purchasing to assure timely supplies, and the reasons for subcontracting. The modal responses for these variables are shown in Table 33.

Table 34 shows the chi-square and binomial test results for the purchasing and materials management nominal variables. Few significant differences were found and no pure regional effects exist. For this section in total, no regional effects were found for any of the variables.

Table 33. Modal responses for the nominal scaled materials management variables

Variable	'nam'	'wem'	'nat'	'wet'
Activities involving purchasing	Make-buy decisions, cost reduction, Material shortages (tie)	Material shortages	Material shortages	Material shortages
Exposure to MRP	Using and benefitting from it	Using and benefitting from it	Never heard of it	Never heard of it
Basis for purchasing decision	Purchase according to prod. plan	Purchase according to prod. plan or purchase only for specific customers' orders (tie)	Purchase according to prod. plan	Purchase according to prod. plan
Two actions to assure timely supply	Long-term contract, multiple sourcing	Long-term contract, multiple sourcing	Long-term contract, multiple sourcing	Long-term contract, multiple sourcing
Reasons for outside purchasing	Production load, lower costs	Production load, lower costs	Production load, production difficulty	Production load, lower costs
Exposure to JIT	Understanding but no necessity to introduce	Understanding but no necessity to introduce	Using and benefitting from it	Never heard of it

Table 34. Significance levels for the nominal scale materials management variable using the chi-square test

Variable	Machine tool	Textile
Activities involving purchasing	0.197*	0.575*
Exposure to MRP	0.120*	0.019*
Basis for purchasing decisions	0.416*	0.937*
Two actions to assure timely supply	0.698	0.894
Reasons for outside purchasing	0.654	0.109
Exposure to JIT	0.830*	0.187*

* More than one fifth of the cells have expected frequencies less than five so these chi-square significance values are suspect. The significance levels obtained by the binomial test are in parenthesis.

SUMMARY

Perhaps the most important observation is that out of the 119 variables analyzed, fewer than 10% were found to be significant using the pure regional effect criterion. Thus there are no surprises, the "siblings" are alike. What differences were found may be important, however, so we summarize them here:

(1) The volume of exports is higher in Eastern Europe.
(2) The proportion of export sales (the export ratio) is higher in western Europe.
(3) Capacity utilization is higher in western Europe.
(4) In western Europe the production planners usually have high school degrees, while the North Americans are college graduates.
(5) The production plan is used for different purposes in North America than in western Europe. In North America the plan is used for operations scheduling combined with manpower or inventory planning. In western Europe, it is used for facility planning combined with inventory planning and subcontracting.
(6) In western Europe a higher portion of the products have costs based on time standards than in North America.
(7) Time standards are perceived to be more accurate in western Europe.
(8) In North America changes in delivery dates have a greater influence on changing production priorities once production has started than in western Europe.
(9) North American firms put a heavy weight on the option of laying off workers to decrease capacity while this is not an useful option in western European companies.

Of the nine variables that describe the differences between these two regions, four come the shop floor control section of the questionnaire. This is the section covering the most detailed aspects of manufacturing practices. Since most of the differences are found in these details, it heightens the sense of similarity of the regions.

Three of the variables that differentiate between the regions come from the general data section. We think that these variables reflect differences in the "business climate" of the two regions.

The final two variables came from the production planning and scheduling section, an area in which differences could be important to firms trying to coordinate their production plans.

There were no differences found in the sales forecasting or purchasing and materials management sections.

REFERENCES

1.* D.C. Whybark and B.H. Rho, "A Worldwide Survey of Manufacturing Practices," Discussion Paper No. 2, Bloomington, IN, Indiana Center for Global Business, Indiana Business School, May 1988.

2.* B.H. Rho and D.C. Whybark, "Comparing Manufacturing Planning and Control Practices in Europe and in Korea," *International Journal of Production Research*, 28, No. 12 (Fall 1990), 2393-2404.

3.* B.H. Rho and D.C. Whybark, "Comparing Manufacturing Practices in the People's Republic of China and South Korea," Discussion Paper No. 4, Bloomington, IN, Indiana Center for Global Business, Indiana Business School, May 1988.

4. L. Wilkinson, *SYSTAT: The System for Statistics*, Evanston, IL, SYSTAT, Inc., 1990.

5. L. Wilkinson, *SYGRAPH: The System for Graphics*, Evanston, IL, SYSTAT, Inc., 1990.

6.* G. Vastag and D.C. Whybark, "An Analytical Framework for Comparing Regions in the GMRG Data Base," Working Paper No. 1, Chapel Hill, NC, Global Manufacturing Research Group, Kenan Institute, Kenan-Flagler Business School, University of North Carolina at Chapel Hill, December 1991.

7. W.W. Daniel, *Applied Nonparametric Statistics*, 2nd. ed., Boston, PWS-Kent Publishing Company, 1990.

* This article is reproduced in this volume.

ACKNOWLEDGEMENTS

The authors would like to acknowledge the Center for Manufacturing Excellence and the Kenan Institute for supporting this research. Gyula Vastag was Citibank International Fellow at the Kenan Institute at the University of North Carolina during the preparation of this paper.

A Comparison of Logistics Management in Hungary, China, Korea and Japan

Robert B. Handfield, Michigan State University, USA
Barbara Withers, University of San Diego, USA

ABSTRACT

Many Western firms are expanding into foreign markets by developing joint ventures, supply and license agreements, and other forms of distributor networks. An important precursor to such ventures is a knowledge of the logistical and manufacturing capabilities involved in foreign countries. This paper explores several performance measures for firms in the textile and machine tool industries in Hungary, China, Korea, and Japan, and distinguishes several major differences with respect to each country's productivity, process technologies, materials management and logistics infrastructure. In particular, the results point to the need for a careful assessment of managerial training requirements in countries such as Hungary and China as a prerequisite to successful implementation of logistics networks.

INTRODUCTION

As increasing numbers of North American firms have felt the need to establish a presence in global markets, coordinated global manufacturing strategies have emerged as a response to both increased competition and the allure of access to global markets [1]. These strategies are sometimes carried out through international alliances that involve a formal long-term relationship between two firms in different countries [2]. Firms may also choose to internationalize through the establishment of supplier linkages in foreign countries. Foreign buying may initially take place solely on need, and may progress to a proactive inclusion of international sources in pursuit of potential new markets, leading to full integration and coordination of global sourcing requirements to maximize buying leverage [3]. In general, it is believed that internationalization is part of a natural process of growth for firms that evolves into an interplay between the development of knowledge about foreign markets and operations on one hand and an increasing commitment of resources to foreign markets on the other [4].

The role of logistics is clearly a critical component of the internationalization process. Nevertheless, several studies have found that many American and European firms continue to de-emphasize logistics capabilities in the global context [5]. For instance, logistical inefficiencies in Europe were particularly prevalent in logistics planning, inventory holdings, use of logistics services suppliers, and the price paid for logistics services [6]. Not only are the transportation and distribution issues complicated by a host of unique regulations in different countries, but the very infrastructure of such environments can be fundamentally different from those found in North America. Before managers enter into global ventures, a clear understanding of the logistical conditions is essential to formulating a logistics strategy and establishing the terms of agreement.

While basic logistics decisions usually involve transportation mode, carrier choice, and the location and size of inventory stocking points, Bowersox [7] points to a more thorough definition of integrated logistics as: "The process of managing all activities required to strategically move and store materials, parts, and finished inventory from suppliers, between enterprise facilities, and to customers." Cohen and Lee [8] expounded on this model to develop an integrated logistics framework consisting of several submodels: material control, production control, finished goods stockpile, and distribution network control, all linked through inventory and scheduling decisions. Whybark [9] further states that a global manufacturing strategy will affect the organization of logistics activities, the evaluation of material flows, the development of new technical skills, and the maintenance of a strategic perspective.

Using these conceptual frameworks as a basis for exploration, this paper examines some of the critical differences found in four countries that represent a diverse set of logistical environments. Moreover, the promise of new markets in Eastern Europe, Asia, and the Far East has attracted many American firms to explore the possibility of manufacturing and distribution sites in these areas [10]. Other reasons for the increasing number of ventures in these areas involve managerial, technological and strategic considerations. Some of these countries are striving to increase export-led growth and are eager for development, while other more developed countries are less inclined. This study focuses on four distinct players from each of these regions, representing a diversity of infrastructural conditions. Hungary and China are relatively new forces in the global market, and are still striving to shed many of the limitations formerly imposed by their centrally planned economies. Many American firms are quickly building plants and distribution networks in these areas in an effort to establish a presence. Korea and Japan, on the other hand, have established themselves as global leaders in many industries, yet their markets have proven to be more difficult to enter. While many American firms have suppliers in these two countries, the logistical and manufacturing conditions are very different from those found in North America. Korea and Japan are also included as a basis for comparing conditions in Hungary and China. Although caution should be used in making generalizations, the data in this study nevertheless demonstrate some important differences in the logistics infrastructure within each of the different areas.

The study first briefly reviews the general political, social and economic conditions within each country. Second, a comparative analysis of several measures of productivity, process technology, materials management, and production and distribution control in the different countries is presented. Finally, the implications of establishing alliances or sources in each of these countries are discussed.

POLITICAL AND ECONOMIC CONDITIONS

A comparison of the economic and demographic differences among these countries is shown in Table 1. A discussion of the political and economic conditions within each country is essential in providing a background for analyzing existing logistical infrastructures.

Table 1. Demographic data (1990)

	Hungary	China	Korea	Japan
Population (millions)	10.5	1,118.0	43.0	123.0
GNP (billions US$)	64.6	292.6*	200.0	1,914.0
GNP/capita (US$)	6,108	261*	4,600	15,600
Privatization started	1988	1979	1960	1950
Exports as % of GNP	29.6%	13.5%*	31.2%	14.1%
Growth rate of GNP	1.3%	4.0%	6.5%	4.8%
Inflation rate	1.3%	4.0%	6.5%	4.8%
Industrial production growth rate	18.0%	19.5%	5.0%	2.1%
Unemployment rate	0.4%	3.0%	3.0%	2.3%
Average wage per month (US$)	150	32	438	1200
Literacy (% of population)	99%	75%	90%	99%

* Data from [11]. All other data from [12].

Hungary

Recent social and political changes in Eastern European countries have introduced an urgency into the transition from their centrally planned economies (CPEs) toward market based economies. The distorting effects of centrally planned economies have ultimately proved devastating and have left a trail of deep and worsening economic crises throughout Eastern Europe [13]. Radical privatization reforms encompass fundamental social, political, and economic changes within each country's infrastructure. The problems that must be overcome in order to achieve these reforms are formidable: not only are there structural hurdles to surmount, but opposition to cuts in subsidies and government programs arising from industrial restructuring will no doubt be strong [14].

Structural problems permeate the social, economic, and political environments of these countries. The socialist ownership structure deprived individuals of the right and motivation to act efficiently. This led to a misallocation of resources that most notably manifested itself as chronic shortages throughout these countries. The bureaucratic structure inherent in the previous governments created a massive managerial elite who are largely unprepared to function in a market driven, sophisticated economy [15]. The industrial structure, skewed toward heavy industry, consists primarily of large, monopolistic firms that are both horizontally and vertically integrated [16]. The centrally planned and controlled economic structure creates a contrived communication system that, when removed, leaves an informational vacuum with respect to the reallocation of resources. All of these structural problems will need to be addressed in the economic reform process.

Hungary, by comparison with the other Eastern European countries of Poland, Romania, Bulgaria, Czechoslovakia, Estonia, Latvia, and Lithuania, is nevertheless comparatively well advanced in its market reform efforts. Prior to the late 1940's when the communist regime took over, Hungary had sound industrial and agricultural bases. During the forty-five years of the communist regime, manufacturing managers in Hungary operated under a controlled system of resource allocation and manufactured output distribution. Since 1968, however, Hungary reverted to a modified centrally planned economy in which managers operated partly in a market and partly in a centrally planned environment. During this period, the country experienced 20 years of "partially liberalizing" prices, and has alternated between periods of reform (1968-1972 and since 1979) and years of reactions (1973-1978) [17]. Generally speaking, the country has enjoyed a relatively high GNP, although growth has been minimal (see Table 1). Until the late 1980s, Hungary enjoyed a single digit rate of inflation, although recent changes in the market structure have increased inflation rates to 18%. Exports are almost a third of GNP, divided approximately evenly between Council for Mutual Economic Assistance (CMEA) Soviet-bloc countries and convertible currency markets [18]. Unlike a market economy, whose output can generally be sold on any market, much of what Hungary and other Eastern European countries produces for Soviet CMEA partners and domestic consumption cannot readily be sold in Western markets for convertible currency, because of the poor quality of much of this output. In turn, this has led to difficulties in convertible-currency balance of payments on increasing imports from non-CMEA countries, and to increasing levels of inflation (see Table 1).

As in many Eastern European countries, the Hungarian government is intent on encouraging export-led growth through the creation of a new middle class of small-and-medium sized businesses. Significant reforms occurred in 1988, including a new Law on Corporate Association that, by permitting joint stock and limited liability companies, set up new operating rules for both foreign and domestic firms, and a Foreign Investment Act that allowed 100% foreign ownership of Hungarian businesses, simplified registration procedures, and conversion of soft currency profits into foreign exchange for repatriation. Nevertheless, existing firms are large and bureaucrat-ridden, and profitability remains an elusive goal. There are few firms in these countries with fewer than 100 employees, a condition that has been described as the "socialist black hole [19]." Unfortunately, a consequence of the centrally planned and authoritative business environment was the evolution of a managerial elite who were deprived of their decision making power [20]. Managerial processes and decisions traditionally motivated by market forces were replaced with an authoritarian system that simply required orders to be executed. Thus, it is not surprising that the managerial elite in these bureaucratic organizations had little reason or opportunity to retain or develop managerial decision-making skills. The results can be seen in Table 1: unemployment is low and industry production growth is minimal. In addition, Hungary's $21.7 billion gross hard currency debt continues to grow.

Despite these problems, East European countries enjoy a reasonable standard of living, and are generally well-educated and have a high rate of literacy [21]. Combined with the fact that their wages are competitive on the world market, there exists a real potential for rapid growth and industrialization within their manufacturing sectors. The procedures to found a joint venture, start a new company, or buy in to an existing enterprise have been simplified, and Hungary's new stock market has become fully operational. It should be noted that all of the data for the Hungarian firms in this study are from 1987, and therefore do not reflect changes that have occurred since.

China

China, although less developed, has been promoting export-led growth for the last decade, with a lesser degree of success. In this sense, China represents a failed attempt at export-led industrialization. The China of the pre-reform period was one of little industrial competition, and was truly a seller's market. Manufacturing capacities and distribution channels were so inadequate that the dependability of market availability of most products was typically low [22]. In this period, many Chinese managers were selected on the basis of their political expertise rather than technical backgrounds. However the reforms developed under the leadership of Deng Xiaoping in the mid 1980s have attempted to reduce planning and move the enterprise closer to the market, stimulate competition, open up the economy to increased foreign presence, and invest the managerial position with considerably more autonomy and accountability [23]. While the reforms in China have successfully privatized agricultural sectors, the market socialization of state-owned enterprises in the manufacturing sector has largely failed.

Nevertheless, a recent figure quoted by the New York *Times* reveals that the Chinese economy is growing at a rate of 6% a year, and is enjoying low inflation, increased foreign investment, and their exports and foreign exchange reserves are at a record high. In this sense, the Chinese economy is enjoying more vigorous growth than the economies of the United States, Japan, India, or most other countries. Foreign investors signed more than 5000 contracts totalling more than $4.5 billion, (a 93% increase over the previous year), with companies such as RJR Nabisco, Gillette, Coca Cola, Procter and Gamble, Heinz, and Avon, which are all manufacturing within China. Xerox, Boeing, and General Electric are all watching their sales grow through exports to China. However, China's economy is still dominated by massive state-owned businesses. The state sector accounts for only about one-third of China's GNP, and 18% of its labor force. For the first time in four decades, China is approaching less than half of its industrial output from state enterprises.

Despite its labor wage advantages (shown in Table 1), the greatest inadequacy of the Chinese manufacturing base has been its inability to respond to unforeseen changes in market environments and its failure to become a dependable producer of manufactured products. This has severely limited the potential for Chinese enterprises to compete effectively against foreign competition without protection in the domestic Chinese market. The government is nevertheless continuing to institute reforms in many areas to encourage a market economy: these include capital creation, housing, pricing for consumer goods, currency valuation, corporate ownership, state subsidies, and foreign investment. The youth market (over two-thirds of the 1.1 billion Chinese are under the age of 35) favors the private sector, since the majority of jobs available are in this area. Wages are also rising faster than prices, as quoted by the Wall Street Journal: basic pay in the first nine months of 1991 rose almost 14 percent, bonuses 18 percent, and allowances 10 percent, while prices in cities rose 8% to 10%.

South Korea

South Korea's economy attained unprecedented levels of growth in recent years, through a policy of export-led industrialization centered on fostering exports. Korea's industrialization policies have been described as "selective intervention," focusing on achieving dynamic efficiency through detailed, industry-specific government policies that promote potential comparative advantages from every possible source [24]. By encouraging exports and promoting infant industries such as cement, fertilizer, steel, chemicals, and consumer durables, Korea to date has made its exports both privately profitable and internationally competitive. The primary variables believed to contribute to export-related growth in Korea include the differential advantages in product uniqueness and price, management's perceptions of the importance of exporting, high rates of return-on-investment, and growth goals [25]. In recent years, these policies have occasionally gone amiss, most notably in the case of heavy engineering in the late 1970s. Recently, selective intervention has lost the support of important segments of the Korean public, who prefer democratic government to economically enlightened dictatorship.

Nevertheless, Korea's growth per capita income was well in excess of 7% over the past three decades. In 1960, the Korean economy was dominated by agriculture and mining, and exports amounted to about 3% of GNP. Today, the economy is dominated by the manufacturing sector, and exports account for more than 30% of GNP (see Table 1), with manufactured products constituting over 90% of this total. Wages remain relatively low, although employee compensation doubled from 1979 to 1984 [26], and the prospects for future growth are excellent. Certain peculiar laws exist within Korea: for instance, Korea denies copyright protection to software, semiconductors, or foreign works.

A common notion, derived from the belief in the primacy of culture in shaping a nation's managerial system, is that Japanese and Korean management styles are essentially identical. This is largely based on the similarity of cultural heritages that stem from East Asian Buddhist value systems. However, major differences exist between these two managerial styles, notably the absence within Korean firms of lifetime employment and seniority-based wages, and Korea's enterprise based union system [27]. In this sense, Korea represents an approach to export-led growth that is somewhat different from Japan's. The importance of culture, in this sense, is diminished by the differences in management.

Japan

Japan is clearly the country of the four most advanced along the path to export-led growth. Its government-initiated efforts began shortly after the Second World War, when the Japanese industrial base was essentially in ruins. This export-oriented drive was particularly emphasized through the "Trade or Die" mentality that became prevalent in many of the Newly Industrialized Countries during the following decades [28]. After a rapid expansion in the 1960s-1980s, exports as a percent of GNP have diminished somewhat, although production growth rate is still strong (Table 1). This is partly due to Japan's increasing stature and diminished labor wage advantage. Through excelling in implementing new technologies and perfecting them through incremental process innovations, Japanese industries were able to successfully enter many American markets by offering products, especially high technology products, having both the best quality and lowest cost [29]. Concurrent with this position, Japan has recently become officially classified as a "wealthy industrialized nation," with high disposable income levels, and the accompanying educational levels, academic and cultural backgrounds, and access to information common to these countries [30]. Despite the perception of strong import restrictions, American consumer products are especially evident. As wages become higher and the labor shortage in Japan becomes more severe, automation is increasingly replacing labor as a means of maintaining price competitiveness. Thus, many Japanese industries are using offshore manufacturing sources such as China and Thailand, and are increasingly anxious to develop their own technology-intensive industries and marketing capabilities. Japanese are also increasingly investing in the United States, in order to increase their political clout and prevent further trade restrictions by creating jobs for Americans, to ensure access to the American market, to become more responsive to the American market, and to hedge against fluctuations in the value of the dollar. Conversely, many American firms are forming strategic alliances with Japanese companies to alleviate the large fixed costs of establishing distributor networks in such markets as shoes, nuclear reactors, pharmaceuticals, automobiles, tires, and glass [31].

METHODS

While the data in Table 1 provide a general illustration of the relative economic status of these countries, they provide little information on the manufacturing and logistical environments within. Attributes of firms with respect to their productivity, process technologies, materials management and control, and production and distribution network infrastructure were compared to assess the major differences for firms operating within these different countries, concurrent with the integrated logistical models proposed by Bowersox [5], Cohen and Lee [8], and Whybark [9]. A description of the sample of firms used precedes this analysis.

Sample

The data used for this study are a subset of a manufacturing data base collected by the Global Manufacturing Research Group [32]. This data base constitutes a worldwide survey of manufacturing practices. Hungary, China, Korea and Japan are four of the countries for which these data are available. The sample includes firms from two manufacturing industries: non-fashion textiles and machine tools. These two industries provide a suitable basis for analysis, in that they represent two diverse points on the process spectrum.

Non-fashion textiles consist of consumer products like towels, sheets, and underwear, as well as industrial goods such as netting, plain woven cloth, and cloth wrapping materials. This industry is largely process-intensive, in that many processes are integrated from spinning through dyeing, weaving, cutting, and final sewing of the products. Because such mass-produced products are characterized by little customization or fashion, competition is largely based on price. As a result, there is often little opportunity for the manufacturer to achieve attractive profit margins. In some of these markets, developed countries are making inroads through intensive automation, often at a lower per unit cost than in the labor-intensive processes of low wage-countries. However, low-wage countries typically rely on the economies of scale inherent in high-volume, low-variety production in order to compete on world markets. Such industries usually rely on a minimally-skilled work force in which there are few opportunities for learning or skill formation to result from the work performed.

The small machine tool industry consists of firms producing industrial products such as lathes, grinders, milling machines, and metal forming equipment. Most processes are batch-oriented job shops, with some firms offering options to be determined by the customer. In contrast to the textile industry, such processes generally rely on a highly-skilled, cross-trained workforce. This industry has received substantial government support in Eastern European countries, largely because of links to defense industry ministries [33]. Although such products compete on the basis of quality and product technology, achievements in Eastern Europe still consist largely of updating current technologies and production of relatively standard machine tools [34]. Once again, products in less-developed countries often suffer from delays in technological diffusion, standardized product mixes, and unreliable quality.

The questionnaire for the study gathered information on general firm data and activities in several areas of manufacturing planning and control. Interviews were carried out by the Korea Productivity Center and the Sogang University on a random sample of firms in South Korea. The Chinese data was collected through the Shanghai Institute of Mechanical Engineering by students. Data from Japan and Hungary were also collected through onsite interviews by members of the Global Manufacturing Research Group.

Overview of data

Because much of the data exhibited skewness that implied departures from normality, a distribution free nonparametric test was used to test for significant differences existing between firms in different countries. In particular, the Kruskal-Wallis H statistic, which is equivalent to the sum of squares for treatments in a one-way analysis of variance, was calculated to determine whether the data from firms in each country was significantly different from those in other countries [35]. This statistic was used in comparing all numerical data, but not for categorical data.

General data on the number of firms from each country and industry, average sales per firm (in dollars), and average exports as a percentage of total sales are shown in Table

2. As a whole, sales for Chinese firms in both industries were significantly less than in the other countries. Although Hungarian firms have a seemingly large percentage of their sales going to exports, most of these sales are to Soviet-bloc countries, and cannot be readily converted into currency. The success of Korea's export-oriented policies are again apparent in this table, particularly in the textile industry where on average 40% of total sales are exports. Japanese textile production, on the other hand, is focused largely on the domestic market, although protectionist measures may be partially responsible for this fact. Exports are also greater for Chinese textile firms than for Chinese machine tools, which produce largely for domestic markets (probably due to the low level of technological proficiency in the machine tool industry). In the remaining sections, various attributes of firm performance in each country are introduced and compared.

Table 2. Overview of firm data

	Machine tool			Textile		
	Number	Sales (US$ millions)	% Export	Number	Sales (US$ millions)	% Export
Hungary	19	41.4	28.6	17	32.7	28.5
China	44	5.3	3.8	56	10.6	22.6
Korea	89	18.3	21.9	33	13.1	39.7
Japan	18	179.4	11.9	36	17.7	1.2
Kruskal-Wallis H statistic		39.5***	23.4***		20.8***	45.8***

*** Significant at the $p=0.01$ level.

PRODUCTIVITY

An overview of some general measures of productivity and growth in Table 3 provide a comparative snapshot of the productive efficiency of firms in these industries. A significant difference in number of employees, sales per employee, and capacity utilization exists between firms in the four countries. In both textiles and machine tools, an obvious difference between the socialist and capitalist firms is the difference in size. The average number of employees in both Hungarian and Chinese firms is significantly larger than Korean and Japanese firms. In socialist economies, the original goal of centralized production and planning was to take advantage of "economies of scale" that might exist. However, these efforts have generally resulted in diseconomies, largely through an inability to take advantage of true capacity. For instance, many Soviet-bloc plant managers are given "quotas" to achieve every year. In negotiating these quotas, different ploys are often used by the factory director:

> He'll try to hide productive capacity or request greater quantities of inputs than needed just to make certain he has enough on hand. . . . They have no urge to set unrealistically high quotas. It is not in their best interest to have those below them fail. The end result is that entire industries, often with bureaucratic connivance, get more than they need while produc-

ing less than they are capable of. Meanwhile, other industries have to halt production for lack of supplies [36].

The results of such tactics are shown in Table 3. Productivity in Hungary and China, measured by average sales per employees, is significantly below the levels attained by Korea and Japan. In the case of Hungary, capacity utilization is also well below normal utilization levels of 80% to 90%. This state of affairs is such that cheap labor, Hungary's and China's principal comparative advantage in the world economy, is rendered ineffective in the face of abysmal productivity and output rates.

Table 3. Productivity

	Machine tool			Textile		
	Employees	Sales/employee (US$ thousands)	Capacity utiliz'n	Employees	Sales/employee (US$ thousands)	Capacity utiliz'n
Hungary	2471	19	71%	2759	13	75%
China	2155	3	76%	1159	8	86%
Korea	468	31	83%	484	36	86%
Japan	660	128	78%	72	258	85%
Kruskal-Wallis H statistic	85.2***	106.8***	10.6**	92.1***	88.2***	7.7*

*** Significant at the 0.01 level, ** at the 0.05 level, and * at the 0.10 level.

Another reason for the productivity gap of large socialist factories has to do with the supply problems that exist in both China and Hungary (an issue discussed later in the study). Factory directors in Eastern Europe and China often try to establish fully integrated operations in order to be as independent of other enterprises as possible. Factories are often equipped to produce items that cannot be done efficiently, rather than depending on a source for the item. Managing such enterprises becomes increasingly complex due to their large size and the wide variety of activities which take place. Such factories lack a central "focus" on a set of critical manufacturing tasks; the result is a general state of chaos which translates to low levels of productivity [37].

PROCESS TECHNOLOGIES

Further evidence of the differences existing between socialist and industrialized manufacturing capability is apparent in examining the state of technology employed within firms in the textile and machine tool industries. Hungary's own experts have testified that the technology gap vis-a-vis the West has probably not narrowed significantly since 1968 [13]. Between 1968 and 1983, Hungary imported a large quantity of new machinery, yet has not succeeded in enjoying any major benefits from this imported technology. The state of technology in Soviet-bloc industries has been compared to a pyramid, with the quality of resources diminishing as one descends to the lower levels [38]. At the base are the ill-equipped enterprises that use labor-intensive techniques and produce at low levels of quality, while the upper branches enjoy a better standing in relation to Western countries. The upper level firms are typically in well-developed industries such as metallurgical industries, machine tools, and electric power generation, which are often critical inputs to defense capabilities. For example, the defense industry builds some ten percent of all metal-cutting tools in the Soviet Union, including about one-fourth of numerically-controlled machine tools. Such new technologies, which result from a deliberate attempt at cooperation between the defense sector and the civilian engineering industry, often receive priority treatment at the expense of consumer-oriented industries further down the "pyramid."

Table 4 provides some crude measures on the state of process technologies in different countries. There is a significant difference in the ratio of equipment dollars per worker for firms in each of the countries. This provides some evidence of a higher intensity of automation in Japan compared with the other three countries, as shown by the ratio of equipment dollars per worker. This higher rate of automation is probably in response to the higher wage rates existing in Japan, particularly in the low-margin labor-intensive textile industry. While many Eastern European firms recognize the importance of intensive development of new technologies, the capacity to respond is restricted to a great extent by ceilings fixed by central authorities. Since capital markets are lacking, many plants can only invest what they can raise by their own efforts.

Table 4. Process technologies

	Machine tool			Textile		
	Equipment (US$1,000/ worker)	Training days/yr	% of firms w/ planners w/ engin'ring degree	Equipment (US$1,000/ worker)	Training days/yr	% of firms w/ planners w/ engin'ring degree
Hungary	8	2.5	53	6	3.3	41
China	3	11.7	18	1	6.7	19
Korea	16	22.9	na	14	5.0	na
Japan	51	4.5	57	106	2.2	40
Kruskal-Wallis H statistic	62.9***	25.6***		60.2***	31.8***	

*** Significant at the 0.01 level.

In terms of the technological knowledge base within these countries, Hungary spends significantly less time training its employees as measured by the number of training days per year, (although the higher levels of "training" days in China may in fact include repeated exposure to communist dogma). Korean machine tool employees receive the greatest intensity of on-the-job training, attesting to their rapid technological development. With respect to the percent of firms that have engineers making production planning decisions, Hungary is clearly not lacking. This figure corroborates the data in Table 1, pointing out that the majority of Eastern Europeans have a solid educational foundation.

In this respect, it appears that the infrastructural requirements for technology transfers from the West to occur in China and Eastern Europe are well-established. However, this does not obviate the need for a well-grounded *technical* education in these countries, where it is generally lacking. Given the size and quality of the Soviet-bloc science and technology establishment, remarkably few technical innovations of any substance have been produced, partly because of a political and economic system that discourages risk taking. Many Soviet technical projects are also limited because they lack good instrumentation, which in turn is a function of the limited ability of the industrial base to produce good instrumentation [39]. This lag is often compensated for by importing foreign technologies, a strategy that unfortunately does not provide many benefits because of a lack of technical learning. To coin a phrase, just as "science is not technology," technology is not manufacturing.

MATERIALS MANAGEMENT AND CONTROL

Another area of critical importance in industrial firms is the management of materials and the associated logistics problems. While manufacturing plants in Newly Industrialized Countries and in the West have developed planning systems such as Material Requirements Planning (MRP) and execution systems such as Just-in-Time to reduce inventory investments and improve efficiency, such systems are by and large unheard of in Eastern Europe and China. A comparison of a simple measure, the number of annual inventory turns (calculated as total sales/total inventory since cost of goods sold was not available), is indicative of the problems of material management in these countries. As shown in Table 5, a significant difference exists between the number of turns in firms from each of the countries.

Table 5. Inventory turns and percent of inventory
held in raw materials and finished goods

	Machine tool			Textile		
	Total inventory turns	Percent of inventory in RM	Percent of inventory in FG	Total inventory turns	Percent of inventory in RM	Percent of inventory in FG
Hungary	3.4	73	5	4.6	62	13
China	2.1	53	14	21.9	39	20
Korea	43.7	51	20	3.5	49	26
Japan	18.5	na	na	4.7	na	na
Kruskal-Wallis H statistic	39.2***	19.5***	6.6**	35.5***	12.9***	3.6

*** Significant at the 0.01 level, ** at the 0.05 level.

In the machine tool industries, Korean and Japanese firms have a very high rate of inventory turnover, signifying an ability to smooth the flow of materials through the plant. Hungarian and Chinese machine tool factories have abysmally low inventory turns.

Further, a significantly larger portion of inventory in Hungarian and Chinese firms is held in Raw Materials (see Table 5). This result suggests that a major stumbling block in socialist economies is the procurement of materials. Hungary, for instance, possesses few natural resources other than agricultural land, bauxite deposits, and some lesser coal, oil, and natural gas deposits. (By comparison, China has a vast array of natural resources). Consequently, most raw materials in Hungary are imported, and as such, are controlled by authorities. In many cases, ministries specify the supplies a factory receives and on what delivery dates. As a result, deliveries are often too small, of substandard quality, behind schedule, or fictitious. The response of factory directors is thus to hoard materials:

> They might not even have any direct use for it. But they might be able to trade it for something else. Sadly, surplus goods stand idle or even rot away in one location while at another production is halted because the input is unavailable [34].

Note that the hoarding of raw materials is a rational response to infrastructural problems within the economy. To compensate for these inherent deficiencies, enterprises often employ *tolkachi* (expediters) to procure materials, whose job is to "beg, borrow or steal" supplies that are otherwise unavailable.

An anomaly in Table 5 is the high inventory turns for Chinese textile firms. In addition, the measure of turns for Hungarian textile firms is not much greater than those of their Japanese and Korean counterparts, and there is not a significant difference in the quantity of finished goods held in inventory. This can partly be explained by the fact that textiles are a consumer good that is badly needed and for which a chronic shortage exists. As such, finished goods levels are minimal, and the factories can sell virtually all that is produced. This is especially true in China, where finished goods are in some cases immediately appropriated by government ministries.

Table 6. Procurement data

	Machine tool			Textile		
	Percent of firms with >4 vendors/ item	Most frequent method of purchasing (% of firms)	Most frequent method of assuring supply (% of firms)	Percent of firms with >4 vendors/ item	Most frequent method of Purchasing (% of firms)	Most frequent method of assuring supply (% of firms)
Hungary	42%	State* (~100%)	Multiple sourcing (74%)	82%	State (65%)	Multiple sourcing (77%)
China	61%	Prod'n plan** (82%)	Hedging (58%)	59%	Prod'n plan (43%)	Hedging (67%)
Korea	18%	Prod'n plan (51%)	Other (88%)	38%	Prod'n plan (43%)	Other (84%)
Japan	7%	Prod'n plan (56%)	Long contract (53%)	9%	Customer orders (47%)	Long contract (56%)

* Raw material is distributed by the State. ** Purchasing according to a production plan or schedule. *** Purchase only for specific customers' orders.

The host of problems associated with procurement of materials is further highlighted by the data in Table 6. In striving to obtain an item or material, Hungarian firms are likely to employ as many sources and suppliers as possible, either through government ministries or through *tolkachi*. Chinese firms are likely to use hedging (i.e. hoarding) in obtaining a dependable source. On the other hand, Korean and Japanese firms are more likely to procure materials from one or two sources, based on production plan requirements, customer orders, or other factors. Japanese manufacturers, in particular, very rarely change suppliers, and expect suppliers to achieve cost reductions as a matter of course [40]. While this has increased the stability of their buyer-supplier relationships, it has made it very difficult for American firms to compete as suppliers within this arena.

PRODUCTION CONTROL AND DISTRIBUTION NETWORKS

In centrally planned economies, the net result of misallocations associated with centralized planning is ultimately felt by the consumer in the form of chronic shortages. The bureaucracies that arise in planning production inevitably lead to poor market linkages and a failure to meet demand. This can be seen through examination of several different facets of firms' production planning and distribution performance.

Whether they explicitly recognize it or not, all manufacturing organizations establish some kind of production plan, which establishes the annual production rate for a product group or other broad category. Although the units can vary (either currency or physical units), the production plan is instrumental in determining the approach to be used in coping with sales activity [41]. As such, key indicators of how "responsive" a production system is to the market are the number of times the plan is revised, the planning increments used, and the assumptions used in determining the plan. A realistic plan is based on actual demand and is revised more frequently according to changes in sales patterns.

A comparison of this data is shown in Table 7. The Hungarian and Chinese firms are significantly less responsive to the market, revising their plans on average three or four times per year, compared with six to ten times per year in Japanese and Korean firms. More importantly, these production plans are more likely to be based on backlog and projected inventory figures as opposed to actual sales data and forecasts. An example of this situation is provided by Desai's [34] description of state-run firms in the Soviet Union. Typically, gross output (*val*) targets are the main elements of production plans, which specify the number of tons to be produced, pages to be printed, etc. A telling cartoon in *Krokodil* depicted a factory manager who produced in fulfillment of a target specified in tons not an assortment of nails but a single gigantic nail! Nevertheless, gross output targets have traditionally dominated all other measurement criteria, including profit and actual demand, and as such have hindered market responsiveness.

This lack of responsiveness to market demand has carried over to the production floor (Table 8). A significant difference in the percentage of late orders exists, and such orders are likely to be delayed by as long as 9 weeks in Hungarian firms. The most common reasons in Hungary for late orders are transportation problems and unavailable materials.

Table 7. Production planning

	Machine tool			Textile		
	Planning increments	Number of revisions/yr	Primary basis of plan (% of firms)	Planning increments	Number of revisions/yr	Primary basis of plan (% of firms)
Hungary	76 days	3.5	Capacity & backlog (84%)	69 days	3.6	Capacity & backlog (86%)
China	66 days	2.4	Backlog & inventory (72%)	47 days	3.1	Backlog & inventory (65%)
Korea	50 days	10.3	Orders & forecast (89%)	39 days	9.1	Orders & forecast (53%)
Japan	29 days	5.0	Orders & capacity (56%)	39 days	5.8	Orders & sales (53%)
Kruskal-Wallis H statistic	21.5***	74.3***		16.7***	23.8***	

*** Significant at the 0.01 level.

Table 8. Shop floor control

	Machine tool			Textile		
	% of jobs released on basis of avail. material	Avg. lateness for late orders (weeks)	Most common reason for lateness (% of firms)	% of jobs released on basis of avail. material	Avg. lateness for late orders (weeks)	Most common reason for lateness (% of firms)
Hungary	58	3 to 9	Trans. prob. (84%)	53	1 to 6	Trans. prob. (77%)
China	20	3 to 9	Mach./lab.cap. (58%)	25	1 to 6	Trans. prob. (57%)
Korea	12	1 to 6	Mach./lab.cap. (66%)	10	< 1	Mach./lab.cap. (61%)
Japan	5	1 to 3	Mach./lab.cap. (56%)	3	1 to 3	Mach./lab.cap. (43%)
Kruskal-wallis h statistic		74.5***			43.1***	

*** Significant at the 0.01 level.

These figures attest to the inability of centrally planned economies to support a logistical infrastructure that keeps materials flowing into factories as required. This statement is corroborated by the figures on delivery performance shown in Table 9. Delivery leadtimes for the Hungarian and Chinese firms are significantly greater than (i.e., more then double!) those of their Japanese and Korean counterparts. The most common reason in Hungary, is, again, material shortages. In a system in which firms can sell all that they produce and where there is no incentive for meeting delivery dates or reducing delivery leadtimes, this vicious cycle of shortages echoes throughout the value chain, from unprocessed materials all the way to the consumer bread line.

Table 9. Delivery leadtime performance

	Machine tool			Textile		
	Avg. promised delivery leadtime (days)	Avg. actual delivery leadtime (days)	Most common reason for leadtime variation (% of firms)	Avg. promised delivery leadtime (days)	Avg. actual delivery leadtime (days)	Most common reason for leadtime variation (% of firms)
Hungary	194	127	Mat.shortage (81%)	123	51	Mat. shortage (86%)
China	134	148	Prod. complexity (72%)	65	27	Prod. load (50%)
Korea	na	55	na	na	46	na
Japan	33	45	Cust.import. (50%)	27	36	Cust.import. (42%)
Kruskal-Wallis H statistic	11.6***	35.9***		24.8***	7.9**	

*** Significant at the 0.01 level, ** at the 0.05 level.

CONCLUSIONS AND IMPLICATIONS

In this paper we have identified several major differences in foreign manufacturing organizations that were evident when comparing performance data from Hungarian, Chinese, Korean and Japanese firms in the same industries. One major problem faced by firms in centrally planned economies is their size, which inhibits management's "focus" on a concise set of manufacturing tasks. The complexity of such organizations hinders the ability of factory directors to effectively manage available capacity. A second problem has to do with the lack of technical knowledge preventing the adoption of new process technologies. Although many managers have a good deal of educational training, much of it is theoretical in nature and is of little help in making capital investment decisions such as retooling and establishing logistics networks.

There is also a lack of managerial skills in Hungary and China with respect to planning production, distribution, and inventories. Because factory directors had only to meet a fixed quota in the past, there is no precedent for matching production and inventory levels with demand. The only rule of thumb common to these managers is the accumulation of raw material inventories whenever possible, regardless of whether or not they are needed. This hoarding of materials not only represents a wasteful use of resources, but serves to worsen the chronic material shortages and distribution problems that exist in the supply chain. This state of affairs is a viscous circle, as production delays caused by material shortages result in longer leadtimes and late deliveries in other sectors of the economy.

As mentioned earlier, many Western firms are seeking to establish international alliances or sources overseas. While a number of potential opportunities for strategic advantages and new markets may exist, managers need to be aware of the existing logistics environments. Conditions in Japan and Korea are likely to be more similar to American planning modes, although new entrants into Japanese home markets may have difficulty in overcoming possible biases regarding the quality of American suppliers. In Hungary and China, however, a significant amount of restructuring may be necessary.

The single greatest resource available to managers in such countries is education in basic management principles. In addition, several infrastructural barriers to effective logistics management will have to be removed. New production planning mechanisms that effectively measure available capacity will have to be devised. Large firms that encompass a variety of operations will have to be broken down into manageable units. Such smaller units will be more prepared to focus on specific tasks, thereby improving the quality of resulting products. In the process, productive capacity can be utilized more efficiently, allowing these firms to compete in world markets initially by taking advantage of lower wage rates. In the long run, technological capabilities can be built up and firms can progress into other types of industries. Increased technical training of the workforce will lead to higher rates of innovation and adoption of new process technologies. In many cases, the building of new distribution and supply networks will be a precursor to the establishment of American joint or sourcing ventures.

The importance of Western direct foreign investment in promoting growth in newly reformed economies cannot be overemphasized. Such ties can encourage the transfer of "hard" and "soft" technologies, for which there is already a solid educational base. In the process, there will be economic hardships, for the labor market will be flooded as unproductive capacity is eliminated. While internationalization offers enticing new markets and forms of increased price competitiveness, managers must enter into such ventures prepared with a logistics strategy to deal with a variety of new and difficult situations.

REFERENCES

1.   S.E. Fawcett, "Strategic Logistics in Coordinated Global Manufacturing Success," *International Journal of Production Research*, 30, No.4 (1992), 1081-1099.

2. L.M. Ellram, "Patterns in International Alliances," *Journal of Business Logistics*, 13, No. 1 (1992), 1-25.

3. R.M. Monczka and R.J. Trent, "Global Sourcing, A Development Approach," *International Journal of Purchasing and Materials Management* (Spring 1991), 2-8.

4. J. Johanson and J-E. Vahlene, "The Mechanism of Internationalisation," *International Marketing Review*, 7, No. 4 (1990), 11-24.

5. D.J. Bowersox, P.J. Daugherty, C.L. Droge, D.S. Rogers, and D.L. Wardlow, *Leading Edge Logistics Competitive Positioning for the 1990's*, Oakbrook, Council of Logistics Management, 1989.

6. J. Cooper, M. Browne, and M. Peters, "Logistics performance in Europe, the Challenge of 1992," *International Journal of Logistics Management*, 1, No. 1 (1992), 28-35.

7. D.J. Bowersox, "The Logistics of the Last Quarter of the 20th Century," *Journal of Business Logistics*, 1, No. 1 (1978), 1-9.

8. M.A. Cohen and H.L. Lee, "Strategic Analysis of Integrated Production-Inventory Distribution Systems, Models and Methods," *Operations Research*, 36, No. 2 (1988), 216-228.

9. D.C. Whybark, "Education and Global Logistics," *The Logistics and Transportation Review*, 26, No. 3 (Sept. 1990), 261-270.

10. M.E. Porter, "Competition in Global Industries, A Conceptual Framework," in: M.E. Porter (ed.), *Competition in Global Industries*, Boston, Harvard Business School Press, 1986, pp. 15-60.

11. 1987 data from Sinhua English Service, as reported in Foreign Broadcast Information Service, *China, Daily Report*, FBIS-CHI-88-035, February 23, 1982, p.11.

12. *The World Factbook*, 1990, Central Intelligence Agency, Photoduplication Service, Washington, DC.

13. *Report of the Central and Eastern European Interparliamentary Seminar*, "The Parliament's Responsibility for Economic Development," published by The Congressional Research Service, U.S. Library of Congress, March 22-24, 1991.

14. D. Lipton and J. Sachs, "Creating a Market Economy in Eastern Europe, The Case of Poland," *Brookings Papers on Economic Activity*, No. 1 (1990), 75-147.

15. A. Aslund, "Principles of Privatization for Formerly Socialist Countries," Working Paper No. 18, Stockholm Institute of Soviet and East European Economics.

16. A. Vahcic, "Capital Markets, Management Takeovers and Creation of New Firms in a Reformed Self-Managed Economy," *European Economic Review*, No. 33 (1989), 456-465.

17. P. Marer, "West-East Technology Transfer, Impact on the USSR and Hungary," *OECD* (September 1985), 18-24.

18. *Foreign Economic Trends and Their Implications for the United States*, U.S. Department of Commerce, prepared by American Embassy Budapest, (July 1990).

19. J.A. Quilch, E. Joachimsthaler, and J.L. Nueno, "After the Wall, Marketing Guidelines for Eastern Europe," *Sloan Management Review*, (Winter 1991), 82-93.

20. J.N. Sajkiewics, "Reconstructing of a Marketing System in Eastern European Economies," *Proceedings of the Northeast Decision Science Conference*, Pittsburgh, PA, 1991.

21. V. Tesar, "The Role and Problems of an Enterprise Director in the Czechoslovak Automobile Industry," *International Labor Review*, 125(4) (July-August, 1986), 435-445.

22. W.A. Fischer, "China as a Player in the World Economy," *China Economic Review*, 1, No. 1 (1989), 9-21.

23. W.A. Fischer, "China's Manufacturing Capabilities," Working Paper, Kenan-Flagler Business School, University of North Carolina at Chapel Hill, 1989.

24. L.E. Westphal, "Industrial Policy in an Export-Propelled Economy: Lessons from South Korea's Experience," *Journal of Economic Perspectives*, 4, No. 3 (1990), 41-59.

25. J. Moon and H. Lee, "On the Internal Correlates of Export Stage Development, An Empirical Investigation in the Korean Electronics Industry," *International Marketing Review*, 7, No. 5 (1990), 16-26.

26. C. Markides and N. Berg, "Manufacturing Offshore is Bad Business," *Harvard Business Review*, 66, No. 5 (Sept.-Oct. 1988), 113-120.

27. J. Lie, "Is Korean Management Just Like Japanese Management?" *Management International Review*, 30, No. 2 (1990), 113-118.

28. M. Kotabe, "Changing Roles of the Sogo Shoshas, the Manufacturing Firms, and the MITI in the Context of the Japanese 'Trade or Die' Mentality," *Columbia Journal of World Business* (1984), 33-34.

29. F.M. Hull, J. Hage, and K. Azumi, "R&D Management Strategies, America versus Japan," *IEEE Transactions on Engineering Management*, EM-32, No. 2 (1985), 78-83.

30. K. Ohmae, "The Triad World View," *The Journal of Business Strategy* (1988), 8-19.

31. K. Ohmae, "The Global Logic of Strategic Alliances," *Harvard Business Review*, 67, No. 2 (1989), 143-154.

32.* D.C. Whybark and B.H. Rho, "A Worldwide Survey of Manufacturing Practices," in: D.C. Whybark and G. Vastag (eds.), *Global Manufacturing Research Group, Collected Papers, Vol. 1*, Kenan Institute, University of North Carolina, 1992.

33. J.M. Cooper, "The Civilian Production of the Soviet Defence Industry," in: R. Amann and J. Cooper (eds.), *Technical Progress and Soviet Economic Development*, Oxford, Basil Blackwell, 1986.

34. P. Desai, *Perestroika in Perspective*, Princeton, Princeton University Press, 1989.

35. R. Larsen and M.L. Marx, *An Introduction to Mathematical Statistics and its Applications*, Englewood Cliffs, Prentice-Hall, 1981.

36. M. Kublin, "The Soviet Factory Director, A Window on Eastern Bloc Manufacturing," *Industrial Management*, 32, No. 2 (1990), 21-26.

37. W. Skinner, "The Focused Factory," *Harvard Business Review*, 52, No. 3 (1974), 113-121.

38. J.M. Cooper, "Technology in the Soviet Union," *Current History* (October, 1986), 317-318.

39. L. Lavoie, "The Limits of Soviet Technology," *Technology Review*, 88, No. 8 (1985), 69-75.

40. J. McMillan, "Managing Suppliers: Incentive Systems in Japanese and U.S. Industry," *California Management Review*, Summer (1990), 38-55.

41. O.W. Wright, *Production and Inventory Management in the Computer Age*, Boston, Cahners Books International, Inc., 1974.

* This article is reproduced in this volume.

ACKNOWLEDGEMENTS

A version of this paper was published in the *Journal of Business Logistics*, Winter, 1992, by the Council of Logistics Management.

Global Manufacturing Practices and Strategies:
A Study of Two Industries

Scott T. Young, University of Utah, USA
K. Kern Kwong and Cheng Li, California State University, Los Angeles, USA
Wing Fok, Loyola University, New Orleans, USA

ABSTRACT

This paper presents an analysis of the machine tool and textile industries in the People's Republic of China, South Korea, Japan, western Europe and the United States. The historical role and importance of these two industries is reviewed for each of the countries and an assessment of their current role is made. The manufacturing practices currently employed in several key areas are also compared for each of the regions. An explanation of the practices and the implications for each of the regions is provided. The discussion of the historical role and current practices serves as a backdrop to determining the manufacturing strategy position for each of the countries in each industry.

INTRODUCTION

Manufacturing strategies in specific industries and markets have been instrumental in the global competitiveness of Japanese firms. Japanese manufacturing has been intensely studied by western researchers in an effort to understand the transformation of the country's industrial capabilities. Schonberger [1] and Hall [2] studied the tactical aspects of the Japanese success, while Wheelwright [3] described Japanese manufacturing firms' strategic use of capacity, facilities, vertical integration, production processes, the work force, quality, and production planning and control.

In recent years, South Korea has threatened the Japanese in several markets, using a labor cost advantage to become the cost leader. Korean manufacturing is now the object of much speculation and interest. Could Korea become the next Japan?

Meanwhile, economic policies in the People's Republic of China loosened to the extent that students of the global marketplace speculated on the future success of a communist entry into selected markets.

Manufacturing in the United States has experienced significant declines in industries once believed sacrosanct, and western Europe, with the possible exception of West Germany, suffered similar declines in the 1980s.

This study examines the manufacturing practices and strategies of China, Korea, Japan, western Europe and the United States in two industries: small machine tools and non-fashion textiles. The objective of this study is to compare international production practices within the context of industry capabilities and manufacturing strategy.

METHODOLOGY

A coordinated effort by the Korea Productivity Center, the Sogang Institute for Economics and Business, the Shanghai Institute of Mechanical Engineering, IMEDE (the international business school in Lausanne, Switzerland), and the Business School at Indiana University provided information on the planning and control practices in China, Korea, Japan, western Europe, and the United States. Questionnaires were administered either by mail or personal interview, translated to the appropriate language and focused upon six key areas of production: forecasting, production planning, scheduling, shop floor procedures, purchasing and materials management [4]. The sample sizes are listed in Table 1.

Table 1. Sample sizes

	China	Japan	Korea	Europe	U.S.
Machine tool	44	18	89	34	45
Textile	56	36	33	24	50

MACHINE TOOLS AND TEXTILES

This research is industry-specific. The results apply only to the machine tool and non-fashion textile industries. Also, vast differences among the countries' capabilities must be considered.

China's machine tools have been described as of inferior quality and technologically primitive. The tools are not carefully engineered and are less durable than those produced in other nations. The machine tool industry has primarily been responsible for providing tools to the Chinese defense industry. Chinese authorities reduced capital investment from 1976 to 1980, resulting in the continued production of outdated tools and the inability to produce higher quality lathes and milling machines. Consequently, little export of Chinese machine tools takes place. The industry is updating technologically and importing advanced technology tools, primarily from Japan, the United States, Switzerland and West Germany [5].

Textiles are another story. China is the world's largest textile producer and textiles are its greatest export. China competes against Korea in the lower quality end of the textiles market and has lower labor costs than Korea. China was fifteenth in the world in textiles in 1973 and sixth by 1983 [6,7].

Textiles are also Korea's greatest export. Sixty-two percent of the textiles produced in Korea are exported. However, structural changes jeopardize Korea's global strength. There have been significant technical innovations in the arts of spinning and weaving, while Korea is burdened with older machines. Rising domestic wage rates have also hurt

Korea's textile industry, while the average Korean works 2833 hours per year compared to 2168 for the Japanese [8]. Quality improvements in the industry have been slow and productivity has diminished.

The Japanese concept of "quality at the source" is not a trademark of Korean manufacturing. Kang commented on how the Korean manufacturing attitude of "get started, see how it goes, and fix it later," differs startlingly from the Japanese fervor about quality control [8]. Despite this different philosophy about quality, Korean products have been noted for quality and innovation.

Japan is the world leader in machine tools. In 1975, Japan ranked fourth in the world and produced half of the U.S. total. By 1982, Japan had captured the world lead and in 1990 possesses over half of the market share in the United States. The Japanese accomplished this by introducing less expensive and easier-to-operate numerically controlled (NC) machine tools. By being at the leading technology edge, Japan quickly gained the world lead in the industry trends of automated and unmanned systems using new materials [10].

Korea, Hong Kong, Taiwan and China threatened Japan's textile industry to the extent that together they took a third of the Japanese market. Japan lost share primarily in the lower cost and quality markets and countered by moving to the production of materials requiring technological capabilities beyond the reach of Asian competitors. Innovative machinery and large investments in research and development are characteristics of the Japanese textile industry [11].

West Germany manufactures 62%, Italy 18.6%, the United Kingdom 9.3% and France 8.7% of the machine tools in western Europe [11]. In the period from 1978 to 1984, these countries suffered an average of 28% reduction in the industry workforce, with the U.K. seeing a 59% reduction in the size of its workforce. These hits were caused by lost business to Far-Eastern suppliers and an overall industry recession in the early eighties. In terms of sales, the U.K. dropped 27% in this seven year period, while West Germany, France and Italy slightly increased sales. Also, there were 10% fewer factories in western Europe in 1984 than in 1975.

With the western European aims of reducing manufacturing costs, achieving more flexibility, and shortening processing times [11], there was an expected increase in the use of CAD/CAM, Numerically Controlled machine tools, FMS (Flexible Manufacturing Systems) and cellular manufacturing. Within western Europe, the ability of many of the United Kingdom firms to meet these challenges was perhaps the most serious concern.

West Germany adopted a textile industry strategy similar to that of Japan. The Germans turned from competition with the low cost Far Eastern countries to production of specialty products.

The West German industry can be described as highly specialized with high productivity. The German banking industry was very involved in the decisions of German textile manufacturing because they had a high equity stake in the industry. Government assistance was minimal. Among other European countries, only Italy increased its global market share in recent years. Italy succeeded because of such factors as geographical production concentrations (which reduced transportation costs), lower labor costs from locations in high unemployment areas, and the use of modern technology and modern management practices, including computerized inventory control. The U.K. textile industry has declined since 1973, a victim of recession, Far Eastern competition, and failed strategies. The U.K. productivity in the textile industry was approximately half of the U.S. textile industry [13].

The U.S. machine tools industry, the world leader from World War II to the early eighties, held a global share of less than 10 per cent by 1989 [14]. Reasons abound for this decline, but one chief reason is the cheap licensing of technology to Japanese machine tool makers in exchange for Far Eastern marketing channels. Numerically controlled machine tools were invented in the United States, but it was the Japanese manufacturers who became the standard bearers for the technology. The U.S. industry has rebounded somewhat from a perilous low in 1982, but a trade agreement was necessary to limit Japanese imports. Japanese firms have since circumvented the agreements by locating plants in the United States, and 30% of the U.S. machine tool capacity was Japanese owned by 1990 [15].

Imports seriously affected labor-intensive and short-run fabric markets in the U.S. textiles industry [15]. In reaction, U.S. manufacturers de-emphasized products that couldn't compete with the imports, and focused their efforts upon domestics, including sheets, towels and pillowcases. In 1990, domestic mills are operating at capacity, primarily supplying domestic markets. The trade deficit in textiles was enormous. In 1977, 743 million pounds of textiles were exported and 1286 million pounds were imported . Ten years later, 913 million pounds were exported and 4417 million pounds were imported.

MANUFACTURING PRACTICES

This section of the paper will discuss production practices in Japan, Korea, China, the United States and western Europe.

Production Plans

When asked what were the two most important uses of the production plan, the responses differed widely across countries. The Chinese machine tool managers listed subcontracting (90%) and material planning (52%), the Japanese listed operations scheduling (61%) and facilities planning (33%), the Koreans materials planning (57%) and operations scheduling (34%), the Europeans operations scheduling (47%) and materials planning (44%), and the U.S. operations scheduling (49%) and manpower planning (47%). (See Table 2.)

Subcontracting was also most important to the Chinese textile industry (84%) while materials planning was most important to the Japanese (58%) and operations scheduling (36%) to the Koreans.

Subcontracting was rarely mentioned by the Japanese and Koreans as a use for the production plan. Operations scheduling was of little concern to the Chinese. Since the Chinese tool industry's largest customer was its own defense industry, there was an obvious absence of market pressures.

The U.S. textile managers had a decided preference for using the production plan for materials planning and operations scheduling, while the European managers found the plan more useful for facilities planning and subcontracting.

Table 2. Most important uses of production plan
(Percent responding. Respondents asked to list two.)

	Machine tool					Textile				
	China	Japan	Kor.	Eur.	US	China	Japan	Kor.	Eur.	US
Budget preparation	32	11	19	38	18	57	19	24	33	10
Subcontracting	91	6	3	35	18	84	14	6	46	2
Manpower planning	0	28	3	15	47	0	3	6	13	38
Facilities planning	20	33	11	35	11	7	22	9	58	8
Operations schedule	16	61	34	47	49	2	44	36	38	58
Materials planning	52	28	57	44	44	30	58	30	38	66
Purchasing	7	28	0	24	24	4	25	0	13	14
Other	7	6	0	0	2	0	3	0	0	0

Computers in Planning, Scheduling and Purchasing

Clearly, computers were slow to gain acceptance for production planning in Korea and China (see Table 3). The Japanese used computers to varying degrees in both industries, but electronic data interchange (EDI) between buyer and supplier was not reported in either industry (Table 4).

The U.S. and European machine tools managers reported much more extensive use of computers for production planning and scheduling than the textile industry managers in their own countries.

Table 3. Use of computers for production planning & scheduling
(Percent responding. Responses may not total 100 percent because of non-response.)

	Machine tool					Textile				
	China	Japan	Kor.	Eur.	US	China	Japan	Kor.	Eur.	US
Not at all	68	17	82	6	9	89	50	79	12	34
Occasionally	20	22	11	21	9	7	25	9	21	34
Moderate	9	39	0	21	38	2	19	0	42	34
Extensive	0	17	0	41	44	0	3	0	17	14

Table 4. How is a purchase order transmitted to a supplier?
(Percent Responding. Responses may total over 100 percent
because of multiple responses.)

	Machine tool					Textile				
	China	Japan	Kor.	Eur.	US	China	Japan	Kor.	Eur.	US
Orally	11	22	13	6	38	27	50	9	25	46
Written	98	89	78	85	80	89	56	79	67	74
Computer	0	0	0	26	22	0	0	0	17	6
Other	9	22	0	6	11	25	11	0	5	14

Deliveries

Lead times varied widely in the machine tool industry. The United States machine tool industry has the longest lead time with an average of 166 days, compared with 42 days for Japan and 54 days for Korea (Table 5). U.S. textiles had more competitive lead times of 26 days, the same as the Japanese, while the Koreans lagged behind with 45 days. However, these figures depend upon the tools being manufactured.

Table 5. Lead time from start of batch to delivery or inventory

	Machine tool					Textile				
	China	Japan	Kor.	Eur.	US	China	Japan	Kor.	Eur.	US
Days	142	42	54	163	166	27	27	45	43	26

The Korean manufacturers were more customer-driven in both industries than the Japanese, Chinese and U.S., according to the determination of delivery dates (Table 6). This was concluded from the response to the question, "How is the delivery date determined: By customer, negotiated or by the supplier?" The Koreans were heavily slanted towards the customer in both industries, while only the Japanese and U.S. textile industries had a similar orientation.

Table 6. How is the delivery date determined?
(Percent responding)

	Machine tool					Textile				
	China	Japan	Kor.	Eur.	US	China	Japan	Kor.	Eur.	US
Mostly by customer	14	33	65	18	24	34	44	45	38	40
Negotiation	70	33	19	59	22	52	42	18	42	20
Mostly by us	14	33	9	24	42	11	6	12	21	38

When listing the causes of late orders (Table 7), the Chinese machine tool industry highlighted lack of labor capacity and material shortages, while their textile industry listed transportation problems and lack of labor capacity. The popular belief has been that Chinese manufacturing had an almost infinite source of labor, but here we discover that they sometimes don't have enough labor to meet demand in these industries. Primitive production scheduling practices could cause these problems. Material shortages were cited most frequently by both U.S. industries as the most frequent cause of late orders.

Table 7. Most frequent causes of late orders
(Percent responding. Respondents asked to list two)

	Machine tool					Textile				
	China	Japan	Kor.	Eur.	US	China	Japan	Kor.	Eur.	US
Lack machine capacity	25	39	39	12	9	11	36	39	8	32
Lack labor capacity	41	28	22	32	29	50	11	15	46	24
Product bottlenecks	9	44	13	47	29	25	36	15	33	38
Transportation	11	0	0	35	0	70	0	0	42	2
Material shortages	34	11	25	26	51	32	28	18	38	48
Quality problems	16	17	44	9	29	16	17	45	17	20
Due date changes	23	33	13	21	9	7	83	9	17	8
Other	20	0	0	0	31	11	28	0	0	10

MRP? What's That?

Considering the lack of computer involvement in production planning in the Far East, it was no surprise that there was little use of Material Requirements Planning (MRP). Korea and China were noticeably more involved with MRP than Japan (Table 8). Seventy-three of U.S. machine tools managers and 50% of the textiles managers reported using MRP. In Europe, 63% of the machine tools managers and 42% of the textiles managers reported some extent of MRP involvement.

Table 8. What is your exposure to MRP?
(Percent responding)

	Machine tool					Textile				
	China	Japan	Kor.	Eur.	US	China	Japan	Kor.	Eur.	US
Never heard of	48	67	49	24	7	55	83	39	46	26
Benefiting from	7	6	9	44	38	0	0	9	38	20
Using, not benefiting	9	17	8	3	2	7	8	18	4	0
No need for	7	0	0	12	20	7	0	0	8	24
Beginning to implement	7	0	0	9	24	7	0	0	0	18
Difficult to implement	20	6	30	3	9	20	6	30	4	12

Sourcing and JIT

A highlight is the high percentage of Far Eastern respondents who never heard of JIT. They may call it Kanban or something else or aren't familiar with the concept (Table 9).

For that matter, the use of JIT by the Chinese and U.S. machine tool industries seemed unnecessary considering their lead times.

Multiple sourcing was also more frequently reported in both industries than single sourcing (Tables 10 and 11.) European and American managers demonstrated a decided preference for multiple over single sourcing.

MANUFACTURING STRATEGY

Several manufacturing strategy areas outlined by Hayes and Wheelwright [16] are pertinent here: quality, production planning, technology, and the work force.

Table 9. What is your exposure to JIT?
(Percent responding)

	Machine tool					Textile				
	China	Japan	Kor.	Eur.	US	China	Japan	Kor.	Eur.	US
Never heard of	48	11	39	9	4	55	47	36	50	22
Benefiting from	0	22	1	18	13	0	0	0	13	24
Using, not benefiting	0	6	0	3	2	0	6	0	0	4
No need for	23	39	0	32	42	5	22	0	25	22
Beginning to implement	0	0	9	12	18	0	6	12	8	18
Difficult to implement	27	11	47	24	18	23	17	42	4	8

Table 10. Important actions to assure timely supply of parts
(Percent responding)

	Machine tool					Textile				
	China	Japan	Kor.	Eur.	US	China	Japan	Kor.	Eur.	US
Long term contract	59	50	9	47	29	66	56	12	75	58
Hedging	48	0	3	9	--	21	11	6	17	16
Single sourcing	2	17	0	15	18	5	6	0	8	16
Large qty. purchase	20	17	0	15	13	11	17	0	13	26
Multiple sourcing	41	56	0	59	67	39	75	0	36	54
Sister plants	39	6	0	15	2	38	3	0	8	4

Table 11. Vendors per purchased part
(Percent responding)

	Machine tool					Textile				
	China	Japan	Kor.	Eur.	US	China	Japan	Kor.	Eur.	US
One	5	17	44	21	13	2	22	30	17	16
2 or 3	34	61	19	68	73	34	67	9	63	74
Four +	61	6	13	6	11	52	8	24	13	10

Japanese quality control was world renowned, with empirical and anecdotal evidence supporting its superiority [1,17]. Korean quality, while not taken as seriously as the Japanese, appeared to meet competitive standards [8,9]. Japan was the world leader in machine tools, with quality and technology being the driving reasons. China has struggled to provide adequate tools for its own domestic needs because of its inferior quality, production planning and technology.

JIT did not appear to be prevalent in these industries. Although the Japanese lead in technology, they appear to be slower to adapt computerized manufacturing information systems than the Koreans.

The Chinese workforce expected lifetime employment in the past, but in 1986, domestic labor reforms instituted labor contracts in which either the employer or employee could decide not to renew at the expiration of the contract [18]. The typical Chinese worker took home $45 per month, including a $10 bonus. Chinese managers appealed to the workers' moral and political motivations rather than material motivations. The state establishes production quotas and profits are appropriated by the government and allocated for units that experience losses. As a result, state enterprises have little motivation to exceed their quotas [19].

Before World War II, Great Britain's spinning and weaving technologies were the world standard. Since that time, U.S.-based technologies dominated the industry and today the U.S. is generally regarded to have the most productive and cost efficient industry in the world [20]. The Far East became strong because much of the textile industry was labor-intensive. West Germany succeeded with a strategy of offshore manufacturing to take advantage of the labor cost differential (labor in West Germany is more expensive than in the United States). Great Britain has not used this strategy as a matter of principle and the U.S. was prohibited from this practice by Item 807 of the Tariff Schedule of the United States, which does not prohibit offshore manufacturing for the apparel industry.

The machine tools industry is one in which the Japanese are in the forefront of technology and innovation with firms in other countries following their lead. In Europe, there were decided differences in the strategic use of the work force between Great Britain and West Germany. West Germany employed a formal system of vocational training which led to a lifetime career. Vocational training was followed by a two or three year apprenticeship.

Such a system contributed to an international reputation for skill in manufacturing. Great Britain's weaknesses in manufacturing were partly attributed to their poor development of human resources [12]. Training was carried out within the plant, so the British plant learning curve could be expected to be longer than that of a West German plant.

Government involvement in industry strategies was extensive. Most governments were protectionist for both industries and involved in recommending global industry strategies. Korea actually prohibited textile imports unless they were re-exported. The United States limited Japanese imports in the machine tools industry. Only West Germany could be described as laissez faire, welcoming a free market.

Firms within these industries varied their strategies somewhat, but industry strategies could be classified into a global competitive scheme. Porter [21] recommended sharply limiting direct cooperation among industry rivals, arguing western governments have misunderstood the role of the Japanese Ministry of International Trade and Industry (MITI). MITI projects work best in stimulating proprietary research and Japanese firms do not contribute their best engineers and scientists to cooperative projects.

Hayes and Wheelwright [16] wrote that firms could be classified into four stages of manufacturing strategy:
- Stage one: Internally neutral. Production simply makes the product and ships it.
- Stage two: Externally neutral. Manufacturing merely meets the standards set by competition.
- Stage three: Internally supportive. Manufacturing attempts to become unique from its competition.
- Stage four: Externally supportive. Manufacturing pursues uniqueness on a global scale, becoming a world class competitor.

According to the Hayes and Wheelwright framework, both Japanese industries could be considered stage four and the Korean industries stage three. The Chinese textile industry was stage three, but the Chinese machine tool industry stage one (see Table 12).

Table 12. Manufacturing strategy stage

Machine tool				Textile			
China	Japan	Kor.	US	China	Japan	Kor.	US
1	4	3	2	3	4	3	3

Because of the divergence of European strategies, it was not possible to determine a unified strategy. The United States textile industry, with its heavy domestic market concentration, could be classified as stage three and the machine tools industry probably fits into stage two.

CONCLUSIONS

The principles of quality and technological innovation which brought the Japanese success in the automobile and electronics industries could be found in the machine tools and textile industries. The survey did not reveal any strategic use of production planning practices other than the fact that the Japanese were faster at producing machine tools than the Chinese, Europeans, and Americans. However, a competitive advantage was gained by simply outperforming the rest of the industry in production planning.

Korea has increasingly become an exporter to be reckoned with. The Japanese success formulas have been well documented and the Koreans seem to be following a similar pattern. The large scale entrance of the People's Republic of China into international trade, clouded by the political events of 1989, was years away in the industries we analyzed. The unification of Europe in 1992 will result in reduced industry costs, but the final impact can not be determined. Presently, national policies and strategies and firm performances differ substantially among the European countries in these industries and it was difficult to speculate on the global impact.

The relative absence of information systems for production planning in any of these countries could produce a source of competitive advantage for the United States. U.S. manufacturers have more information available for production planning than their Far Eastern competitors.

This research described global manufacturing practices and strategies in the machine tools and textiles. A lesson to be learned from this research is that studying international operations at the industry level is an approach that will yield information quite different from what is expected. Micro studies of individual firms are also needed to provide an in-depth understanding. However, the industries selected for this study were very influenced by government policy. Cultural and language barriers have made international empirical study difficult, but the only way we can go develop a proper global perspective of operations is for researchers to visit plants in all corners of the world. This research, with the collaborative efforts of the aforementioned groups, is a step in that direction.

REFERENCES

1. R.J. Schonberger, *Japanese Manufacturing Techniques*, New York, The Free Press, 1982.

2. R.W. Hall, *Zero Inventories*, Homewood, IL, Irwin, 1983.

3. S.C. Wheelwright, "Japan--Where Operations Really Are Strategic," *Harvard Business Review*, 59(4) (1981), 67-74.

4.* B.H. Rho and D.C. Whybark, "Comparing Manufacturing Practices in the People's Republic of China and South Korea," Discussion Paper No. 4, Bloomington, IN, Indiana Center for Global Business, Indiana Business School, May 1988.

5. A. Davenport, "Forging a Modern Machine Tool Industry," *The China Business Review,* May-June, 1988, 38-44.

6. The World Bank, *Korea: Managing the Industrial Transition.* Washington, D.C., The World Bank, 1987.

7. B. Johnstone, "Textiles Diversification Helps to Protect Profits," *Far Eastern Economic Review,* 142 (October 13, 1988), 54-56.

8. T.W. Kang, *Is Korea the Next Japan?* New York, The Free Press, 1989.

9. I.C. Magaziner and M. Patinkin, "Fast Heat: How Korea Won the Microwave War," *Harvard Business Review,* 67(1) (1989), 8393.

10. R. Sarathy, "The Interplay of Industrial Policy and International Strategy: Japan's Machine Tool Industry," *California Management Review,* 31(3) (1989), 132-160.

11. "Holding the Lead in the Machine Tool Industry Proves Tough," *Business Japan,* 32 (September 1987), 95-104.

12. Commission of the European Communities, *The Social Aspects of Technological Developments Relating to the European Machine-tool Industry,* Luxembourg, Office for Official Publications of the European Communities, 1986.

13. F. Ghadar, W. Davidson and C. Feigenoff, *U.S. Industrial Competitiveness,* Lexington, MA Lexington Books, 1987.

14. M. Holland, *When the Machine Stopped,* Boston, Harvard Business School Press, 1989.

15. *Standard & Poors Industry Surveys,* New York, Standard & Poors, 1990.

16. R.H. Hayes and S.C. Wheelwright, *Restoring our Competitive Edge,* New York, John Wiley & Sons, 1984.

17. D.A. Garvin, "Quality Problems, Policies, and Attitudes in the United States and Japan: an Exploratory Study," *Academy of Management Journal,* 29 (1986), 653-673.

18. J.P. Horsley, "The Chinese Workforce," *The China Business Review,* May-June 1988, 50-55.

19. S.C. Zhuang and A.M. Whitehill, "Will China Adopt Western Management Practices?" *Business Horizons,* March/April 1989, 58-64.

20. B. Toyne, J. Arpan, A. Barnett, D. Ricks, T. Shimp, J. Andrews, J. Clamp, C. Rogers, G. Shepherd, T. Tho, E. Vaughn, and S. Woolcock, *The Global Textile Industry*, London, George Allen & Unwin, 1984.

21. M.E. Porter, *The Competitive Advantage of Nations*, New York, The Free Press, 1990.

* This article is reproduced in this volume.

ACKNOWLEDGEMENTS

A version of this paper has been published in the *International Journal of Operations and Production Management*, 12, No. 9 (1991), 5-17.

Cross-National Comparison of
Production-Inventory Management Practices

Attila Chikán, Budapest University of Economic Sciences, Hungary
D. Clay Whybark, University of North Carolina-Chapel Hill, USA

ABSTRACT

This paper is based on the results of surveys on manufacturing practices conducted in South Korea, western Europe, Hungary, and the People's Republic of China. Selected questions are evaluated from the standpoint of relating to the market, managing production and organizing for production decisions. There are clear differences among regions in the way these activities are conducted. These differences stem from the market orientation (economic system) and cultural differences in each of the regions. For the market related factors the Chinese and Hungarian firms seem to form one group, while those factors influenced by culture seem to group the Hungarian and western European firms together.

INTRODUCTION

This paper presents the results of an evaluation of a set of surveys carried out in South Korea, China, western Europe and Hungary. The survey was initiated by Professors Boo-Ho Rho of Sogang University, South Korea (who first directed the preparation of the survey in South Korea) and D. Clay Whybark of Indiana University, USA (who did the same in China). Since then it has been conducted in several other parts of the world [1]. The survey covers production-inventory practices in two industries: nonfashion textiles and small machine tools. The results have provided interesting insights within the countries themselves [2] and cross national comparisons promise further important lessons.

Why is it important to be familiar with the practices of other nations, especially in a field like production and inventory management, which is basically considered the internal affair of a company? Today's widening cooperation between geographic regions and nations of different cultural backgrounds provides an important reason for making such comparisons. The results can be applied in the case of cooperation or joint ventures: If we want to be successful we have to know about the management practices of the countries with which we wish to cooperate. On the other hand, if we wish to understand the differences in output, both in quality and quantity, or simply the differences in productivity and want to find ways of improvement, the practices of other countries can provide useful information and lessons. Finally, we think that such studies deepen our knowledge of the general nature of production and inventory management: if we see common features or similarities in the practice in several different countries then we can say with high probability that those features issue from the general nature of the system.

The four regions included in this analysis are very different. This means that the international comparison provides a many-sided picture which can help realize the above-

mentioned objectives and advantages quite well. Other regions of the world are being surveyed and will be included in the project in the future [1].

SURVEY METHODOLOGY

The method of conducting the survey was similar in the case of South Korea and China,
namely, questionnaires were first distributed by mail. The response to the mailed survey
yielded an unsatisfactory number of completed returns. The next step therefore was to
send interviewers to the companies where they gathered enough additional questionnaires
to provide a total of approximately fifty responses from each industry. In western Europe
the survey started with a mail distribution and this was followed up with telephone
canvassing involving 35 companies in the machine-tool industry and 34 textile firms. In
Hungary, interviewers were sent directly to companies, and they collected responses from
79 companies. The Hungarian survey included companies from several industries (since
there are not enough machine tool or non-fashion textile companies to provide a solid
basis for evaluation). Thus the survey in Hungary shows the general picture of the
manufacturing industry.

The difference in the Hungarian sample population doesn't really disturb the cross
national evaluations since there were only a few examples where the differences between
industries were greater than between countries. This finding corroborates one of the
outcomes of some of the early comparative research [3,4].

The questionnaire consisted of 65 questions in the following categories:
　　I. Company Profile
　　II. Sales Forecasting
　　III. Production Planning and Scheduling
　　IV. Shop Floor Control
　　V. Purchasing and Materials Management

The limits of this paper do not allow a detailed evaluation of all the survey results.
Furthermore, we cannot attach the questionnaire itself, which is quite lengthy (see
reference [1]). We shall provide some of the results which we found most interesting and
draw some general conclusions. Our comparisons involve all the data but the illustrative
Figures use only the data of the machine-tool industry for China, South Korea and western Europe, since the majority of firms in the Hungarian survey were from the machine
industry.

We introduce the results of our analysis of the survey by analyzing how production
relates to the market. Considered are tasks like planning, forecasting and other aspects
of matching production to consumption. We then look at some of the techniques used for
production and inventory management and, finally, turn to reviewing some of the organizational roles in production-inventory decisions.

RELATING TO THE MARKET

One of the key differences in the production-inventory systems of the different regions is the manner in which they respond to the market (dictated, in part, by the degree of central planning in the local economy). Determining this response means, among other things, defining whether production is to order or to stock. Although production to order is dominant in all four areas of the world, there are still important differences among the areas.

In Hungary and South Korea over 70% of the companies have more than 90% of their production to order. In western Europe only 36.4% in the machine industry and 21.7% in the textile industry produce 90% to order but the numbers increase to 66.4% and 47.8% for 60% production to order. In China, the picture is quite different. Free market customer orders play a rather small role since only 4.7% of the machine and 12.0% of the textile industry firms produce over 90% of their production to order. The state however, comes in and substitutes for the consumer so that 66% of the textile industry and 29% of the small machine tool industry production is to state order.

The proportion of production to order is a very important aggregate indicator of production system design policy for the particular industry. In general, the more a company produces to order, the less speedily it can satisfy consumer demand. So, as a rule, one can say that the stronger the market requirements for speedy delivery the more a firm produces to stock. The most demanding market in this sense seems to be in western Europe. The "luxury" of producing to order in the three other areas of the world have quite different roots, varying from state planning in China to the global supplier role of the firms in Korea.

The ratio of raw material and finished goods inventories is another major design factor in responding to the market. This ratio is shown in Table 1 for three of the geographical areas. (We do not have sufficient data for South Korea).

Table 1. Percent of Companies in the Survey Having Ranges of
Ratios of Raw Material to Finished-good Inventories
(r=ratio of raw material to finished goods)

	$r \leq 2$	$2 < r < 5$	$r \geq 5$
W. Europe	54.0%	26.0%	20.0%
China	35.0%	31.6%	33.0%
Hungary	6.4%	32.1%	61.5%

In Hungary, companies have the highest proportion of total inventories in raw materials among the three areas as seen in Figure 1. This is due to the raw material scarcity in the shortage economy. This is like China, but in Hungary there is no state intervention to ensure supply as there is in China. The ratio of finished goods is the highest in western

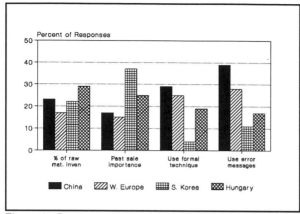

Figure 1. Raw material inventory and forecasting

Europe, which, as we've already said has the most demanding market in terms of speed of supply. The data we do have from South Korea indicates that they keep somewhat higher raw-material stocks overall than west Europeans, but still much less than Hungarians.

Forecasting in some form is done everywhere, but the use of forecast information is different from region to region. In western Europe the forecast is used mostly for financial budgeting, indicating the importance of financial planning in that market. In China and South Korea the main use of the forecast is for production purposes. In contrast, there is no clear picture in Hungary of either a dominant use or user group.

Important differences arise in production planning. In China production plans are used for operations scheduling more than anything else. The use of production plans is uniformly distributed among budgeting, subcontracting, manpower planning, operations scheduling, inventory control, and purchasing in western Europe and Hungary--with one important difference. Each of these uses is mentioned by almost all Hungarian companies (which shows the domination of production over other activities) while in western Europe all uses are mentioned by only about half of the companies .

There are interesting findings in the production planning details as well. For example, west Europe is the only area where the production plan is connected to manpower planning. In the other regions, manpower levels are "rigid." Hungary is the only region where production plans are not used for purchasing. In Hungary the delivery times are so long that supplies must be ordered long before the production plan is completed.

In China, purchase orders are released mostly on the basis of the production plan, while in western Europe the customer order is mostly used. This is surprising, given western Europe's higher make-to-order percentages but consistent with a more market oriented policy and pull type inventory systems. Both in China and Hungary inventories play an important role in releasing purchase orders, reflecting the shortage economy impact.

Decisions to start production (which means the start of replenishment of finished goods inventories--just as purchasing is the replenishment of input inventories) are based on factors very similar to those for purchase order release. Inventories play a much less important role in China and Hungary in this regard, since production to stock is much less frequent, i.e., the role of "indicator" is not played by inventories.

The causes of lateness in delivery of orders to customers are very different between regions and there are practically no differences between industries. Material shortage is the most important reason behind late deliveries in Hungary (none of the other reasons come close to it), while in South Korea capacity problems are the main cause. In China transportation problems are the most important, although a difference does occur between industries. Material shortages play an important secondary role in China's textile industry but not much at all in the machine tool industry. There is no leading reason in western Europe, the causes are uniformly distributed, which in our interpretation means that the causes are random, individual events.

TECHNIQUES USED IN PRODUCTION MANAGEMENT

As can be seen in Figure 1, the role of past sales in forecasts is highest in South Korea and Hungary. On the other hand the use of formal techniques and error measures is lowest in these two regions. The dominant sales forecasting technique in west Europe is time series analysis. Very few west European companies use consensus techniques while there is greater use of consensus in other regions. We suspect that this refers to cultural differences where western European companies rely more on positive, objective measures, while in other regions connections among people play a more important role.

It was surprising to learn that the forecasting period is similar in all four regions. In general, there is an annual forecast updated quarterly. This is interesting because in Hungary there is a tendency to think that this timing is a remnant of the classical planned economy when the use of this periodicity was mandatory for all companies. Given the universality of our finding, one is tempted to think that this periodicity is closer to a "natural" periodicity than a result of state planning convenience.

It is surprising that South Korean firms have the lowest level of use of computers for forecasting and production planning of the four areas. In order of increasing use, South Korea is followed by China, then comes Hungary, and western Europe. The position of Hungary is a particular surprise since it is generally felt that Hungary is in a poor position as far as computer hardware is concerned. In western Europe there is slightly

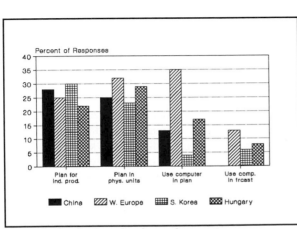

Figure 2. Production planning and computer use

more use of computers in the machine tool industry than the textile industry. The use of computers in forecasting and production planning, is shown in Figure 2.

Despite the disparity in the use of computers between the four regions, there is not much difference in the elaborateness of the plans produced. There is not much difference in the proportion of plans made for individual products (as opposed to groups of products) or the proportion of planning in physical units (as opposed to planning in values). At this level of detail, however, the two European regions are closer to each other than to the Asian ones, as can be seen in Figure 2.

The production planning area is another area where one can see well the impact of cultural differences. In production planning, there are greater differences, in many respects, between the European and the Asian countries than between the planned and open markets. That is, the production planning approaches used in Hungary are most like those used by the west European companies and least like the Chinese firms. This is despite the fact that Hungary and China are "socialist" countries with a high ratio of state-owned companies. We take this to indicate that for certain operational activities cultural differences sometimes can be more important than the forms of ownership.

Advanced methodologies are used practically only in western Europe. Figure 3 shows that MRP is applied in Hungary and South Korea to some extent, while in China there was no application reported. Just-in-time can be found almost exclusively in west Europe. Results are similar in the case of other advanced methodologies: up-to-date concepts of operation of the production inventory systems still have not penetrated the industries of Hungary and the Asian countries.

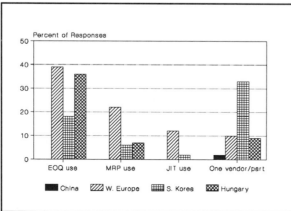

Figure 3. Techniques and methods used

ORGANIZATIONAL ROLES IN PRODUCTION DECISIONS

There are several ways in which production-inventory decisions are affected by various people inside and outside the production organization. We will look at two aspects of that in this section: functions that influence production-inventory decisions and the organizational level of the persons involved.

Functional Units' Influence

Analyzing the influence of various functional units on production-inventory decisions provides very important information on the differences in corporate culture between geographical regions. In this section we will describe the roles played by various departments in several activities.

The general pattern of organizational participation in forecasting is rather similar in all regions. The sales department dominates the forecasting task, although substantial participation of planning is experienced and production takes part to some extent. The finance department is rarely involved. As for production planning, the roles that various departments play in South Korea, Hungary and western Europe are very similar. In these regions production has somewhat more influence than planning, the participation of sales is substantial, and again, finance plays a negligible role. China is different however. There the planning department exercises the predominant role, with substantial participation of production. Sales plays a minor role, and finance has essentially no influence at all.

There are very important differences both by industry and by geographical region in the practice of shop floor control. The planning department plays the leading role in ordering the start of production in the textile industry in China, while in the machine industry, planning's role is much reduced in favor of production. The west European textile industry behaves like its counterpart in China, while in the west European machine industry the role of production is much less important than in the same industry in China. In Hungary, on the other hand, production plays the dominant role in both industries, while in South Korea marketing is the most influential in both. This may help make the South Korean firms much more sensitive to market needs, despite their high percentage of make-to-order production.

In the shop floor control activities in general, however, one can see smaller differences between the various countries than in forecasting and planning. As seen in Figure 4 companies in Hungary use more formal methods of shop floor control (when it is measured by the proportion of written purchase orders or priority changes as opposed to verbal direction). Initiations of changes in priority in western Europe mostly come from marketing or administration while in South Korea and China they come from production. In Hungary, we note a very mixed situation with respect to priority changes.

As for initiating a purchase order, the situations in China, South Korea and Hungary are somewhat closer to each other than to the west European one.

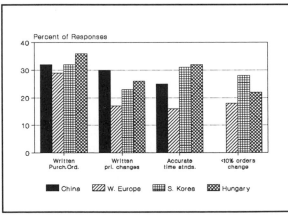

Figure 4. Shop floor conditions

In all four regions production plays the leading role, but the margin of difference in western Europe is small, while in the other regions the difference between the influence of production and the other areas is large.

It's interesting that the response by Hungarian firms to questions regarding the influence of various departments on purchasing were characterized by multiple answers. This may indicate some uncertainty about the actual influencing areas or that everybody has something to say about purchasing.

Another important finding is that finance plays a major role in production decisions, only in western Europe. Finance's role is much less important in the other three regions where production and planning are the departments most concerned with decisions related to the physical processes.

Organizational Level of Influence

The level of executive attention paid to forecasting is highest in western Europe, where top management participation is the most frequent. Forecasts are most often prepared at the middle management level in Hungary and China, while about 40% of the South Korean companies produce forecasts at the lowest managerial level.

In production planning there are inconclusive industry and geographical differences except in China where middle managers predominate in the process. Clear differences exist among the users of the production plan. The users are people in production in western Europe, Hungary and in the Chinese machine industry. In the textile industry in China, the production plan is mainly used by the planning department. Both the producer and user of the production plan is at a somewhat lower level in the organization than those who prepare and use the forecast. (Interestingly enough, in western Europe and Hungary the user of the production plan is characteristically at a higher level than the producer of the plan).

As a generalization, the greatest differences are between China and the other regions and the Chinese textile industry especially seems to be very much "planned." The responses from Hungary are characterized by having several managerial levels marked on the same questionnaire, perhaps again reflecting some uncertainty about the responsibility for decisions or involvement of many people. In South Korea the involvement of lower organizational levels is more pronounced than in any other region.

CONCLUSIONS

The results of the survey presented here provide enough evidence to establish the statement that the production practices of the various regions are quite different. One cannot say, however on the basis of the individual findings here, which set of practices is superior. Also it must be considered that these practices (and others not discussed here) together comprise a system of operations where we cannot single out one of the elements and judge it by itself. However, we think that the survey supports the concept that the practices of the companies reflect the conditions and requirements of the environment both from the economic/political and social/cultural points of view. The majority of the characteristics which were only briefly discussed here can be well explained by causes stemming from the market situation and/or the cultural background of the different regions.

The most general conclusion, then, is that production inventory systems are basically influenced by two major groups of factors, namely, the market situation and the cultural background.

Considering the activities influenced by the market factors, western Europe and Korea seem to be in one cluster, while Hungary and China are also close to each other. As for the cultural background, western Europe and Hungary are in one cluster while China and Korea are in another.

It seems that in issues concerning forecasting, planning, inventory positioning, and so on, market factors play the greatest role, while in organizational related concerns, the cultural background is more important. It is harder to make such a distinction in terms of the techniques employed in the different regions. These conclusions correspond to common sense, but, according to our view, they are far from evident when performing an analysis of a field of a single company's internal operations. The groupings we found, however, come through clearly in the comparison of the practices across companies and regions.

REFERENCES

1.[*] D.C. Whybark and B.H. Rho, "A Worldwide Survey of Manufacturing Practices," Discussion Paper No. 2, Bloomington, IN, Indiana Center for Global Business, Indiana University Business School, May 1988.

2. A. Chikán, "Characterization of Production-Inventory Systems in the Hungarian Industry," Paper presented at the Fifth International Workshop on Production Economics, Igls, Austria, 1988, to be published in *Engineering Costs and Production Economics*.

3.[*] B.H. Rho and D.C. Whybark, "Comparing Manufacturing Practices in the People's Republic of China and South Korea," Discussion Paper No. 4, Bloomington, IN, Indiana Center for Global Business, Indiana University Business School, March, 1988.

4.[*] B.H. Rho and D.C. Whybark, "Comparing Manufacturing Practices in Europe and South Korea," Discussion Paper No. 3, Bloomington, IN, Indiana Center for Global Business, Indiana University Business School, March, 1988.

* This article is reproduced in this volume.

ACKNOWLEDGEMENTS

A version of this paper appeared in *Engineering Costs and Production Economics*, 19, 1-3 (1990), and in A. Chikán (ed.), *Inventories: Theories and Applications*, Amsterdam, Elsevier, 1990.

SECTION FOUR

STUDIES OF OTHER RELATIONSHIPS

The papers in this section represent quite different approaches to the analysis of the Global Manufacturing Research Group data. They vary from the comparisons among regions, industries or size presented in the previous sections and they differ one from another in this section. All look at different relationships between elements of the data, generally on a multi-region basis. Three of the papers look at questions concerning inventory, lead time, delivery performance and factors affecting their behavior. Two of the papers raise issues of productivity while another is concerned with the influence customers have on manufacturing practices. The final paper in the section traces the changes in manufacturing practices in Hungary over a five year period. All papers show the diversity of insight that is possible when as many people as possible are involved in the analysis of the data.

Gyula Vastag and D. Clay Whybark
Global Relations Between Inventory, Manufacturing Lead Time and Delivery Date Promises

This paper looks at the relationships between manufacturing lead time, delivery date promises and inventory. It tests hypothesis concerning the delivery time promises compared to the manufacturing lead time and tries to account for any anomalies by the presence of inventory. In addition the paper evaluates some of the results of the implementation of JIT programs. The data used were taken from China, Hungary, Japan, North America and western Europe.

Curtis P. McLaughlin, Gyula Vastag and D. Clay Whybark
Statistical Inventory Control in Theory and Practice

Does it matter if there are computers, MRP or JIT systems, or whether companies know about EOQ? Do the presence of these technologies and techniques change inventory levels or delivery times in any substantive way? These are the questions that this article addresses. Using data from China, Hungary, Japan, Korea, North America and western Europe, this paper investigates the relationship between the techniques and performance. The paper also covers some concerns about inventory theory and presents a description of how actual practice deviates from the assumptions required by the theory.

Karen Brown and Gyula Vastag
Determinants of Manufacturing Delivery Reliability: A Global Assessment

As customer satisfaction has become more important in the market place, delivery reliability has become a competitive concern for firms around the world. Though many theoretical and mathematical studies have been done on the relationship between scheduling and delivery performance, very little empirical work has been conducted. This study evaluates some of the potential causes of lateness using the GMRG data. Furthermore, a comparison is made of differences in delivery performance between several western countries and three countries experienced with centrally planned economies.

Xiao Cheng Zhong and D. Clay Whybark
A DEA Analysis of Sales Output and Comparison of Manufacturing Practices in China, Hungary and the United States

The data envelopment analysis (DEA) technique provides a method for evaluating the relative performance of individual firms. In this article DEA is applied to the sales output per worker and capital investment dollar in China, Hungary and the United States. The analysis discloses, not surprisingly, that Chinese firms are least effective at transforming their labor and capital investment into sales outputs. Some of the manufacturing practice differences that account for differences in the relative performance are described as well.

John G. Wacker
A Comparison of the Relative Productivity in the Machine Tool Industry for the U.S. and Japan

This paper addresses the important question of relative productivity differences between the Japanese and U.S. firms in GMRG data base. After describing the difference in basic production philosophy and the data used for the study, the data are used to fit equations that relate capital and labor to output. The coefficients are then used to test hypotheses related to labor and capital productivity and issues of differential use of resources. Conclusions concerning the differences in productivity are drawn.

D. Clay Whybark
Are Customers Linked to Manufacturing?

The notion of "customer driven" manufacturing implies that customers should have linkages to manufacturing. It's impossible to think about customers having an impact on manufacturing if there is no effective communication of customer desires and interests. This paper evaluates the ability of the customer to impact forecasts, production plans or schedules using data from Korea, China, North America and western Europe. Data on questions that reveal the extent to which customer information is used in forecasts, the degree to which customer plans are a basis for production planning, or how customer pressure influences shop priorities are used to assess the degree to which customers are linked to manufacturing.

Attila Chikán and Krisztina Demeter
Manufacturing Practices in a Transition Economy

This article presents one of the first temporal comparisons of the data from the Global Manufacturing Research Group. It reports on the changes in manufacturing practices from the questionnaire administered in 1986 to one administered in 1991. The motivation was the transition in the Hungarian economy and the article clearly reflects this transition in the changes in manufacturing practices that have already occurred. The article is an important precursor to the studies that will follow the worldwide administration of the second GMRG questionnaire starting in 1993.

Global Relations Between Inventory, Manufacturing Lead Time and Delivery Date Promises

Gyula Vastag, Budapest University of Economic Sciences, Hungary
D. Clay Whybark, University of North Carolina-Chapel Hill, USA

ABSTRACT

This paper reports a study of the inventory, lead time and delivery promise data from the Global Manufacturing Research Group. It finds, as theory suggests, that manufacturing lead time increases as work-in-process inventory increases. Also, many companies promise delivery in less time than it takes to produce the product, but the difference is not explained by finished goods inventory. The research also shows the chief benefit of JIT, for those few firms that are achieving benefits, seems to be reduced raw materials.

INTRODUCTION

This paper describes an analysis of some of the data collected by the Global Manufacturing Research Group (GMRG) on manufacturing practices in two industries: non-fashion textiles and small machine tools. The data gathering was begun in 1986 in South Korea and the People's Republic of China, and has been extended to Australia, Bulgaria, Chile, Finland, Hungary, Japan, Mexico, North America, the former USSR and Western Europe. A common questionnaire was used in order to be able to compare findings between the regions. Information obtained from the surveys is contained in a data base housed in the Kenan Institute at the University of North Carolina.

The data covers areas of practice from forecasting to materials management, principally from the materials planning and control functions. The industries were chosen because they are found virtually everywhere in the world. In addition, the machine tool companies mainly use batch processing, while the textile companies are closer to a process form of production. A comprehensive description of the project, questionnaire and data formats is provided in [1].

A variety of studies have been performed on the data base. Regional differences have been studied in bilateral (e.g., Korea and Western Europe [2]), to multilateral comparisons [3]. Specific issues like the linking of the market to manufacturing [4] have also been investigated. The analysis described in this paper is also concerned with a specific issue, that of the relationship between inventories, manufacturing lead times, and promised delivery dates.

The discussion of the analysis starts with a description of the data used in this study. The next section describes the relationship between manufacturing lead time and inventory. Following that, promised delivery times are compared to the manufacturing lead times and the effect of variables that might mitigate the manufacturing lead time, customer promise date relationship is explored. Finally, the impact of JIT is evaluated and some conclusions drawn.

THE DATA USED FOR THIS STUDY

This study uses a subset of the Global Manufacturing Research Group data base. Specifically, data on typical manufacturing lead times, delivery times promised to the customer, and levels of inventory, as well as values for variables that relate to the management of inventories were investigated. To provide insights on developed and less developed economies, data from the People's Republic of China, Hungary, Japan, North America, and Western Europe were used. Both the machine tool and textile industry data were included.

Table 1 shows the distribution of companies sampled by type of economy, region and industry. Even though the minimum number of companies for any subgroup is 17, the sample size for any single experiment could be less, due to non-responses.

Table 1. Distribution of companies in the sample

Economy	Region	Machine tool	Textile	Total
LDE*	China	44	56	100
	Hungary	19	17	36
LDE total		63	73	136
Developed	Japan	18	36	54
	North America	45	50	95
	Western Europe	34	24	58
Developed total		97	110	207
Grand total		160	183	343

* Less developed economy

MANUFACTURING LEAD TIME AS A FUNCTION OF RELATIVE INVENTORIES

The first set of experiments on the data had to do with manufacturing lead time and inventory relationships. Is there a general relationship (i.e., one that can be seen for the data as a whole) between lead time and any of the three inventory types; raw materials, work-in-process, or finished goods? The lead time was reported as the typical manufacturing time for a batch or product. The inventory data consisted of total inventory (in monetary terms) and the percentage distribution between raw materials, work-in-process, and finished goods.

Some theory exists to help frame the hypotheses. Shop floor control theory indicates that as work-in-process (WIP) increases, the manufacturing lead times should increase. While there is no direct theory concerning raw material inventories, their existence might reduce manufacturing lead times. The argument is that there would be no need to wait for raw materials to be delivered before starting manufacturing. Finally, finished goods inventory levels should not affect manufacturing lead times.

The first step was to normalize the inventory data for company size. The data base provides annual sales (in US$) and total inventory was also expressed in US$. Using sales as a measure of company size, inventory values were normalized by computing the number of days of supply of each type of inventory. The manufacturing lead times were compared to the number of days of supply.

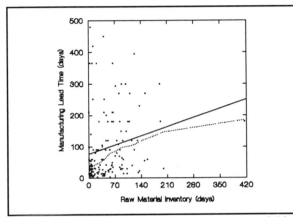

Figure 1. Manufacturing lead time versus raw material inventory

The analysis was started by preparing scatter plots of the manufacturing lead time as a function of the days of supply of the different types of inventory. These are reproduced in Figures 1-3. To help determine any general relationships that might exist, linear regressions were developed for each of the sets of data. The regression lines are the straight lines shown on each figure.

To detect any underlying relationships that might be more complex than linear, the LOWESS regression technique of Cleveland was applied [5]. This robust method does not impose any specific model on the data but weights the dependant variable points (the manufacturing lead time) in successive narrow regions around the independent variable (the inventory) to develop a smooth curve through the data. The data are allowed "to speak for themselves." The LOWESS regressions are the curved dotted lines also shown in Figures 1-3.

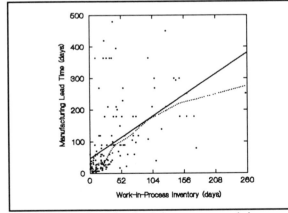

Figure 2. Manufacturing lead time versus work-in-process inventory

Despite the conjecture about the relationship between raw material and lead time, there is a significant positive slope to the linear regression line in Figure 1. The slope indicates that, roughly, for every day of added raw material inventory, there is slightly less than a half day added to the manufacturing lead time. The relationship is a weak one, however, explaining only about 4% of the overall variance.

For the work-in-process data, Figure 2, the slope is positive and significant. Approximately 25% of the overall variance is explained by the WIP inventory and, on the average, for every additional day of WIP there will be approximately one more day of manufacturing lead time. That suggests that the additional day of WIP means one more day a batch must spend in queue on the factory floor.

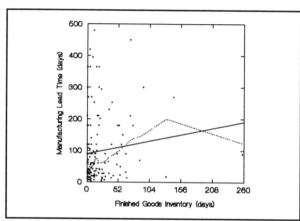

Figure 3. Manufacturing lead time versus finished goods inventory

The regression line for finished goods, shown in Figure 3, is not significant. This supports the contention that there is no relationship between lead time and finished goods. This result could be due, in part, to the use of the linear model when it is not appropriate. Indeed, even though the LOWESS regression lines lie close to the linear regressions in all three figures, they still suggest that the relationships may be curvilinear. The clearest conclusion, however, is that there is still a lot of scatter in each of the figures.

To address some of the possible reasons for the scatter, the relationships for the subgroups of Table 1 were explored, still using linear regression and scatter plots for general understanding. For raw materials there was a significant relationship only for the textile industry in the two less developed economies and for both industries combined in the LDEs. The slope was still positive (e.g., the more raw material, the longer the lead time), but the maximum amount of variance explained was still only about 25%. At the country level, the slopes turned negative (in line with the original conjecture) but there were no significant relationships. The conclusion from these experiments is that there is not a clear, strong relationship between raw material levels and manufacturing lead time, even at the country level.

The results of the detailed studies for the relationship between finished goods inventories and manufacturing lead times were again not consistent, but were even less supportive of relationship than in the raw material case. There were significant relationships for the textile industry in all regions and for both industries combined for the LDEs, but the maximum variance explained was less than 15%. At the country level, there were no significant relationships and the slopes were both positive and negative, indicating no consistent relationship.

The WIP relationship was significant and positive for all sub-groups except the machine tool industry in the LDEs. In all cases the slopes were positive, but substantial scatter still remains (the maximum variance explained was for the textile industry at 35% in the developed economies and over 50% in the LDEs). The differences in lead times and WIP levels was not significant between the two economies, but was between the two industries. Figure 4 shows the average lead times and WIP inventory levels for the textile and machine tool industries. In both cases, the greater values are in the machine tool industry, which may help explain why the greatest amount of scatter is in that industry. The overall conclusion remains, however, that there is a strong positive relationship between WIP and manufacturing lead time in the data.

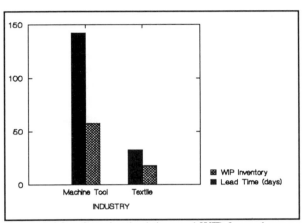

Figure 4. Manufacturing lead time and WIP for each industry

PROMISED DELIVERY TIME VERSUS LEAD TIME

Data is available in the data base on delivery promises made to customers by the companies in the sample. It comes in two forms. One form of the data is for firms that make fixed delivery time promises that remain constant for quite some time. This can occur when there is a standard delivery time set by the industry. The fixed time is reflected in a comment like, "we always tell our customers we'll deliver in three weeks." For those firms, the data consists of the fixed promised delivery times.

The other form of promised delivery times is for companies that vary the time as conditions change. The minimum and maximum times used in promising delivery to their customers is available for these firms. The times can vary because of local conditions, like the amount of business, or external factors, like delivery of raw material. In this study, the causes aren't analyzed, but the times are related to the manufacturing lead time and other factors.

Theory indicates that the manufacturing lead time should be a lower bound on the delivery lead time. That is, the sales force should not promise delivery before manufacturing can build the product. In circumstances where the promise times are less, the explanation should be found in finished goods inventory, or poor delivery service.

The relationship between the fixed delivery time promises and the manufacturing lead time is shown in Figure 5. The 45 degree (dashed) line divides the companies into those with "feasible" promise times and those whose promises are "infeasible." Companies which lie above the line promise to deliver the product in more time than it takes to produce it. Companies below the line promise delivery times that are less than the manu- facturing lead time. The com- panies which lie on the line are those which promise a delivery time equal to the time it takes to produce the product.

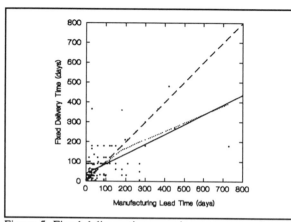

Figure 5. Fixed delivery time promise versus manufactur- ing lead time

Forty-five (30%) of the com- panies are in the infeasible region (below the line), some- what less than the 65 compa- nies that are above it. Thirty- eight companies lie on the line. To help visualize any tenden- cies that might exist, both a LOWESS (the dotted curved line) and a linear regression line (solid) are plotted on the chart. Both lines indicate that, it is more likely that a company with a long manufacturing lead time will promise deliv- ery in less than the manufacturing lead time.

To search for the reasons why some companies that promise deliveries in less time than it takes to make the product, the finished goods inventory, safety stock and order lateness data were studied. The mean values and significance levels, using the Kruskal-Wallis test (for details, see Daniel [6]) of those data are provided in Table 2. Only the finished goods inventories are significantly different. They are higher, as might be expected, for those companies that have promise dates less than the manufacturing lead times (those in the infeasible region).

Table 2. Mean values of finished goods inventory, lateness and safety stock and significance levels of the differences in the different feasibility regions

Variable	Feasible region	Infeasible region	On the line	Kruskal-Wallis test
Finished goods inventory (days)	11.4	40.2	25.4	0.026
Average lateness per order (days)	5.3	5.6	4.7	0.133
Finished goods safety stock (weeks)	6.0	7.3	5.4	0.866

Surprisingly, the companies with shorter promise times than manufacturing times don't have any greater lateness than those who make promises longer than the manufacturing lead time. Likewise, safety stock, which could have been used like regular finished goods inventory to improve delivery times, is also similar for each group. Only finished goods inventory is left to account for the differences between the time to manufacture and the promised delivery time.

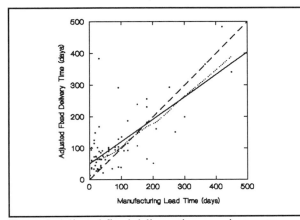

Figure 6. Adjusted fixed delivery time promise versus manufacturing lead time

The promised delivery time was adjusted to account for the days of supply of finished goods inventory and was plotted against manufacturing lead time. This is shown in Figure 6. There was a reduction in sample size from Figure 5, since some of the firms did not report finished goods inventory. Nevertheless, 13% of the firms are still in the infeasible region after the adjustment. The difference in finished goods inventories is not great enough to account for the difference between the promise times and manufacturing lead times. For the companies in the infeasible region, the average promise times are 77 days and the average manufacturing lead time is 172 days. The 31 days of average finished goods inventory does not make up this difference.

The relationship between manufacturing lead time and customer promise time was checked for the companies that have variable promise times. If the *maximum* promise times are still shorter than the manufacturing lead times, that could be even more severe than in the case of fixed lead times. The maximum promise times were adjusted for finished goods inventory and checked against the manufacturing lead times. Even under these conditions, 7% of the firms were in the infeasible area. For those firms, the finished goods inventory levels (an average of only 12 days supply) could not make up the differences between promise times of 103 days and manufacturing lead times of 201 days. The gap is very nearly the same as in the fixed delivery promise time case.

THE IMPACT OF JIT

Another set of inquiries in this study looked at the impact of JIT on the relationships between inventories and delivery promise times to customers. JIT is one of the contemporary approaches to inventory and production management being implemented in many companies. Among the benefits claimed for JIT are reduced manufacturing lead times and reduced inventories. These claims are investigated as was the suggestion that the use of JIT would enable firms to be better at promising delivery.

The firms were divided into one group that had never heard of JIT or felt no need to introduce it and another that is using and benefiting from JIT. Despite the global publicity for JIT, less than 10% of the respondents fall into the latter group. The performance for the firms in each of the groups is summarized in Table 3 along with the Kruskal-Wallis significance levels for the differences.

Table 3. Mean performance measures for firms using JIT and those not

Variable	No JIT	JIT	Kruskal-Wallis test
Raw material inventory (days)	59.0	27.4	0.009
Work-in-process inventory (days)	38.9	36.2	0.585
Finished goods inventory (days)	18.7	23.9	0.589
Manufacturing lead time (days)	82.2	88.7	0.496
Fixed promise time (days)	78.7	49.0	0.029
Firms in infeasible region (%)	12.9	16.1	*
Firms in adjusted infeasible region (%)	3.0	6.5	*

* Not significant using the binomial test.

Of the inventory reductions expected from JIT only the raw material inventories are significantly lower. The other inventory categories were not significantly different, nor was the manufacturing lead time reduced. In fact, the customer promise times were less for those firms with JIT, even though the manufacturing lead times were not significantly different (indeed the average lead time is higher). This difference may be reflected in the higher percentage of infeasible delivery promises given by the JIT firms, even though the binomial test did not show the differences to be significant [6].

It seems that JIT may engender a false feeling of confidence that leads to promising much shorter delivery times than without JIT. Moreover, the raw material reductions give credence to the complaints of suppliers who claim that JIT simply means moving the inventory back further in the supply chain.

CONCLUSIONS

This study strongly supports the theory that relates work-in-process inventories to manufacturing lead time. The theory suggests that manufacturing lead time will increase with work-in-process inventory. The raw material inventory has very little effect on manufacturing lead times, surprisingly, and finished goods inventory seems to have no effect at all. A high proportion of firms promise delivery within the manufacturing lead time, more so than can be explained by finished goods inventory. This is especially true for those firms with long manufacturing lead times.

For those firms that are using and benefiting from JIT, only the raw material inventories are reduced (perhaps by being pushed back on to the supplier). Also, reductions in the promised delivery time to customers is not matched by reductions in the manufacturing lead times for the JIT firms.

There are few differences between the less developed countries and the developed countries. In some cases the relationships seem stronger in the LDEs, especially for the textile industry, but no other difference stands out. The two LDE countries chosen, China and Hungary, are both reforming centrally planned economies. The data suggest that central planning has not on left a differential legacy on the tested aspects of material management.

REFERENCES

1.* D.C. Whybark and G. Vastag, eds., *Global Manufacturing Research Group Collected Papers, Vol. 1*, Chapel Hill, NC, Global Manufacturing Research Center, Kenan Institute, University of North Carolina, March, 1992.

2.* B.H. Rho and D.C.Whybark, "Comparing Manufacturing Planning and Control Practices in Europe and in Korea," *International Journal of Production Research*, 28 No. 12 (Fall 1990), 2393-2404.

3.* G. Vastag, and D.C. Whybark, "Manufacturing Practices: Differences that Matter," *International Journal of Production Economics*, 23, Nos. 1-3 (1991), 251-259.

4.* D.C. Whybark, "Are Markets Really Linked to Manufacturing?" Chapel Hill, NC, Center for Manufacturing Excellence, Kenan Institute, University of North Carolina, Paper, November 1990.

5. C.W. Cleveland, "Robust Locally Weighted Regression and Smoothing Scatterplots," *Journal of the American Statistical Association*, 74, No. 368 (December 1979), 829-836.

6. W.W. Daniel, *Applied Nonparametric Statistics*, 2nd ed., Boston, PWS-Kent Publishing Company, 1990.

* This article is reproduced in this volume.

ACKNOWLEDGMENTS

The authors would like to acknowledge Citibank, the Center for Manufacturing Excellence, and the Kenan Institute of Private Enterprise for supporting this project. A version of this paper is scheduled for publication in the *International Journal of Production Economics* in 1993.

Statistical Inventory Control in Theory and Practice

Curtis P. McLaughlin, University of North Carolina-Chapel Hill, USA
Gyula Vastag, Budapest University of Economic Sciences, Hungary
D. Clay Whybark, University of North Carolina-Chapel Hill, USA

ABSTRACT

In this paper the authors report on three areas where statistical inventory control (SIC) expectations diverge from reality. First, actual inventory performance seems immune to the use of modern techniques like MRP or JIT. Second, simulation studies seem to provide higher than expected customer service levels. Finally, dynamic organizational actions appear to change both the rules of the game and the way that it is scored. These observations suggest that the lack of effectiveness of SIC models in practice can not be blamed exclusively on the scientists or the practitioners. The paper suggests that practitioners have not done well in applying the models that are available to them. It also points out that theory and practice are still far apart and suggests some research to remedy this. Finally the paper concludes that there is still a lot of research and understanding to be gained in the SIC area. It is not and should not be a dead area of exploration.

INTRODUCTION

Even though scientific papers continue to accumulate in the area of statistical inventory control (SIC), SIC has not enjoyed the attention or impact in recent years that it has in the past. This is partly due to other initiatives in industry and academia like material requirements planning (MRP), just-in-time (JIT), total quality management (TQM), and so forth, and somewhat due to the "bad name" that SIC has picked up in some circles. The SIC problem, even for a single item at a single location, can be extremely complicated as defined by the scientific community. Nevertheless, the problem is "solved" in industry every day for multiple items in multiple locations.

This paper will explore three broad areas where the scientific expectations for SIC and reality differ. The first area is concerned with empirical observations of industrial performance. They indicate that actual inventory and order lateness results don't seem to be affected by the new initiatives mentioned above among the small percentage of firms that have implemented them. By extension, the number of and payoffs from industrial applications of SIC techniques are well below the potential expected by the scientific community. In the second area, scientific research, many research projects have been based on simulation models which provide levels of customer service that are much higher than the parameters chosen by the researchers would suggest. This suggests a technical problem in the simulation models. The final area, actual practice, discloses that levels of service provided to customers are often substantially higher than would be predicted from models of the inventory control system used, implying that the models don't capture reality.

Each of these three areas, actual inventory and lateness performance, statistical inventory control research, and industrial practice will be addressed in this paper. After exploring each of the areas, some observations and suggestions for the future of statistical inventory control work will be presented.

ACTUAL INVENTORY AND LATENESS PERFORMANCE

The data used to evaluate the impact of various management techniques on inventory levels and lateness performance comes from firms in the small machine tool and non-fashion textile industries. In a project described by Whybark and Vastag [1], the Global Manufacturing Research Group has gathered data on manufacturing practices around the world over the past few years. Among other information, the data base includes finished goods, work-in-process and raw material inventory levels, finished goods safety stock and delivery lateness.

In an earlier paper, Vastag and Whybark [2] used some of this data to explore the impact of JIT (just-in-time) on manufacturing lead-time, inventory and delivery performance. One of their findings was that JIT techniques are not widely practiced, despite the length of time they have been around and the publicity JIT has gotten. In evaluating JIT methods to help manage inventories, they found that the benefits were limited to reductions in raw materials, something that suppliers have suspected for a long time.

The data base on manufacturing practices does not have any direct information on the use of statistical inventory control (SIC) techniques. Consequently, in this paper, other examples of "modern" methods of inventory management are considered. Specifically the use of computers in forecasting and in production planning and scheduling; and the use of the economic order quantity (EOQ) technique, material requirements planning (MRP) systems and just-in-time (JIT) approaches will be investigated. These methods will be analyzed as to their effectiveness for reducing inventories and decreasing lateness.

The evaluation will be based on a comparison of the inventory levels, safety stock levels and lateness performance of the firms using the techniques or computers with those that do not. The data used comes from more than 400 respondents in the People's Republic of China, Hungary, Japan, Korea, North America and Western Europe. There are no less than 236 respondents for any single question. (Of course, the number of responses for any single choice on a catagorical question can be much fewer.) All data are standardized to permit valid comparisons. The inventory data are expressed in days of supply of raw material, work-in-process (WIP), and finished goods inventory, while safety stock is expressed in weeks of protection. The lateness performance is expressed as the average number of days late in delivery of an order to the customer.

The statistical significance of the comparisons are computed using the Kruskal-Wallis non-parametric test. This test determines whether the factors (use of computers or techniques) are significant in accounting for differences in inventory and lateness performance. It is based on the ranks of the individual responses for each of the factors.

Table 1 presents the data for the question of whether a firm uses the computer in developing forecasts. The response was simply yes or no. The first observation is that only 20% of the respondents said that they used computers in forecasting. There is an improvement in inventory levels (accept for safety stock) for those firms that use the computers although the differences are not statistically significant. Similarly, there is an improvement of average lateness for those firms using computers, but not strongly significant. While there is some evidence that using computers for forecasting could be helpful, their use is not widespread.

Table 1. Impact of using computer in forecasting

	Response	Raw matl.	WIP	Finished	Safety	Lateness
No use*	80%	61	41	28	6	10
Use	20%	40	33	19	7	7
K-W significance		.109	.992	.351	.712	.058

* Values are: percent of respondents; average days of inventory, weeks of safety stock, days of lateness; and Kruskal-Wallis significance levels.

The responses to the question on use of computers in production planning and scheduling is presented in Table 2. A higher percentage of the respondents use computers for these tasks than for forecasting. Overall, however, there are still fewer than half that do. The results are not highly significant for these data, except for WIP, and even for WIP there is no consistent pattern of change. Only with moderate use of the computer in production planning and scheduling do improvements in inventory levels and delivery lateness occur.

Table 2. Impact of computer usage in production planning and scheduling

	Response	Raw matl.	WIP	Finished	Safety	Lateness
No use*	54%	60	40	28	6	9
A little	16%	59	42	21	8	9
Moderate	18%	54	27	23	6	6
Extensive	12%	49	52	22	6	12
K-W significance		.259	.003	.125	.117	.219

* Values are: percent of respondents; average days of inventory, weeks of safety stock, days of lateness; and Kruskal-Wallis significance levels.

Table 3 presents the data for analyzing the use of the economic order quantity (EOQ). The question asked simply whether the EOQ was used for determining production or purchasing quantities. As with computer usage, an overwhelming majority, 73%, make no use of the technique. The significance values aren't particularly strong, although there is some evidence that the use of the EOQ helps in reducing finished goods inventories. There is a slight (not statistically significant) increase in lateness occasioned by the reduction in finished goods inventory, however. The major conclusion from Table 3 is that the EOQ technique is not widely used.

Table 3. Impact of using the economic order quantity

	Response	Raw matl.	WIP	Finished	Safety	Lateness
No use*	72%	54	39	26	7	9
Use	28%	62	38	25	6	10
K-W significance		.246	.584	.116	.815	.548

* Values are: percent of respondents; average days of inventory, weeks of safety stock, days of lateness; and Kruskal-Wallis significance levels.

The question that was asked about the use of material requirements planning (MRP) was more complicated than that for EOQ, as seen in Table 4. The question asked about the intensity and success of MRP as well as whether it was being used. Indeed, the results did reflect some of the frustration of implementation. Although more than half had heard of MRP, fewer than half were trying it, some without benefits.

Some of the responses in Table 4 represent very small sample sizes and, again, the significance values are not very strong. Even with these caveats, there is no discernable pattern of benefit from the application of the technique except for lower finished goods inventories. The group that knows MRP, but doesn't need it, seems to be right. The group that says they are not benefiting from MRP (although the sample size is very small) is a mystery since they seem to have some of the best results. The key result here, again, may be that nearly 50% of the firms haven't heard about MRP.

Table 4. Impact of material requirements planning (MRP)

	Response	Raw matl.	WIP	Finished	Safety	Lateness
Never heard of MRP*	45%	72	45	35	7	9
Know it, but don't need MRP	13%	47	36	21	6	7
Just starting to use MRP	7%	44	55	19	7	13
Starting, but have problems	18%	41	24	15	6	11
Using, but not benefiting	1%	17	14	3	8	3
Using MRP and benefiting	16%	52	37	25	6	9
Kruskal-Wallis significance		.599	.118	.053	.259	.164

* Values are: percent of respondents; average days of inventory, weeks of safety stock, days of lateness; and Kruskal-Wallis significance levels.

The last approach to be evaluated here is just-in-time (JIT), again with some information on intensity and success of use. Surprisingly, a slightly higher percentage of the respondents have heard of JIT than MRP, as can be seen in Table 5. Overall, however, fewer than 50% are using it. As with the MRP results, there are some responses that have quite low sample sizes. This is particularly true of those that say they are not benefiting, but seem to be doing well. Unlike the MRP case, the group that feels they don't need JIT may be wrong. They have some of the highest inventories among the groups.

Table 5. Impact of Just-in-Time (JIT)

	Response	Raw matl.	WIP	Finished	Safety	Lateness
Never heard of JIT*	37%	65	32	22	8	9
Know it, but don't need JIT	19%	77	57	39	6	8
Just starting to use JIT	8%	32	28	29	5	9
Starting, but have problems	27%	46	43	24	7	11
Using, but not benefiting	2%	16	5	2	6	10
Using JIT and benefiting	7%	28	32	21	4	6
Kruskal-Wallis Significance		.010	.117	.217	.337	.035

* Values are: percent of respondents; average days of inventory, weeks of safety stock, days of lateness; and Kruskal-Wallis significance levels.

There is a little more evidence that there is value in implementing JIT than there was for MRP, both for inventories and for lateness. The picture is not totally consistent, however, across the stages of implementation. Full implementation does seem to provide reductions in inventory and decreases in lateness, however. These data support the finding that inventory reductions are primarily in the raw material category under JIT. That is something that suppliers have felt for a long time.

In this evaluation of the application of computers and techniques for managing inventories around the world, only one conclusion can be strongly drawn. There are many fewer firms using these resources than are not. The data on the benefits from their use are very mixed. There is not a consistent pattern of inventory reductions or lateness decreases shown. Thus, there is not a compelling argument for increasing their use. By extension, the same can probably be said for statistical inventory control procedures as well.

STATISTICAL INVENTORY CONTROL RESEARCH

There is a long history of research in statistical inventory control. Even recently, as shown in a study by Meredith and Amoako-Gyampah [3], 20% of a sample of doctoral dissertations were on inventory control and that increases to more than 40% when scheduling and forecasting are included. The most frequent topic (nearly 25%) of articles published from 1982-1987 was inventory control. The topic is literature driven, however, not industrially defined. There seems to be an inexhaustible ability to extend the work of other researchers, without having to validate the models, parameters or approaches in the field. This has lead to SIC research getting a bad reputation from many practitioners, consultants and some academics.

The lack of application of SIC techniques, the studied indifference of business managers to the ever expanding literature of SIC, and the lack of testimonials of benefits from SIC models have contributed to the deprecation of the field. On the other hand, the real problem of trying to manage thousands of individual items, facing uncertain independent demand, in several locations still exists. The inability of MRP or JIT to address some of the key issues associated with this problem still leaves the field wide open for appropriate research. How to direct that research should be a key priority for the scientific community.

There are some key concerns about SIC research that will not be covered in this section. There are significant issues of whether the research models capture reality well enough to draw valid conclusions. Some aspects of this concern will be discussed in the last section of the paper. There are issues of the technical comparability of research studies, the lack of attempts to draw simple general principles for practitioners, and too little emphasis on solving problems "given constraints" instead of developing helpful insights into the value of and means to remove the constraints. These latter concerns won't be covered here.

This section will look at two technical issues arising in past SIC research. The first of these has to do with the definitions of customer service that are used in defining safety stock levels for models of independent inventory systems. The second has to do with the generation of random numbers for simulation experiments, a popular way to conduct research on inventory systems. Both of these issues stem from experience with and observations of simulation research over several years. Many of the simulation results can be characterized as having much lower stockout levels than would be expected from the parameters used. This leads to much higher fill rates (customer service levels) than might be anticipated.

As an example of such observations, consider the recent work of Jacobs and Whybark [4]. In a simulation experiment that compared reorder point (SIC) systems with MRP approaches, they developed plots of service level as a function of average inventory. A set of their runs was made with a mean absolute forecast error of 20% and an order quantity to demand ratio of about four. This meant that the average cycle stock was approximately 100 units (e.g., when demand averages 50 units per period, order 200 units every four periods). When the service level was on the order of 90% for these runs, the average inventory was about 65 units. This implies the use of *negative* safety stock. Many other such examples of such unexpectedly high service levels can be found in the literature. In this section we consider two possible reasons for this, the definition of customer service and the generation of random numbers.

The Customer Service Criterion

Technical discussions of customer service levels almost always start out with a description of demand uncertainty and the use of "safety stock" to protect against stockouts except for the rare occasions when demand is very high during the lead time (i.e., the time during which a replenishment order is outstanding). Some portion (often "k") of the standard deviation of the demand during lead time (SDDLT) is used to calculate the safety stock. For continuous review policies, the reorder point (ROP) is then calculated by adding the safety stock to the average demand during lead time (DDLT):

$$ROP = DDLT + (k \times (SDDLT))$$

Assuming a well understood distribution, like the normal, appropriately selecting "k" can provide a specified level of protection, e.g., a 95% coverage. The implication of this logic is that there will be a stockout during only 5% of the reorder cycles. Unfortunately, that is not the same as having 5% stockout levels or 95% fill rates. In addition, several other service level criteria might be used for SIC as pointed out by Schneider [5].

Vollmann, Berry and Whybark [6] provide a description of the theory and the formulas for calculating safety stock values and fill rates. In order to determine the fill rate, the expected number of stockouts per reorder cycle associated with the "k" value must be found. This requires calculating the expected number of stockouts for each time demand exceeds the safety stock coverage and weighting that by the probability of having a stockout. The expected number of stockouts per reorder cycle divided by the demand per reorder cycle will give the expected shortage rate and the complement is the expected fill rate. Clearly, large order quantities that mean long reorder cycles will increase the fill rates for a given uncertainty distribution and k value.

Using a k value to establish a probability of stockout and equating that to the shortage rate is a possible explanation for the high levels of customer service that get generated in simulations of SIC systems. Uniformly high service levels means it is hard for researchers to discriminate between the policies that they are evaluating in their studies. As a result they use low or negative safety stock values to reduce service levels in order to discern the differences between them.

Generating Random Numbers

Another possible explanation for the high levels of fill rates observed in simulation models of SIC systems has to do with random number generation. Some approaches base the generation of other distributions on the uniform distribution (although different technologies do exist). In early simulations, for example, the generation of normally distributed random numbers was based on the sum of twelve uniformly distributed random numbers. The estimation of the normal this way can lead to differences in the probabilities in the tail of the distribution. This has been calculated by Yang [7] and is shown in Table 6.

Table 6 shows that demands in the range of one standard deviation are overestimated, while those in the "tail" are underestimated. The differences are not great, but it takes demands from the tail to have shortages in any reasonable inventory system. Since these high demands are low probability events anyway, any underestimation of demand in the tails would overstate the fill rates relative to the theoretical value. In a variation on the theme, Silver and Rahnama [8] indicate that when statistical estimates of demand parameters are used to calculate reorder points, the reorder points should be biased upwards. In

Table 6. Approximate and exact normal probability values

Number of standard deviations above the mean, Z	Probability of a value exceeding Z		Difference in probability values
	Normal	Approximate	
1	.1587	.1607	.0020
2	.0228	.0223	-.0005
3	.0013	.0010	-.0003

other words, the estimates of the demand distribution parameters underestimate the negative consequences.

INVENTORY CONTROL IN INDUSTRIAL PRACTICE

There are a number of organizational and situational responses that can explain some of the differences between actual fill-rates and those predicted by theoretical and simulation models. The customer-oriented organization has a number of ways, some subtle and some not, to respond to the demands and due dates initially expressed by customers.

These responses can occur at all stages of the process of fulfilling a customer's request: quotation, order entry, stock status checking, order acknowledgement, production planning, production scheduling and dispatching, sales service and expediting, and shipping. At each of these stages there are steps that tend to increase the experienced fill rate. Each stage is discussed here.

Quotations

Quotations, whether formal or informal, include both price and delivery. If the delivery date (sometimes immediate) is unsatisfactory based on stock status checks or lead times, the customer will not place the order, biasing the orders actually placed toward higher fill rates. On the other hand, sometimes salespersons will be dishonest in estimating lead times to stimulate sales. This will bias the system toward lower-than-anticipated fill rates.

Order Entry

A savvy sales service or order entry group reviews orders as they are received to identify potential outliers in terms of large quantities, unusual delivery lead time, special tooling requirements, batching requirements for infrequently cycled product groups, and in interpretation of ambiguous instructions such as ASAP (as soon as possible). Properly handled, such interventions can lead to significantly improved fill rates by providing special treatment for probable outliers.

Stock Status

Once the order is entered into the information system, the next step is a stock status check. The nature of this check depends on the specific inventory system used, but the intent is the same. Orders are netted out against available inventory either on the shelf or in existing production schedules. The stock status check produces information, however, beyond whether or not the item is in stock or on order. Information about the quantity of shortages or displays of run-out times can lead to changes in batch sizes already scheduled and to modifications of the cycling of families of products through the plant. The information from the stock status check can also lead to major adjustments in the stages that follow: order acknowledgement and production planning.

Order Acknowledgement

Most simulation studies assume demands occur according to some specified distribution independent of inventory or production status. This is equivalent to having due dates set by customers and demands occurring against inventory on the due date without any company knowledge. In actuality, the manufacturer knows that the customer-specified due date may or may not conform to the quoted delivery times and the quoted delivery times usually reflect aggregated backlogs. Thus, the manufacturer seldom accepts the customer delivery request directly, but respecifies the delivery based on the stock status results. Furthermore, the company typically computes its fill rates on this later promised date, not on the customer-specified due date. This adjustment of due dates virtually assures a high fill rate, if the information in the inventory system is accurate.

The customer either accepts the revised date or lodges a complaint on receipt of the acknowledgement. If the customer complains, the order is moved out of the routine system that the models represent and into an exception handling mode.

Production Planning

Most inventory theory assumes that a company operates with one planning system or another (e.g., SIC, MRP or JIT). But that is seldom the case. Assembled items may be assembled to order on a JIT basis out of components that are planned by an MRP system. Original equipment manufacturer (OEM) customers may be on a JIT system, while spare parts are managed by SIC. Major customers may have ordering patterns or special requirements that move their orders from one type of system into a different system.

An example of mixed systems is provided by a manufacturer of replacement recording charts for electronic instruments which are normally shipped from stock managed by SIC. However, its circular charts are printed in green and two of its major customers produce photographic film under red lighting. Green cannot be read under red lights. Therefore, the charts for these two customers are printed quarterly in special colors and shipped in separate lots. Obviously, the fill rates for these special items are 100%.

The same manufacturer has some large government customers for chart paper who order their annual requirements on a single bid which calls for a single shipment against a relatively long scheduled lead time. Again, these orders would have unusually high fill rates, because they were not processed through the SIC system through which a high percentage of the orders, but a lower percentage of the unit volume proceeded.

Production Scheduling and Dispatching

Once production planning is done, production scheduling and dispatching respond in a number of ways to out-of-stock situations and to the special needs of key customers. For example, one of the situations where an automated SIC system often has difficulty is the new stock item, since there is little sales history. However, given the potential importance of such items, production planning often overrides the SIC system to make sure that the new item is available to fill pipelines and to sample new customers. Similarly, production planning meetings emphasize any shortages of high volume or key items and these are pushed ahead in the dispatching process. Where products are batched in families, inventory status and internal politics can determine which items go first within the family sequence.

Sales Service and Expediting

A continual set of negotiations and responses goes on between the customer, customer service, production planning, the shop floor, and shipping to maximize fill rates by mechanisms seldom duplicated in models or simulations. Delivery dates are moved forward and backward in response to the needs of key (or noisy) customers. The lagging items are expedited, or in some cases stolen from other orders having longer lead times calling for the same items.

As an example, consider a firm that receives an order for an item on March 14 for shipment on June 1. The item is in stock, but this order drops available inventory below the reorder point triggering production of a batch of the item to be delivered May 15. A second order is received on March 21 for delivery April 30 which calls for a quantity in excess of the available inventory. The system would treat the second order as out-of-stock and ship it on May 15 when the new batch is physically available. However, an expediter could note that the material for the June 1 delivery could be taken from the batch arriving May 15 and the order due out on April 30 could be filled from the existing stock, avoiding any loss of fill rate. Alternatively, if the customer uses the item continuously, the sales service organization could call the customer and arrange to split the order, shipping part now from stock and the rest on May 15. Then, instead of the failure to fill initially reported, there are now two orders, both of which are filled on time.

Shipping

For the customer the operative fill rate concerns the arrival of the material at their point of use, not the time of departure from the warehouse or plant. Therefore, customer-observed fill rates can be affected by choices of modes of transportation and by enhanced coordination with transportation firms.

Most academic studies treat each order as if it were only for a single item. More often, however, orders are received for multiple items. This is especially true in distribution systems that have truly independent demand from wholesalers and retailers. With multiple-item orders one has to deal with discrepancies between line item and total order fill rates. In logistics systems, due dates are often as much determined by the shipping schedule as by the customer's needs or the completion of the order. Increasingly economies of scale in transportation have pushed distribution into periodic review systems, which theoretically call for larger safety stocks than continuous review systems. However, periodic shipment systems are ideal for the shipment of partial orders, since the incremental transportation cost of splitting the order is nil.

The addition of electronic data interchange (EDI) systems for the rapid transfer of order data, when combined with periodic review and shipment systems, also serves to speed up cycle times and provide better coverage in the field.

The combination of periodic shipments, EDI, order splitting, multiple items per order and customer needs means that the fill rate calculation must be defined to meet the customers' needs, not the producer's system. For example, if split line item orders mean high order fill rates, but low item fill rates, the real question must be how well the customer was served. All this takes negotiation--something that inventory research does not currently encompass.

Summary of Industrial Practice

Many real situations involve tactics by inventory managers that are not well modeled by our current scientific approaches. These practices fall into several broad categories. Managers may *change demand* by influencing customers' orders when stocks are low (i.e., asking a customer to postpone an order or to order less). In some instances the *measures are changed* in order to make the reports look better (i.e., counting a partial order as "meeting the customer needs"). In many instances there is *extra safety stock* in the drawer of a salesperson's desk or in an inventory bin known only to a few people. In numerous cases the managers *override parameters* in an effort to improve their service measures.

The models and simulations of inventory theory are intended to be simplified representations of the real world. However, the real world is itself both complex and dynamic. Given the existence and use of these organizational and situational responses, it is not surprising that what is reported empirically seems to work out somewhat better than what would be expected theoretically.

CONCLUSIONS

This paper accepts that there is a need for statistical inventory control (SIC) in industry and notes three areas where expectations diverge from reality. First, actual inventory performance seems immune to the use of modern techniques like MRP or JIT. Second, simulation studies seem to provide higher than expected fill rates. Finally, dynamic organizational actions appear to change both the rules of the game and the way that it is scored. These observations suggest that the lack of effectiveness of SIC models in practice can not be blamed exclusively on the scientists or the practitioners.

On the side of the practitioners, it is disappointing that there is so little evidence of impact of use of techniques or computer hardware. There does seem to be a shortfall of ability to use the systems that have been put at their disposal. Perhaps there is a need to work harder at training the industrial community. In the real world of multi-item orders, numerous adjustment mechanisms within the producing units, and the will to respond to customers rather than to due dates, the managers use a broad spectrum of responses to cope. This complication can sometimes frustrate the attempts to meet customers' needs.

On the other hand, there may be some technical issues still outstanding in the research of the scientific community. This pales, however, compared to the problems induced by the distance between the theory and reality. Existing models do not seem to address the issues that concern practicing managers. They have trouble with the complexity of the SIC system situation as the organization copes with it. If the theory is to gain greater acceptance, it must address more of the issues of concern to managers.

The implications for research are clear. The models must permit the balancing of supply and demand on a short term, negotiated basis. This might call for research using open simulations with interventions by decision makers in two "organizations" that use various information support systems to negotiate with each other. More field research may be required in order to establish the nature of demand, due date setting and the other parameters that feed the models of SIC situations. In any event there is certainly no evidence that SIC is dead (or should be) as a field of study.

REFERENCES

1. D.C. Whybark and G. Vastag (eds.), *Global Manufacturing Research Group, Collected Papers Volume I*, Chapel Hill, NC, Kenan Institute, University of North Carolina, 1992.

2.* G. Vastag and D.C. Whybark, "Global Relations Between Inventory, Manufacturing Lead Time and Delivery Date Promises," *International Journal of Production Economics* (forthcoming).

3. J. Meredith and K. Amoako-Gyampah, "The Genealogy of Operations Management," *Journal of Operations Management*, 9, No. 2 (1990), 146-167.

4. F.R. Jacobs and D.C. Whybark, "A Comparison of Reorder Point and Material Requirements Planning Inventory Control Logic," *Decision Sciences*, 23, No. 2 (March/April 1992), 332-342.

5. H. Schneider, "Effect of Service-levels on Order-Points or Order-Levels in Inventory Models," *International Journal of Production Research*, 19, No. 6 (1981), 615-631.

6. T.E. Vollmann, W.L. Berry and D.C. Whybark, *Manufacturing Planning and Control Systems*, 3rd ed., Homewood, IL, Irwin, 1992.

7. S. Yang, "Standardized Normal Distribution, N(0,1) vs. Approximate Standardized Normal Distribution," personal correspondence, August 1992, based on Uspensky, *Introduction to Mathematical Probability*, New York, McGraw-Hill, 1937, pp. 277-278.

8. E.A. Silver and M.R. Rahnama, "Biased Selection of the Inventory Reorder Point when Demand Parameters are Statistically Estimated," *Engineering Costs and Production Economics*, 12 (1987), 283-292.

* This article is reproduced in this volume.

ACKNOWLEDGEMENTS

A version of this paper is scheduled for a special issue of the *International Journal of Production Economics* containing the Proceedings of the Seventh International Symposium on Inventories, Budapest, Hungary, August 1992.

Determinants of Manufacturing Delivery Reliability:
A Global Assessment

Karen Brown, Seattle University, USA
Gyula Vastag, Budapest University of Economic Sciences, Hungary

ABSTRACT

Issues of due date setting and due date performance have received a reasonable amount of attention in the literature. However, most research has involved mathematical investigations focused on artificial problems, and very little work appears to have explored the causes of late deliveries in the field. The entire issue is important, however, and is worthy of this sort of examination: late deliveries have ill effects on customer satisfaction, and, in the worst case can lead to loss of future business. The Global Manufacturing Research Group (GMRG) data base was used to examine a set of variables hypothesized to predict order lateness. A causal model was initially tested with a sample of firms from countries with established market economies. Results indicated that engineering changes and the method of due date setting were significant determinants of order lateness. Moreover, manufacturing lead time and order lateness were highly related, validating the recent emphasis on the strategic advantages of manufacturing speed. Results from these initial analyses were compared with those from emerging market economies of Eastern Europe, where different manufacturing strategies were hypothesized to exist. Comparative findings indicate that western models may not hold in planned economies, possibly because delivery reliability is a market-driven performance measure that is less relevant in these environments. Given some of the limitations on the types of data available in the GMRG data base, future research will be needed to more rigorously test the relationships explored here.

INTRODUCTION

Delivery reliability is a critical factor for manufacturers, and has been demonstrated to influence customer satisfaction, as well as return on investment [1,2]. A firm's ability to deliver a product reliably can be represented by a number of measures, including, number of on-time deliveries, percent of orders that have later than promised delivery dates, average lateness overall, or average lateness of late orders, to name a few [3]. Regardless of the measure used, if a customer is told that a product will arrive on the first of the month, but it is not delivered until sometime later, the supplier should expect to lose customer goodwill, and perhaps to lose the customer altogether [4]. The potential problems are magnified when we consider the propensity for unhappy customers to relate their dissatisfaction to other existing customers or potential customers, thus creating a ripple-effect of deteriorating market share or restricted growth [5].

Although delivery reliability may be one of the easiest performance dimensions to measure, it is one of the most complex to manage because of its cross-functional nature [6,7]. On the demand-management side, we have marketing and sales personnel who may feel compelled to offer an early due date and numerous customer features in order to make a sale. These promises may be infeasible, yet manufacturing personnel have traditionally been placed in a "can't say no" position when it comes to the receipt of these kinds of orders [8]. If those who make the promises do not work in cooperation with the engineering function (where the custom features must be designed), or with the operations function (where capacities and loads must be assessed), their promises are unlikely to be realistic. Furthermore, they may fail to consider the uncertainties associated with the transportation function, leading to an even greater gap between the promised delivery date and the probable delivery date.

Two recent trends, the proliferation of just-in-time systems, and escalating customer expectations, promise to make delivery reliability an increasingly important issue. Just-in-time (JIT) purchasing systems rely on frequent, timely deliveries of small batches; production will halt if they do not arrive when scheduled--there is simply no slack in the system. Regardless of the presence or absence of a JIT system, customers in nearly all settings are expecting more from their suppliers and the definition of what is "on time" seems to be a moving target. Given the specifications of JIT inventory systems, it could be argued that early delivery is still less likely to be costly than late delivery [9]. For the purpose of this paper, therefore, we will restrict our analysis to the issue of late delivery. We begin with a survey of the literature on delivery reliability and follow that by presenting a conceptual model that frames our empirical analysis.

LITERATURE REVIEW

Although delivery reliability is an important strategic variable, there appears to have been no empirical, field-based research investigating its causes. Much of the scheduling research has involved due date issues, and some of this work has viewed due date performance as a dependent variable (see Graves [3] for a review). This research has been helpful to us in formulating an understanding of due date performance dynamics and outcomes, but it has been somewhat limited to variables that are within the grasp of the operations function, and it has been almost entirely based on mathematical formulations. An additional limitation to previous research is that it has been based on assumptions and models prevalent in western market economies, and it may not be generalizable to planned economies, or to the emerging market economies of central Europe.

Graves [3] provides a useful framework for describing the research on operations scheduling, and some of its elements are relevant to our concern with delivery reliability. First, he classifies problems as to the manner in which requirements are generated--by customers (i.e., make to order) or by internal inventory replenishment decisions (i.e., make to stock). For purposes of examining delivery reliability, the make-to-order perspective may be most relevant. Second, he characterizes problems according to the criteria used in comparing the effectiveness of various scheduling methods: schedule cost versus schedule performance. The latter criterion includes level of resource utilization and order tardiness. Order tardiness, a surrogate for delivery reliability, in many cases must be traded off with resource utilization; it is difficult to optimize both. However, we must acknowledge that "world class" organizations have been described as those, which find ways to optimize variables that have been viewed as being in competition with other [10]. In spite of the significance of delivery reliability to the topic of production scheduling, it appears to have received less attention than cost or resource utilization criteria. We should note, however, that there are exceptions. One example of a study using due date performance is that of Bookbinder and Noor [9]. They assessed schedule performance as a function of the percentage of time that the due date is honored. In another study, Bertrand [11] recognized mean lateness and the standard deviation of the lateness as performance criteria for a scheduling model. The use of the standard deviation is a useful departure from other studies because it is a truer measure of delivery reliability than mean lateness alone. Additionally, the Critical Ratio decision rule incorporates the problems associated with due dates that are imposed on the manufacturing system by customers or by the marketing function [12].

An examination of research on due date issues reveals a bias toward variables that are internal to the production function (e.g., processing time, congestion) with little mention of customer preferences as input variables. This bias may limit the realistic application of these models in field settings, and is symptomatic of broader concerns about the relevance of scheduling research in general [3]. One notable exception to the bias toward manufacturing variables is found in a paper by Vepsalainen and Morton [4]. These authors stress the importance of strategically differentiating customers--that is, recognizing that some customers are more important to the firm than others. They acknowledge that not all customers are equal, and suggest the use of a priority index that incorporates a rating of each customer's relative importance. This is a useful addition to the literature, but we are still left with questions about actual causes of late delivery in the field.
In summary, the concept of due date performance has been recognized as being an important issue in the literature on manufacturing scheduling. However, research that considers customer-related variables has been limited, and there appears to have been no work which investigates the possible causes of late deliveries. It would seem that research of this nature should be a prerequisite to further studies of scheduling and due date setting.

CONCEPTUAL MODEL

Cross-functional empirical field research on delivery reliability may benefit practitioners by suggesting priorities for improvement, and it would surely provide practical direction for the research efforts of those in the management sciences. Global comparisons will allow us to understand the applications of western models in developing market economies, and to suggest modifications that may lead to a better fit in these environments. In this paper, we examine delivery reliability using a global data base. The sequential process model in Figure 1 provides the framework for a discussion of our hypotheses, which focus on western models.

Figure 1 is a logical representation of a well-understood sequence of events. As such, it is not intended to provide new insights into the order process; it is presented simply to guide our discussion of the factors that can lead to delivery reliability or to order lateness. In this section, we review the issues at each step and discuss related hypotheses.

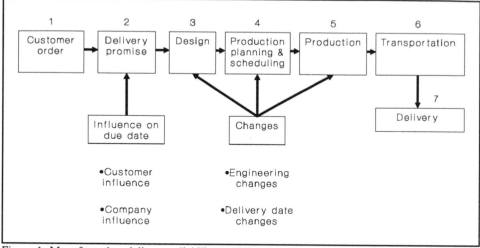

Figure 1. Manufacturing delivery reliability model

Customer Order

The customer order is a universal element of the process, although its form may vary from one economy to the next (e.g., state plan orders versus free market customer orders). In a free market economy, if a firm follows a make-to-stock policy, an internal "customer" determines when inventory levels should be replenished, often based on inventory carrying policies and set-up cost assessments. Make-to-stock policies are established when the manufacturing strategy emphasizes capability for rapid response to customer demand, and a relatively standard product. In contrast, make-to-order implies a situation in which customers are willing to wait for a product that meets their specifications. The reliability of due date performance can be jeopardized in these situations when enthusiastic sales people make unrealistic promises.

> Hypothesis 1: When a firm operates on a make-to-stock policy, there
> will be fewer late orders than there will be when the firm follows a
> make-to-order policy.

Delivery Promise

The delivery promise that follows the customer order will be based on several possible processes: it may be demanded by the customer as a condition of doing business, it may be specified by manufacturing based on an analysis of capacities and loads, or it may be negotiated based on the interests of all involved parties [9]. When the manufacturing organization decides on the production schedule, this can occur independently of any interaction with sales personnel. Under these circumstances, the due date promised to customers may or may not have been considered in the due date used internally for manufacturing [7]. At the other extreme, when sales personnel make promises in order to clinch a sale, they may not have given consideration to manufacturing or engineering capacities. Negotiation, a compromise between these two ends of the spectrum, is likely to produce a workable solution for the customers and the plant.

> Hypothesis 2: Delivery reliability will be best under circumstances in which the due date is negotiated with the customer.

Design

Much has been written lately regarding the importance of integrating the customer, the engineering function, and the manufacturing function during the design process [13,14]. Such interdisciplinary efforts can reduce the likelihood or magnitude of late engineering changes. Engineering changes can lead to the development of a product with different manufacturing requirements than the one that was considered at the time the delivery date promise was made. These changes will affect the delivery date for a product and can have a ripple-effect on production schedules and delivery reliability for other products. Some cross-cultural differences are likely to exist; Vollmann, Berry and Whybark [15] have suggested that the propensity toward engineering changes will be most prevalent in market economies because of their orientation toward meeting customer needs.

> Hypothesis 3A: There will be more late orders in plants which allow extensive engineering changes, particularly if these changes occur after the start of production.

> Hypothesis 3B: In companies which tend to negotiate due dates with customers, there will be fewer engineering changes because of a closer degree of attention to customer needs.

Production Planning and Scheduling

We include capacity planning here, as well as detailed production scheduling. This stage is directly affected by the previous three steps, and operates in a dynamic fashion to accommodate a continuous series of order arrivals. Although some customers are of greater strategic importance than others, the extent to which an "important" customer can preempt an earlier arriving order may be disruptive to the scheduling for all remaining customers [4,7]. The tendency for altering schedules through expediting or "hot lists" has been noted as a characteristic of U.S. firms; Japanese firms, in contrast, are known for adhering to schedules once they have been established [10]. The power and authority of a state planning system also may cause schedules to remain more rigid, thus reducing the number of late orders caused by changes. So, we may expect cross-cultural differences here, as well.

> Hypothesis 4: In organizations that tend to change due dates in order to accommodate new jobs entering the system, delivery reliability will be the lowest.

Production

Once production has commenced, it is still possible for changes in schedule or design to disrupt the flow. These issues have been addressed above. Additionally, however, unexpected obstacles can interfere with the production process, and create further delays [16]. For example, materials may not arrive or they may be of insufficient quality; machinery may be inoperable, or the capacity assumed by the sales department is simply not available. Wide-scale survey research by Schmenner [6] has demonstrated that many of these obstacles can be diminished if efforts are made to reduce manufacturing lead times. The reason for the improvement is that the reduced lead time goal forces managers to improve quality, reduce work-in-process inventory, and make other changes that generally make plant operations more reliable. However, cross-cultural differences may exist. Problems with materials and capacity have been proclaimed by observers to be far more significant in eastern European planned economies than they are in western market-type economies [17,18,19]. Additionally, Vollmann, Berry and Whybark [15] have noted that it may be easier to adjust capacity levels in the United States than in other locations. Again, we see the potential for cross-national differences in the relationships among the variables in the sequence of events from order placement to order delivery.

> Hypothesis 5: Plants with the shortest manufacturing lead times will have the best delivery reliability.

Transportation

Transportation delays can occur for a variety of reasons, most of which are similar to those for production. Additionally, when there is a production delay and a late delivery seems imminent, firms will tend to resort to costly rush shipping as a means to make up for lost time [4]. Transportation has been included here as a means of making our sequential model complete, but it is outside the scope of our current analysis.

Delivery

The difference between the promised delivery date and the actual delivery date is the focus of our analysis. We provide more discussion of the way this variable is operationalized in later sections.

Summary of Conceptual Model

Delivery reliability is an important success factor in manufacturing, but its causes have not been empirically tested in the field. In the sequence of events from order placement through order delivery, a number of variables may influence the probability that an order will be late. Many of these are within managerial control, or, at the very least, may be anticipated from a probabilistic point of view. Most of our assumptions about the sequential causal process are based on western models and processes, but they may not apply when transferred to other cultures and economic systems. The study described below tests our hypotheses using an international data set.

METHODS

The Global Manufacturing Research Group (GMRG) data base was used to test our hypotheses about delivery reliability. This data base was created as a part of a large research project aimed at describing manufacturing practices world-wide. Full details about its development and the administration of the survey are available in Whybark and Vastag [20]. Briefly, though the survey was conducted in 665 machine tool and textile plants in several countries during the years 1986 to 1990. The GMRG survey was created for descriptive purposes, and, as such, was not designed to test any specific hypotheses. Methodology for the present study involved an exploration of the existing data base. Our task was to select the appropriate sample groups for making the comparisons implied in the hypotheses, and to choose from the data base the items that most closely approximated the variables described in the preceding section.

Sample Selection

There were two central research questions for this study. First, can the hypotheses based on current operations management literature be supported by empirical evidence? Second, do western models of delivery reliability apply in other economic and social environments? The second question required a segmentation of the data base into two groups. The first group was composed of plants from the market economies of western Europe and North America. The second group was composed of plants from centrally planned economies in central and eastern Europe including Bulgaria, Hungary and the USSR. Summaries of the sample characteristics are displayed in Tables 1 and 2.

Table 1. Market economy sub-samples selected for testing delivery reliability hypotheses

Region	Machine tool	Textile	Total
Western Europe (Ten countries)	56	24	80
North America (U.S. and Canada)	45	50	95
Total	101	74	175

Table 2. Centrally planned economy sub-samples selected for testing delivery reliability hypotheses

Country	Machine tool	Textile	Total
Bulgaria	15	13	28
Hungary	19	17	36
USSR	18	32	50
Total	52	62	114

Free Market Economies

Data from free market economies were gathered in three separate but coordinated efforts in western Europe, the U.S. and Canada, and Finland. Data gathering circumstances varied slightly across these three efforts, but all used the same survey instrument.

The data gathering activities in western Europe started with a mail questionnaire in 1988. Eight hundred companies were chosen by taking a random sample from the industrial listing data base maintained by COMPASS (through the Zurich office). To accommodate the languages spoken in these companies, the questionnaire was translated into French, German and Italian. The response rate was very low, less than 1%. After the limited response to the mail questionnaire, a telephone solicitation was conducted using directories of trade associations in the countries where data was gathered; i.e., Scandinavia, the Benelux countries, the United Kingdom, France, Germany, Austria, Switzerland and Italy. This involved direct telephone calls to the appropriate people in the selected companies and asking for their commitment to respond to the survey. This produced a substantially higher response rate (about 25% of the firms having promised to return the survey actually did so). A total of 34 usable responses were obtained for the machine tool industry, and 24 usable responses were received for the textile industry.

The Finnish survey was carried out in 1990, independently of the western European survey. It was mailed to 100 firms in the small machine tool and the electrical industries. Forty-three companies responded; 21 from the electrical and 22 from machine tool industry. The 22 machine tool companies were selected for inclusion in analyses for this paper. The number of responses from Finland is the largest of any European country, therefore it is over represented.

Data gathering efforts for the U. S. and Canada took place in 1989. Directories of trade association members were used to select a random sample of firms to be contacted from each industry. A manufacturing executive from each of selected companies was then contacted by telephone and was asked to participate in the study. Follow-up phone calls were made to answer questions and remind respondents to complete and return the survey. About 50% of the companies that agreed to participate actually returned the completed form (after three to four follow-up phone calls).

Centrally Planned Economies

The second data set consisted of former centrally planned economies of central Europe, such as Bulgaria, Hungary and the Soviet Union. These countries represented various forms of centrally planned economies [21].

The Bulgarian data were collected during the second half of 1990, but respondents gave answers based on 1989 activities. Central planning was still active in Bulgaria during 1989 and production was more or less stable. The data were gathered in on-site personal interviews with deputy directors or experienced department heads in 15 machine tool and

13 textile companies. These constituted 6.3 and 9.2% of the total number of enterprises in the Bulgarian machine tool and textile industries, respectively. In general, the comparison between the industry and the sample data indicated that the samples were quite representative, although the companies included in the sample were slightly exceeding the average industry performance indicators.

The Hungarian survey was carried out in 1988 and involved on-site interviews in 79 companies. The original sample included several industries but only those from the machine tool and textile industries were included in the analyses for the present study [22]. The sample represented 5.4 and 19.3% of the companies in the machine tool and textile industries, respectively. The sample differed from industry averages with respect to sales and number of employees because it was biased toward large state-owned companies. However, these companies were the primary candidates for joint ventures and privatization efforts, and developing a knowledge of their manufacturing practices had been considered a high priority in the research project.

The Soviet data were gathered in early 1991 and included machine tool companies and non-fashion textile companies from all major regions of the country. The sample was randomly selected from the Center for Opinion Research database which consists of more than 160,000 records. Initially, Russian-language questionnaires were sent to 400 companies. However, because of an extremely low response rate to the mailed survey, teams of researchers were sent to interview key managers in the selected enterprises. This method was more successful, and the final sample included 18 machine tool companies and 31 textile companies [23].

Variables

The variables we chose, their names and the interpretation of their meanings, are displayed in Table 3. These included two measures of delivery lateness, as well as answers to questions about manufacturing lead time, percent of orders for which there are engineering changes, frequency with which due dates are changed after the start of production, and the nature of the due date setting process.

Table 3. Variables selected for testing delivery reliability hypotheses

Variable name(s)	Interpretation
LATEDAY	Average lateness per order expressed in days. This variable was derived from other variables in the GMRG data base.
LATEPCT	This is a measure of relative delivery lateness, calculated as the average absolute lateness divided by the promised delivery time.
LT	Average manufacturing lead time of a typical batch, expressed in days.

(continued)

Table 3, continued

Variable name(s)	Interpretation
ECNPCT	The percentage of manufacturing orders for which engineering changes occurred after processing had begun. Possible responses included four categories: 1. Greater than 60% (80%). 2. 30-60% (50%). 3. 10-30% (20%). 4. Less than 10% (5%). In the analyses we used the values in parenthesis.
ORD, STK, PLN	The percentage distribution of make-to-order, make-to-stock and make-to-state plan products. The last category was used primarily in planned economies. However, some firms from western Europe used that response to describe work done for the government.
NEG	An indicator of the way in which delivery due dates are determined. This was a categorical variable with three possibilities: 1. determined mostly by customers. 2. determined by negotiation. 3. determined mostly by the company.
CHGPR4	An assessment of the influence of delivery date changes on shop floor priorities after the start of production on an order. Possible responses included: 1. little. 2. moderate. 3. heavy.

RESULTS

Results of the data analysis are presented in this section. We begin with an overview of descriptive statistics, and follow that with tests of hypotheses on both subsamples.

Descriptive Statistics

Prior to testing hypotheses, we analyzed the data within and between the subsamples to assess the general nature of the variables under consideration and the validity of grouping these variables. The results of these analyses appear in Figures 2 and 3, and Tables 4-9.

Figures 2 and 3 show scatter plots and histograms for all variables in both regions [24]. The histograms, based on the non-missing values of the appropriate pairs, show that the distribution of these variables is highly asymmetric. We used the Lilliefors test, which is based on the non-parametric Kolmogorov-Smirnov test, to assess the normality of all ratio scale variables in both subsamples. All of these variables (LATEDAY, LATEPCT, LT, ECNPCT, ORD, STK) must be considered non-normally distributed at the $p=0.000$ significance level.

The most visible difference between the groups regarding the distribution of the individual variables was in the due date determination (the NEG variable). The three options are

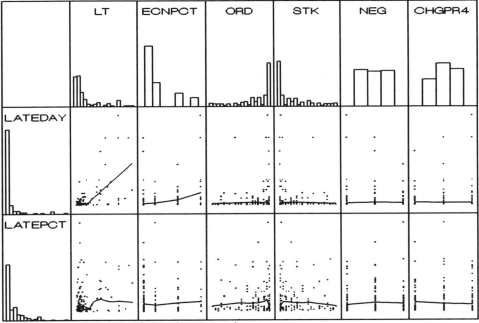

Figure 2. Data description for market economies

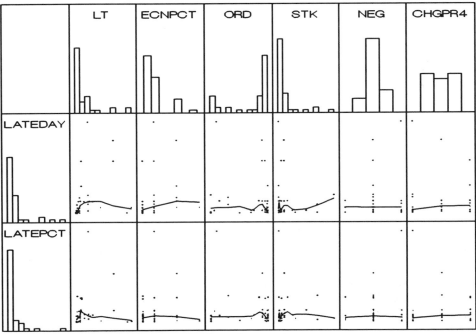

Figure 3. Data description for centrally planned economies

almost evenly distributed in the market economies but in the centrally planned economies the customer had the least influence in determining the delivery date (an expected outcome) and negotiation was the most popular option. In the market economy subsample, delivery date changes had a greater influence on shop floor control priorities (CHGPR4 variable) than they did in the centrally planned economies.

To show the relationships between the dependent and independent variables we used LOWESS smoothing (for details see [25]), which is a robust locally weighted regression. Results show that the average delivery lateness (LATEDAY) is positively related to manufacturing lead time (LT) and engineering changes (ECNPCT) in the market economy group. In contrast, these relationships do not appear to exist in the planned economy subsample. The general conclusion from these plots is that the market economy subsample is more homogeneous than the group of the planned economies and the normality assumption does not hold for any ratio scale variable.

Table 4 presents means and standard deviations for the data sets in the market economy group. Finland is shown separately from western Europe because the data from that country were gathered at a later time and under somewhat different circumstances.

Table 4. Mean ratings for research variables for Finland,
North America and western Europe

Variable (Kruskal-Wallis p value)	Finland	North America	Western Europe
LATEDAY (0.097)	1.101 (1.06)*	6.48 (13.19)	3.86 (9.89)
LATEPCT (0.011)	.035 (.07)	.061 (.07)	.053 (.09)
LT (0.154)	40.91 (33.03)	87.97 (112.63)	113.96 (152.94)
ECNPCT (0.239)	.207 (1.82)	.226 (.260)	.161 (.199)
ORD (0.003)	84.18 (29.14)	75.17 (29.39)	66.68 (32.42)
STK (0.001)	11.27 (27.33)	24.73 (29.47)	33.86 (32.78)
PLN (NA)	5.00 (13.801)	000 (000)	000 (000)

* Standard deviation in parenthesis

Table 5 presents the same information for the central and eastern European group, with Hungary, Bulgaria, and the USSR shown separately--again, because of different data-gathering circumstances.

We used a nonparametric test, the Kruskal-Wallis one-way analysis of variance by ranks to test the assumption that the different regions in the subsamples could be grouped together [26]. There were some differences between countries and regions within our two market economy subsamples but, in general, there were more differences between the two economic groups that there were within them. The differences in the market economy subgroup can be explained by the Finnish sample which contained only the machine tool industry with high make-to-order percentage and less relative lateness.

Table 5. Mean ratings for research variables
for the planned economy countries

Variable (Kruskal-Wallis p value)	Bulgaria	Hungary	USSR
LATEDAY (0.000)	.075 (.592)*	5.30 (10.17)	2.27 (5.81)
LATEPCT (0.843)	missing (missing)	.060 (.099)	.049 (.104)
LT (0.000)	missing (missing)	92.49 (104.90)	29.41 (45.48)
ECNPCT (0.010)	.071 (.053)	.171 (.179)	.160 (.185)
ORD (0.000)	60.36 (15.98)	96.91 (3.99)	52.64 (25.49)
STK (0.000)	13.75 (5.87)	3.086 (3.99)	17.51 (25.49)
PLN (0.000)	25.89 (17.27)	000 (000)	36.61 (41.39)

* Standard deviation in parenthesis

As we indicated earlier, the differences within the centrally planned economies subgroup were greater than those within the market economy subgroup. In Bulgaria and the USSR the state planning system was still operating at the time of the survey, but in Hungary economic reforms had been underway since 1968. The low level of lateness in Bulgaria can be explained by the fact that delivery dates are determined by the manufacturers. An order was considered to be delivered on time if it was shipped in the period (usually a quarter) for which the delivery date was specified.

Tables 6 and 7 show the distribution of due date setting in the two environments. The differences between the countries or regions in the same subgroup were tested with the chi-square test [26]. In both groups, the differences were significant (at p=0.000), but more than one fifth the cells were sparse for the planned economy subsample. However, we felt that there is enough theoretical and empirical evidence to consider these three countries different regarding due date settings.

One issue worth noting here is that customers have a greater role in due date setting in the market economies, which we would expect. Generally, the more planned an economy is, the more uncertain the delivery of raw material and component part supplies. Therefore, companies cannot promise a fixed delivery time to their customers because it depends on the arrival of the purchased materials.

Table 6. Method of due date setting for market economies

Method of due date setting	Finland	North America	Western Europe
Customer	54.55*	34.83	25.86
Negotiation	45.45	22.47	51.72
Company	0.00	42.70	22.41

* Percent of non-missing responses by category
Note: the differences between regions are significant at the p=0.000 level

Table 7. Method of due date setting for centrally planned economies

Method of due date setting	Bulgaria	Hungary	USSR
Customer	0.00	27.60	4.35
Negotiation	42.86	48.28	86.95
Company	57.14	24.12	8.70

* Percent of non-missing responses by category
Note: the differences between countries are significant at the p=0.000 level

Tables 8 and 9 show the differences in the influence of delivery date changes on shop floor priorities.

Table 8. Influence of delivery date changes on shop floor priorities
for market economies

CHGPR4	Finland	North America	Western Europe	Total
Low (%)	23.81	16.67	36.36	24.10
Medium (%)	57.14	42.22	30.91	40.36
High (%)	19.05	41.11	32.73	35.54

Table 9. Influence of delivery date changes on shop floor priorities
for centrally planned economies

CHGPR4	Bulgaria	Hungary	USSR	Total
Low (%)	0.00	14.71	53.49	32.56
Medium (%)	0.00	23.53	25.58	22.09
High (%)	100.0	61.76	20.93	45.35
High (% of all companies)	32.14	58.33	18.00	34.21

The Kolmogorov-Smirnov test was used to detect the differences in the distribution patterns in the market economy subsample. In North America, the delivery date changes had a greater influence on shop floor priorities than in either Finland (at p=0.000 significance level) or western Europe (at p=0.002 level). In the centrally planned economy subgroup, the regional patterns were not directly comparable since the Bulgarian respondents indicated only whether the delivery date changes had a "high" impact. In this case, the countries were compared using the binomial test [26]. Bulgaria was not different from the expected pattern (p=0.155), but Hungary and the USSR were different (p=0.01).

Overall, we felt that considering the difficulties and problems of multi-country surveys, our two subsamples were reasonably homogeneous and representative of the populations from which they had been drawn. Survey method variance does represent a possible concern here, however, and should be given more consideration in subsequent administrations of the GMRG survey.

Tests of Hypotheses

Hypotheses were centered around questions about the causes of delivery lateness in manufacturing settings. Before these questions could be addressed, it was necessary to determine how the criterion variable was to be operationalized. This led to a comparison of the absolute measure of order lateness (LATEDAY) versus the relative measure of order lateness (LATEPCT). It was our bias that the absolute measure would be best for two reasons. First, when an order is late, it is late--if parts are not available, production cannot commence, or if the product is not available for the end-user, that customer may go elsewhere. The later it is, the more unhappy the customer is. Second, there are psychometric problems associated with using the proportional measure in that it requires mathematical combination with another variable. Each time we combine variables by dividing or multiplying, we run the risk of magnifying the extent of error associated with the variables [27]. Thus, we are likely to have a more reliable variable if we choose one that has been drawn straight from the survey, or that has entailed the least amount of mathematical manipulation. On the other hand, it makes intuitive sense to give a relative definition of lateness and define it as a proportion of the promised delivery time. The main advantage of this approach is that in a way it gives standardized values--it compares the absolute lateness to the promised delivery time--and eliminates the differences between technologically different industries. We did test all hypotheses with both dependent variables as a means of examining the validity of these competing perspectives on the issue.

Hypothesis 1

This hypothesis had predicted that the nature of a firm's manufacturing environment (make-to-stock versus make-to-order) would influence order lateness. This proposition was tested using regression analysis. The results confirmed the findings of Figures 2 and 3 that overall there were no significant linear relationships between the manufacturing environment and the two measurements of delivery lateness.

One possible explanation for these weak results is that a make-to-stock policy, in itself, does not necessarily guarantee reliable delivery. The effects of the policy will be modified by the firms's policies regarding finished goods inventory levels and the nature of its due date promises to customers. Ideally, customer due date promises in make-to-stock environments would be based on information about on-hand inventories and existing production schedules. The limitations of the GMRG survey restricted our ability to test these interactions, however. Thus, although Hypothesis 1 was not fully supported, the trends in the data suggest that it may represent a reasonable proposition.

Hypothesis 2

This hypothesis had suggested that delivery performance would be best when due dates were negotiated with the customer. This hypothesis was tested using MANOVA with the three methods of due date determination as predictor variables and the two measures of lateness as dependent variables. Overall, the method of delivery date determination did not have a direct significant influence on either measure of delivery lateness.

Hypothesis 3

This hypothesis predicted that engineering changes after the start of production would serve as obstacles to delivery reliability. Using the Kruskal-Wallis test this hypothesis proved to hold true for the absolute delivery lateness variable in both subsamples (p=0.000 for the market economy subsample; p=0.024 for the planned economy subsample).

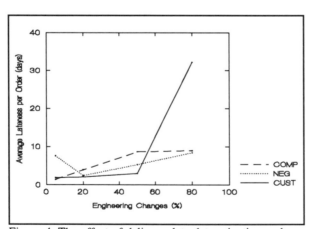

Figure 4. The effect of delivery date determination and engineering changes on delivery lateness

Figure 4 shows the effects of delivery date determination and amount of engineering changes on average absolute delivery lateness for the market economy subsample. This figure indicated that there may be an interaction effect present. Based on two way analysis of variance there was a significant main effect for engineering changes (F=7.204, p=0.000) and a significant interaction effect between due date setting and engineering changes (F=4.136, p=0.001) indicating that, through customer involvement, delivery lateness can be kept low even with a high level of engineering changes. Because of the high number of missing values we could not perform this analysis for the centrally planned economy subsample.

Figure 5. Engineering changes and delivery date setting in market economies

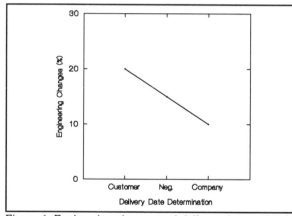

Figure 6. Engineering changes and delivery date setting in centrally planned economies

Using the Kruskal-Wallis test, we examined the relationship between due date setting and engineering changes to assist us in developing an under-standing of the interaction effect described above. Al-though it was not significant, there were clear and very different patterns within the two subgroups. Figures 5 and 6 show these relationships. In market economies if the cus-tomer is involved in delivery date setting there are few engi-neering changes. In centrally planned economies, the rela-tionship may be a reflection of the market situation: the com-pany will not make engineering changes unless it is "forced" to do so. If the customer sets the due date, the customer may have a considerable power over engineering, too.

Hypothesis 4

Hypothesis 4 had to do with the relationships between due date priority changes and deliv-ery reliability. No relationship was found to exist but we feel this may be an artifact of the survey design. The question was somewhat ambiguous in that it did not specify the cause of the due date change. For example, it may have been interpreted as a change in due dates coming from parts vendors, or it may have meant that customers requested changes in due dates. The range of possible interpretations probably washed out any effects for what appeared to be a logical relationship. We should point out that the new version of the GMRG survey instrument has been designed to reduce these types of interpretation problems.

Hypothesis 5

This hypothesis linked delivery reliability with lead time, a relationship, which proved to hold true for the absolute measure of delivery reliability in both subsamples as pre-viously indicated in Figures 2 and 3. The shorter the lead time, the better the delivery reliability and the relationship can be considered linear (F=92.493, p=0.000 for the market economy subsample and F=6.780, p=0.011 for the centrally planned economy subsample). This relationship did not hold for the relative measure of delivery lateness in either subsample.

DISCUSSION

This study represents a first attempt at uncovering the variables that cause problems with delivery reliability. Given the growing strategic importance of delivery reliability, this represents a worthwhile endeavor that is deserving of further effort. Mathematical modeling has been helpful to researchers in developing an understanding of relationships under controlled conditions, but it has not taken us into the "real world" to examine the validity of assumptions that have been held, untested, for decades.

A comparison was made between two different types of economic systems and two different measures of delivery reliability. Although the number of centrally planned economies has decreased considerably their successors have inherited many features of the old system and company practices cannot be changed in a short period of time. Models for determining delivery reliability may differ between market and planned economies. Although our results were limited, probably because of some weaknesses in survey design and data-gathering, they do indicate that engineering changes and the nature of the due date setting process may influence absolute delivery lateness. Additionally, there is some evidence that customer involvement influences the number of engineering changes and it appears to work differently in different types of economies. Our analysis has underscored the importance of manufacturing lead time reduction by revealing its relationship to delivery lateness. The results also confirmed that an absolute measure of delivery reliability is better than a relative one. Considering the difficulties with multi-country surveys, further analyses be needed to assess the nature of delivery reliability models in various economic environments.

REFERENCES

1.   W.L. Berry, V.A. Zeithaml and A. Parasuraman, "Five Imperatives for Improving Service Quality," *Sloan Management Review,* 1990 (Summer), pp. 29-38.

2.   A.V. Roth and J. G. Miller, "Success Factors in Manufacturing," *Business Horizons,* 1992 (July-August), pp. 73-81.

3.   S.C. Graves, "A Review of Production Scheduling," *Operations Research,* 29, No. 4 (1981), 646-675.

4.   A.P.J. Vepsalainen and T.E. Morton, "Priority Rules for Job Shops with Weighted Tardiness Costs," *Management Science,* 33, no. 8 (1987), 1035-1047.

5.   K. Albrecht and R. Zemke, *Service America!* Homewood, IL, Dow Jones-Irwin, 1985.

6. R.W Schmenner, "The Merit of Making Things Fast," *Sloan Management Review,* 1988 (Fall), pp. 11-17.

7. A. Kumar and G. Sharman, "We Love Your Product, But Where is it?" *Sloan Management Review,* 1992 (Winter), pp. 93-99.

8. T. Hill, *Manufacturing Strategy: Text and Cases,* Homewood, IL, Irwin, 1989.

9. J.H. Bookbinder and A.I. Noor, "Setting Job-Shop Due-Dates with Service-Level Constraints," *Journal of the Operational Research Society,* 36, no. 11 (1985), 1017-1026.

10. R.J. Schonbergen, *Japanese Manufacturing Techniques,* New York, The Free Press, 1986.

11. J.W.M. Bertrand, "The Effect of Workload Dependent Due-Dates on Job Shop Performance," *Management Science,* 29, no. 7 (1983), 799-816.

12. W.L. Berry and V. Rao, "Critical Ratio Scheduling, an Experimental Analysis," *Management Science,* 22, no. 2 (1975), 192-201.

13. J.D. Blackburn, *Time Based Competition, The Next Battleground in American Manufacturing,* Homewood, IL, Business One Irwin, 1991.

14. B. King, *Better Designs in Half the Time,* Methuen, MA, GOAL/QPC, 1989.

15. T.E., Vollmann, W.L. Berry and D.C. Whybark, *Manufacturing Planning and Control Systems,* Homewood, IL, Irwin, 1988.

16. K.A. Brown and T. R. Mitchell, "Performance Obstacles for Direct and Indirect Labour in High Technology Manufacturing," *International Journal of Production Research,* 26 (1988), 1819-1832.

17. F.H. Jellinek, "Go East Young Man," *Inc.,* Sept. 1990, pp. 84-86.

18. J. Morcom, "First Europe, Then the World," *Forbes,* Oct. 29, 1990, pp. 134-135.

19. T.M. Rohan, "East Germany, Another European Comeback," *Industry Week,* Sept. 17, 1990, pp. 66-69.

20. D.C. Whybark and G. Vastag (eds.), *Global Manufacturing Research Group Collected Papers, Vol.1,* Chapel Hill, NC, Kenan Institute, Global Manufacturing Research Center, University of North Carolina, 1992.

21. J. Kornai, *The Socialist System, The Political Economy of Communism*, Princeton, NJ, Princeton University Press, 1992.

22. A. Chikán, "Characterization of Production-Inventory Systems in the Hungarian Industry," *Engineering Costs and Production Economics*, 18 (1990), pp.285-292.

23.* A. Ardishvili and A.V. Hill, "Manufacturing Practices in the Soviet Union," *International Journal of Operations & Production Management* (forthcoming).

24. L. Wilkinson, *SYGRAPH: The System for Graphics*, Evanston, IL, SYSTAT, Inc., 1990.

25. W.S. Cleveland, "Robust Locally Weighted Regression and Smoothing Scatterplots," *Journal of the American Statistical Association*, 74 (1979), 829-836.

26. W.W. Daniel, *Applied Nonparametric Statistics*, 2nd ed., Boston, PWS-Kent Publishing Company, 1990.

27. F.L. Schmidt, "Implications of a Measurement Problem for Expectancy Theory Research," *Organizational Behavior and Human Performance*, 10 (1973), 243-251.

* This article is reproduced in this volume.

ACKNOWLEDGEMENTS

The authors would like to acknowledge the support of the Kenan Institute, the Global Manufacturing Research Center, and Citibank for supporting this research.

A DEA Analysis of Sales Output
and Comparison of Manufacturing Practices
in China, Hungary and the United States

Xiao Cheng Zhong, Shanghai Institute of Mechanical Engineering, China
D. Clay Whybark, University of North Carolina-Chapel Hill, USA

ABSTRACT

This paper presents a comparison of the manufacturing planning and control practices among firms in the machine tool industry in the People's Republic of China, Hungary and the United States. The analyses are based on data collected in factories in those countries using a common questionnaire covering sales forecasting, production planning and scheduling, shop floor control, purchasing and materials management.

First, we use the multi-set data envelopment analysis (DEA) model to evaluate how well the three countries use three input resources to produce sales. The results indicate that the U.S. machine tool factories have the best performance, followed by Hungary, then China. Secondly, we present the significant differences in manufacturing practices among Chinese, Hungarian and U.S. firms. The key differences are in the use of labor and market influences. These findings are quite important for the economic reforms being carried out in Hungary and China, and are helpful for planning cooperative business or joint ventures in the future.

INTRODUCTION

A joint project involving a worldwide survey and comparison of manufacturing practices was started in 1985 by many researchers around the world. Many of the findings contain important information for wide cooperation between companies in different geographic regions and nations. Other findings provide insights into ways of improving management and enhancing productivity from one country to another. In June of 1990, the Global Manufacturing Research Group (GMRG) was set up for further promoting and coordinating these joint research efforts. The work will be beneficial to all countries involved in attaining manufacturing excellence and achieving economic objectives.

This paper compares the manufacturing practices among firms in China, Hungary and the United States in the machine tool industry. The People's Republic of China is a developing socialist country still dominated by a centrally planned economy. Since 1978 there has been an economic reform campaign trying to establish what the Chinese call "a planned commodity economy with market regulation mechanisms." In Hungary, the deep economic reform and dramatic change in the last decade has been termed the "New Growth Path." The United States is one of the most developed countries with free market economy but now facing serious international competition. Currently all three countries are striving to upgrade their manufacturing capability, to globalize their industry and to make themselves more competitive in the world.

This analysis is based on data collected from machine tool factories in each of the countries. A common questionnaire containing 65 questions with over 100 data items was used. The questions covered the areas of general information, sales forecasting, production planning and scheduling, shop floor control, purchasing and materials management.

In this paper, we first use multi-set data envelopment analysis (DEA) to make a comprehensive assessment of the relative productivity of the factories in the three countries [1]. This analysis ranks the U.S. as the most productive and China as the least productive, with Hungary in between. After applying DEA, we use statistical analysis, mainly the chi-square test and Tukey method of analysis of variance to determine the significant differences in manufacturing practices between the countries. The combination of the DEA assessment and significant differences analysis in this paper deepens the insight over using either analysis alone. It is especially helpful in highlighting areas of manufacturing practice that need attention in the less developed regions or countries.

DEA ANALYSIS

A basic problem in industrial management is to become more productive over time. Increasing productive efficiency is generally considered a desirable goal for enterprise managers in most societies. The more productive each firm is, the wealthier the country will be. But the problems of definition and measurement of productive efficiency are extremely difficult.

Economic productivity models have been studied beginning with Adam Smith in 1776. A generally accepted definition is productivity (P) equals output (O) divided by input (I), (i.e., P=O/I). If outputs and inputs can be measured in the same units, it is straightforward process to measure productivity directly. This is often possible in engineering or natural sciences. In economics and business, measurement is complicated by the fact that the inputs consist of several different production factors and outputs also consist of a variety of useful goods and services as described in [2]. The multi-set DEA method enables us to combine a variety of variables to do a comparative evaluation. It has already proved useful as a method for cross-national comparison in previous work [3,4,5].

We used data gathered by the GMRG members for this analysis. It came from 41 factories in China, 29 factories in the U.S. and 16 factories in Hungary. The questionnaire had limited performance measurement data, as the focus was on manufacturing practices. This analysis considers sales as the only output, and labor and equipment as the only inputs. These are the same variables used in previous DEA analysis studies using the GMRG data.

> *input variables*
> X_1 = total number of employees
> X_2 = number of factory workers
> X_3 = investment in production equipment

output variables
$$Y_1 = \text{annual sales}$$
$$Y_2 = \text{annual export sales}$$

Before applying the DEA model to the data, we provide a brief general description of the procedure. First, we use model (M1) to evaluate the relative productivity (producing sales from labor and equipment investments) of each of the firms within a country using all firms as a reference set.

$$
\begin{aligned}
\min \quad & -e^T s^{+\alpha} - e^T s^{-\alpha} \\
\text{s.t.} \quad & Y^\alpha \lambda^\alpha - s^{+\alpha} = Y_o^\alpha \\
& -X^\alpha \lambda^\alpha - s^{-\alpha} = -X_o^\alpha \\
& e^T \lambda^\alpha = 1 \\
& \lambda^\alpha, s^{+\alpha}, s^{-\alpha} \geq 0
\end{aligned}
\tag{M1}
$$

where, $\alpha = 1,2,3$, respectively, indicates the three countries of interest. The e^T is a unit vector and each λ $(0 \leq \lambda \leq 1)$ forms a linear combination of inputs and outputs. The S values are slacks and (X_o^α, Y_o^α) is the input-output vector of the firm being tested. The firm's relative performance measure h_o^α is defined as:

$$
h_o^\alpha = \left(\sum_{i=1}^m \frac{x_{io} - s_i^{-*}}{x_{io}} + \sum_{r=1}^n \frac{y_{ro}}{y_{ro} + s_r^{+*}} \right) / (m+n)
\tag{1}
$$

where there are m outputs and n inputs. For this measure, we have $0 <= h_o^\alpha <= 1$, with $h_o^\alpha = 1$ if the firm being tested is on the frontier (e.g., has the highest level of sales for the given inputs). If $h_o^\alpha < 1$, then (X_o^α, Y_o^α) is not on the frontier.

For firms not on the frontier in their countries, we adjust the data by projecting them onto the frontier. The input-output vector after adjustment $(X_o'^\alpha, Y_o'^\alpha)$ is on the frontier, where $X_o'^\alpha = X_o^\alpha - s^{-\alpha*}$, $Y_o'^\alpha = Y_o^\alpha + s_o^{-\alpha*}$. Then we use model (M2) to evaluate the relative productivity of each firm with all the adjusted firms in the three different countries as a reference set.

$$
\begin{aligned}
\min \quad & -e^T s^+ - e^T s^- \\
\text{s.t.} \quad & Y'\lambda - s^+ = Y_o'^\alpha \\
& -X'\lambda - s^- = -X_o'^\alpha \\
& e^T \lambda = 1 \\
& \lambda, s^+, s^- \geq 0.
\end{aligned}
\tag{M2}
$$

where $\quad X' = (X'^1, X'^2, X'^3) \quad Y' = (Y'^1, Y'^2, Y'^3)$

A summary of the DEA analysis is provided in Table 1. Note how close to the frontier the U.S. firms lie (high average value for h) compared to the Chinese firms. An analysis of variance shows the differences among these three countries are significant. The U.S. firms have the highest sales output for a given level of labor and equipment. The Hungarian firms are next and the Chinese firms are the lowest. The question of why there are such sharp differences among different countries has long been of concern. Many comparative management studies have shown that company specific practices, environmental factors (such as education, sociology, and natural resources), political-legal institutions and economic policies all have a significant bearing on managerial activity and effectiveness in any society [2].

Table 1. Summary of DEA Analysis

	China	US	Hungary
Number of units on the frontier	0	18	4
Fraction of units on the frontier	0	62%	25%
Average performance value (h)	0.4534	0.9852	0.7131

In this study we will only consider manufacturing practices as collected by questionnaires to analyze the significant differences among them.

DIFFERENCES IN MANUFACTURING PRACTICES

In this analysis we are doing comparisons among three different countries so we use the F-test in analysis of variance and Tukey method [6] to test the significance of differences for quantitative data, and use the contingency table in a chi-square test [7] to test the homogeneity for nominal (or categorical) data. We discuss here some of the results--those we found most interesting in light of the DEA analysis. The presentation follows the five categories of the questionnaire.

General comparisons

The Chinese and Hungarian firms have a much higher employment level than their U.S. counterparts (see Figure 1). There is no significant difference between Chinese and Hungarian firms, however. The

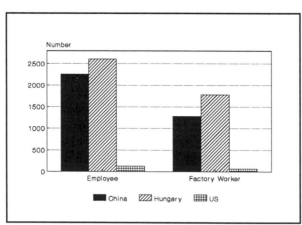

Figure 1. Employment

size differences may be a reflection of socialist ownership since these countries tend to build big firms. The investment in production equipment shows a different picture (see Figure 2). In absolute value there is no significant difference between the Chinese and the U.S. firm investments. But the overall employment in Chinese firms is about 20 times that in U.S. firms, so the Chinese firms have substantially smaller investments in equip-

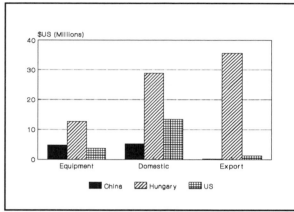

Figure 2. Equipment and sales value

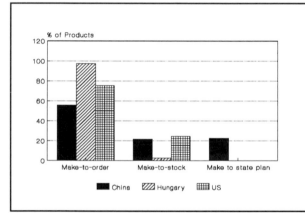

Figure 3. Production category

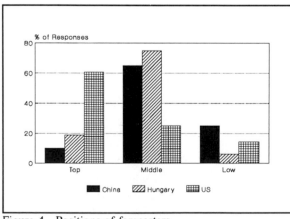

Figure 4. Positions of forecasters

ment per employee than the U.S. firms. This low level of investment could be a main reason for China's low DEA performance values.

While the Hungarian firms have the highest investment, the per employee value is between that of the U.S. and China. The Hungarian firms have the highest sales and portion of export sales which contributes to their effectiveness values being higher than those of the Chinese firms.

As seen in Figure 3, the Hungarian firms have virtually all production to order, while about a quarter of the U.S. production is to stock. These reflect different market responses [8]. The Chinese still have about a quarter of their output to state plan.

Forecasting

The forecasts are prepared by personnel from the same functional groups in the three countries. The management positions, however, differ between the Chinese and Hungarian firms and those of the U.S. firms. About 60% of the people involved in the U.S. are in top level positions. More executive attention is paid to forecasting in the U.S. than in Hungary and China (Figures 4 & 5).

There is no significant differ-

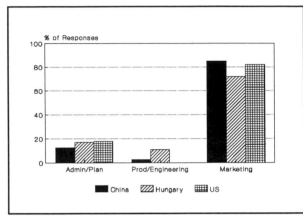

Figure 5. Function of forecasters

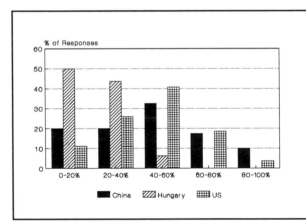

Figure 6. Weight of subjective factors in forecast

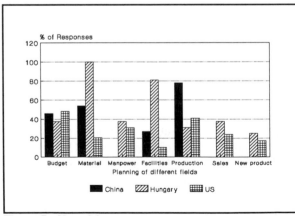

Figure 7. Use of forecasts

ence between Chinese and U.S. firms in the weight given to subjective factors in forecasting. Fewer Hungarian firms, however, give much weight to subjective factors in forecasting (see Figure 6).

On average, the Chinese firms modify their forecasts 2.3 times per year, the Hungarian firms 3.2 times per year, and the U.S. firms about 5.2 times per year. The U.S. production activities can be more closely connected to the fluctuation of the market this way.

Differences occur between the countries in the use of forecasts. The Chinese firms use forecasts most heavily for production planning, but almost not at all for sales planning, manpower planning or new product development. The Hungarian firms make more use of forecasts for material and facility planning than the others (see Figure 7).

Production planning and scheduling

All three countries have yearly production plans but most of the Chinese and Hungarian firms divide the yearly plan in quarterly increments, while U.S. firms divide the plan into months.

As with forecasting, there are differences in the positions of the people who prepare these

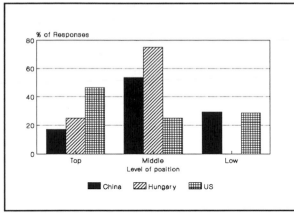

Figure 8. Position of production planning

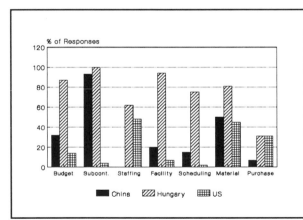

Figure 9. Uses of production planning

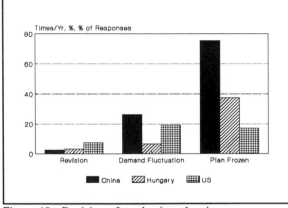

Figure 10. Revision of production planning

plans. The Chinese firms have the lowest portion of the top management involved in production planning while the U.S. firms have the most. In Hungary, the production plans are prepared entirely by the top and middle level management with little low level involvement (see Figure 8). Again the level of management attention and planning detail show the importance of meeting the market.

There are some significant differences in the uses of the production plans. The Chinese firms mainly use the production plan for administrative tasks like subcontract planning, material planning and budget preparing. The Hungarian firms also make heavy use of the production plan for such tasks (facility planning and staffing). The U.S. firms, however, use the plans more for scheduling, material planning and other tasks more closely related to customers (see Figure 9).

Again, mirroring the forecasting activities, the Chinese firms revise their plan about 2.4 times per year, the Hungarian firms 3.2 times per year and the U.S. firms about 7.5 times per year. In addition, about three-fourths of the Chinese responders said that their plans are frozen. This makes them very difficult to change, and, on average, it takes a demand fluctuation

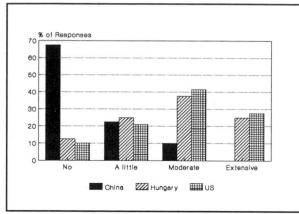

Figure 11. Use of computers in production planning

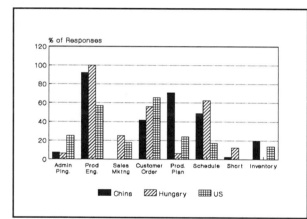

Figure 12. Starting production

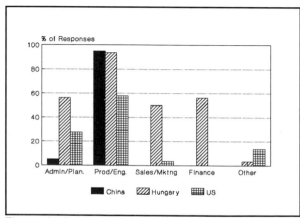

Figure 13. Responsibility for purchasing

more than 26% to do so. The U.S. firms are much more flexible and the Hungarian firms are in-between (see Figure 10).

As an example of lack of investment to support people, the Chinese firms show the lowest level of use of computers in production planning. About two-thirds of all firms do not use computers at all. The Hungarian and U.S. firms are about at the same level of use of computers (see Figure 11).

Shop floor control

There are a number of significant differences among China, Hungary and the U.S. firms in the area of shop floor control. In Chinese firms the sales/marketing persons have no authority to start production, while around a quarter of the Hungarian and U.S. firms let them do so. The basis of production orders in China is mainly the production plan, while U.S. firms give most weight to customer orders and Hungarian firms use the schedule (Figure 12).

Chinese firms have the administration/planning and production/engineering groups exclusively responsible for the purchase of material. In Hungary the sales/marketing and finance groups play an important role, as well. The basis of purchas-

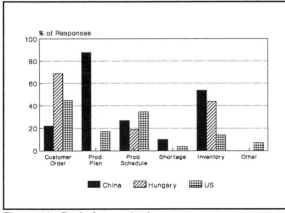

Figure 14. Basis for purchasing

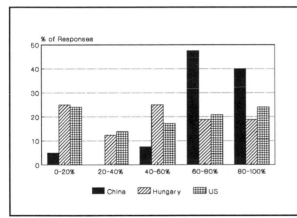

Figure 15. Self-fabricated parts

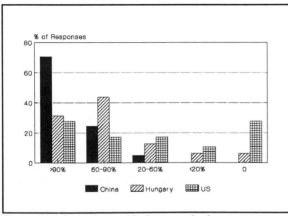

Figure 16. Operations with time standards

ing is quite different too. The Chinese firms' purchases are mainly based on the production plans, the inventory position and the customer order. The Hungarian firms give high priority to customer orders but do not use the production plan at all. The U.S. firms use a variety of bases (see Figures 13 and 14).

The Chinese firms show a much higher fraction of self-fabricated parts than their counterparts. Almost 90% of the Chinese responders produce more than 60% of the fabricated parts in their plants but only 37% of Hungarian firms and 44% of U.S. firms are doing so. This drive for high levels of internal fabrication may contribute to the low level of performance found by DEA for the Chinese firms (see Figure 15).

The Chinese firms show a higher use of time standards than the other firms. Over 70% of the responders have time standards for more than 90% of the operations, and no firms have less than 20% of the operations using time standards. The U.S. firms have the fewest operations with time standards (see Figure 16). The use of time standards in costing presents a similar picture for the countries.

Figure 17 shows the factors used for changing the priorities

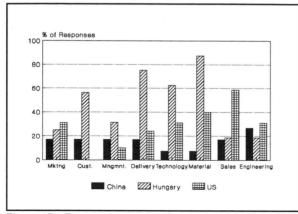

Figure 17. Factors that change production priorities

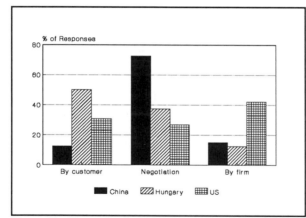

Figure 18. Determination of delivery dates

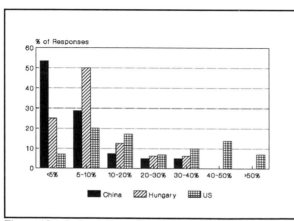

Figure 19. Order delivery lateness

on the production orders in process. In Hungary, material shortages, delivery date changes and manufacturing problems have the greatest influence on the priorities. The U.S. firms give the highest priority to changes in the sales plan. But in China, the greatest influence comes from engineering problems.

There are differences in who determines delivery dates for customer orders. The U.S. firms themselves mostly determine the delivery dates. About half of the Hungarian firms report that the delivery dates are determined by the customers. More than 70% of the Chinese firms said they determine the delivery date by negotiation (see Figure 18).

The percentage of the late orders is quite different. No firms in China report more than 40% of the orders are late. But about 20% of U.S. firms do so (see Figure 19). However, the average lateness in China is about 2 months, while it is about 1 month in the U.S.. The situation in Hungary is in between.

Figure 20 shows several alternatives for changing the capacity required to meet a peak schedule in the three countries. Overtime and subcontracting are used by all to increase capacity. The U.S. firms lay off extra workers and reduce

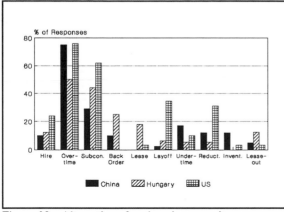

Figure 20. Alternatives for changing capacity

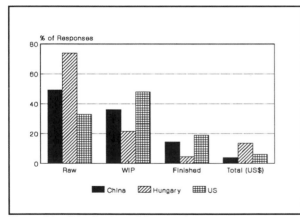

Figure 21. Distribution of inventories

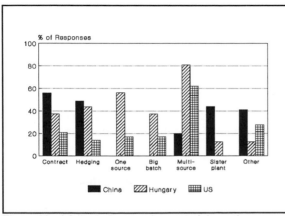

Figure 22. Methods for assuring material supply

the work time to decrease capacity. Thus, the U.S. firms more effectively match capacity to need, as picked up by the DEA analysis.

Purchasing and materials management

The inventory patterns are different between the countries. Hungary has the highest amount and about two-thirds of their inventory investment is in raw materials and purchased parts, with only about 5% in finished goods. The U.S. firms hold about 50% of their inventory in work-in-process and only 33% in raw materials. The Chinese firms hold about 50% of their inventories in raw materials and about 15% in work-in-process inventory (see Figure 21).

The Hungarian and U.S. firms mostly use multiple sources of supply to assure timely supply. This may help reduce raw material costs, as well. In contrast, the Chinese firms use contract, hedging and sister plants; practices left over from centrally planned economy procurement methods (see Figure 22).

CONCLUSIONS

There are many important differences in manufacturing practices among China, Hungary and the United States in the machine tool industry.

Generally speaking, the Chinese firms and the U.S. firms are at the two extremes and the Hungarian firms are in between, as indicated in the performance measured by DEA. Many of these differences reflect the two different economic systems: centrally planned and free market. Though China has carried out economic reform for more than 10 years trying to introduce the market mechanism into her planned economy, there still are big differences between China and the U.S. in almost every aspect. The Hungarian firms have made great changes in the past decade, as seen by their position between China and the U.S..

The clearest indication of the reason for the performance differences in the machine tool industry is the low sales productivity by labor in China. It is only about one-tenth of its Hungarian counterpart and one-forty-fifth that of the U.S. sales/employee. A part of this problem is the low equipment investment in China. Another aspect is the flexibility and responsiveness of the U.S. systems, and the extent of the market influences on production. Many other contributing factors reflect the residual effects of state planning. Though make-to-state-plan production is only used for 25% of the production in China, the residual effects are much bigger than that. For example, the production plan is not as flexible as others, the forecasts have little impact on production and don't get high level attention. The sales and marketing group has much less authority and influence on production than elsewhere. Paying attention to these factors could help the Chinese industry go a step further in economic reform and comparative productivity.

REFERENCES

1. A. Charnes, W.W. Cooper, B. Goland, L. Seiford and J. Stutz, "Foundations of Data Envelopment Analysis for Pareto-Koopman Efficient Empirical Production Functions," *Journal of Econometrics*, 30 (1985), 91-107.

2. R.N. Farmer and B.M. Richman, *Comparative Management and Economic Progress*, Bloomington, IN, Cedarwood Publishing Co., 1970.

3. L.F. Wu and C.Z. Xiao, "Comparative Sampling Research on Operations Management in Machine Tool Industry Between China and the Countries in Western Europe," *Journal of Shanghai Institute of Mechanical Engineering*, 11 (1989), 61-67.

4. C.Z. Xiao, "Cross-National Comparative Study of Manufacturing Practices with DEA Method," Proceedings of Pan-Pacific Conference VII, 1990, 13-15.

5. C.Z. Xiao and W. Xu, "Comparative Research on Manufacturing Practices with DEA Method," presented at the International Manufacturing Practices Workshop, Shanghai, 1990.

6. J. Neter, W. Wasserman and M.H. Kutner, *Applied Linear Statistical Models*, Homewood, IL, Irwin, 1985.

7. C.T. Clark and L.L. Schkade, *Statistical Analysis for Administrative Decisions*, 4th ed., Cincinnati, Southwestern Publishing Co., 1983.

8.* A. Chikán and D.C. Whybark, "Cross-National Comparison of Production-Inventory Management Practices," Discussion Paper No. 8, Bloomington, IN, Indiana Center for Global Business, Indiana Business School, April, 1989.

* This article is reproduced in this volume.

A Comparison of the Relative Productivity in the Machine Tool Industry for the U.S. and Japan

John G. Wacker, Iowa State University, USA

ABSTRACT

The productivity of U.S. and Japanese firms in the machine tool industry is compared in this paper. The output considered is sales, adjusted for the percentage of parts manufactured in the firm in order to estimate the value-added. Labor and capital are the inputs. A Cobb-Douglas production function is fitted to the data and used to evaluate capital and labor productivity. In addition a regression analysis is used to explore the relationship between direct and indirect workers. The overall productivity of Japanese firms is higher than that of their U.S. counterparts as is the productivity of Japanese non-factory workers.

INTRODUCTION

The economic success of Japan has captured world attention. Dubbed an economic miracle, Japan now accounts for a substantial and growing proportion of the United States trade deficit. Japan is the envy of the many developing nations that have failed to achieve similar growth.

Various propositions have been made to account for Japan's impressive success. One of the most popular, at least in the eyes of Western observers, has been the thesis that Japan's cultural heritage tends to facilitate economic growth. The values of thrift, discipline and commitment to work entailed in this heritage have made possible rapid growth while eliminating many of the traditional malaises of developing manufacturing economies, such as high labor turnover and a nondisciplined work-force.

The reasons for the popularity of the cultural thesis are quite obvious. First, the cultural explanation reflects Western fascination with an area that is still a mystery to many. Second, it supposedly explains why Japan has made it economically while other countries have not. Third, it is supported by empirical findings showing differences in work values and management styles between East-Asian and Western countries [1,2].

While the role of culture differences in manufacturing performance may provide an explanation for part of the divergent patterns of manufacturing growth between Japan and the U.S., it is not a complete explanation. Furthermore, focusing on the cultural thesis alone could distract attention from other explanations. Therefore, it is essential to examine the validity of existing models of manufacturing growth before jumping to any conclusions regarding the possible role of culture in that process.

The recent success of the Japanese in gaining a growing share of global markets has triggered a quest for the roots of Japan's impressive growth in manufacturing output. The present paper suggests that along with the cultural explanations for the Japanese manufacturing success, the role played by major production factors such as labor, capital and technology must also be taken into account. A model incorporating these factors is presented and examined for Japan and United States. This paper examines the differences in the management of labor, capital and technological resources in the manufacturing industries compared with those in the U.S.. The overall basic hypothesis is that with the same level of resources, Japanese firms have more output than U.S. firms.

BASIC PHILOSOPHICAL DIFFERENCES

One primary difference between Japan and United States is the degree to which managers and workers interact to achieve production quotas. Working relationships between laborers and management are so close that managers and other non-factory workers often help factory workers with their tasks [3,4]. Japanese managers are reluctant to hire workers who do not directly help manufacturing goals. This philosophical difference is reflected in the lower factory overhead costs [5]. The difference also leads to two additional hypotheses. Hypothesis Two: Japanese firms are expected to have a lower percentage of non-factory workers than U.S. firms. Hypothesis Three: Japanese firms are expected to have higher productivity from the non-factory employees with respect to factory output than U.S. firms.

The three basic hypotheses are:
- Hypothesis One:
  H_0: There is no significant difference in the productivity between Japan and the United States. H_a: Japan is more productive than the United States with the same set of resources.
- Hypothesis Two:
  H_0: There is no significant difference in the amount of non-factory workers between Japan and the United States. H_a: The United States has more non-factory workers relative to factory workers than Japan.
- Hypothesis Three:
  H_0: There is no significant difference in productivity for non-factory workers in Japan and the United States. H_a: Japanese non-factory workers are more productive than United States non-factory workers.

DATA

To analyze Hypotheses One and Three, production functions are the most traditional method. For production functions estimation, the output should be measured in exactly comparable units, capital should reflect the exact amount of capital expended (or the equivalent capital rent) should be utilized [6], and labor should be measured by number of workers. The data utilized for the empirical estimates in this study are directly comparable since output (value-added) and capital were gathered in the same time frames and were adjusted for differences in exchange rates. However, to estimate output it is necessary to make some adjustments for the amount of value not produced inside the corporation. The data gathered [7] do not have any direct estimates of the value-added. This study, therefore, adjusted the total sales by the percent of internally fabricated parts. The number of factory workers and the number of other workers represent the amount of labor inputs [8]. The capital expenditure for new equipment will represent the amount of capital rent. This measure presents some theoretical problems since the new capital is generally not utilized strictly as replacements for the amount of capital equipment utilized. Consequently, the productivity of new capital would not correspond to the amount expended for equipment in the current time period, but rather to a later time period. However, this has been a traditional problem with estimation of production functions [9].

METHODOLOGY

The statistical methodology utilizes a modified generalized Cobb-Douglas production function. The traditional estimation procedure calls for a logarithmic transformation of the explanatory resource variables and the output dependent variable. The specific estimation form is:

$$\ln(OUTPUT) = \beta_0 + \beta_1 \ln(LABOR) + \beta_2 \ln(CAPITAL) + \ln(\varepsilon_i) \qquad (1)$$

However, since there are inter-country differences in the production function, a modified form of the production function will be utilized to test for statistical differences in the production functions and productivity. The form suggested here is:

$$\ln(OUTPUT) = \begin{aligned}&\beta_0\\ &+ \beta_1 \ln(LABOR_1)\\ &+ \beta_2 \ln(LABOR_2)\\ &+ \beta_3 \ln(CAPITAL)\\ &+ \beta_4 (JAPAN)\\ &+ \beta_5 \ln(LABOR_1)(JAPAN)\\ &+ \beta_6 \ln(LABOR_2)(JAPAN)\\ &+ \beta_7 \ln(CAPITAL)(JAPAN)\\ &+ \ln(\varepsilon)\end{aligned} \qquad (2)$$

This form of estimation is the natural logarithmic equivalent to using qualitative binary variables in traditional estimation procedures [10]. The estimated constant (β_0) of the regression is the technology coefficient which represents the level of the technology. $LABOR_1$ represents the factory workers; $LABOR_2$ represents the amount of non-factory labor, and capital represents the new capital expenditures. Differences in the base level of technology are estimated by the binary variables: Japan=1 and U.S.=0.

The interpretation of the estimated coefficients for labor and capital (β_1, β_2 and β_3) are output elasticities with respect factory labor, non-factory labor, and capital respectively. Because they are output elasticities, they are unit free and alleviate some of the problems that occur because of data incompatibility between the U.S. and Japan. The coefficient for the binary variable Japan (β_4) represents the statistical difference between Japan and the U.S. in technological base. If it is positive and significant, it means that Japan is more productive with any combination of resources. However, the coefficients for the cross products of the binary Japan variable and the resource variables (β_5, β_6 and β_7) represent the statistical differences in the output elasticities with respect to that resource. If any of the cross product coefficients is positive and statistically significant, it is interpreted that Japan is more productive in the use of that resource. Alternatively, if any cross product coefficient is negative and statistically significant, it is interpreted to mean that the Japan is less productive using that resource.

The summation of the labor's and capital's coefficients is an estimate of returns to scale. If it is greater than one there are increasing returns to scale (meaning a doubling of all inputs will produce more than a doubling of outputs) and a summation of less than one means that there are decreasing returns to scale (doubling inputs will less than double output). The resource coefficients can be interpreted in the traditional manner of neo-classical economic theory [11].

For Hypothesis Two, an statistical regression with total workers as the dependent variable and the factory workers as the explanatory variable will give a relative comparison of the propensity to add non-factory workers. Again a qualitative binary variable will be introduced to estimate the relative differences (Japan=1 and U.S.=0). The specific estimation form is:

$$TOTAL = \beta_0 + \beta_1\ FACTORY + \beta_2\ JAPAN + \beta_3\ JAPAN(FACTORY) + \varepsilon \quad (3)$$

where *TOTAL* represents the total workers, *FACTORY* represents factory workers, *JAPAN* is the binary qualitative variable for Japan, and *JAPAN(FACTORY)* is the interactive variable. The estimated coefficients β_0 and β_1 represent the absolute and relative propensity to add workers in the U.S.. The estimated coefficients β_2 and β_3 represent the relative and absolute differences for Japan compared to the U.S.. (The ε represents the error term for the regression.) The statistical significance of β_2 and β_3 estimate the differences between the two countries.

RESULTS

The following results were estimated for the modified Cobb-Douglas production function of equation (2) with a resulting $R^2 = 0.557$ for n = 40.

			t values
	-	2.80	
	+	0.826 ln($LABOR_1$)	2.360
	+	0.102 ln($LABOR_2$)	0.417
	+	-0.015 ln($CAPITAL$)	-0.087
ln($OUTPUT$) =	+	(-0.792) ($JAPAN$)	-0.266
	+	(-0.423) ln($LABOR_1$)($JAPAN$)	-0.553
	+	(0.742) ln($LABOR_2$)($JAPAN$)	1.453
	+	(0.089) ln($CAPITAL$)($JAPAN$)	0.233
	+	ln(ε)	

The statistical results were disappointing because the data was collected on individual firms rather than aggregated across many firms, causing lower explanatory power through disaggregation effects. But some conclusions can be made from the estimates. The results indicate that the U.S. and Japan have similar technological bases, since the Japan variable was statistically insignificant. Not surprisingly, the productivity of the factory workers is most statistically significant in explaining output variations. There is no statistically significant productivity difference between factory workers or between new capital equipment in Japan and the U.S.. However, the non-factory workers have a statistically significant higher productivity ($p<0.100$), supporting Hypothesis Three. Additionally, there are differences in returns to scale between the two countries (Japan had 1.321 and the U.S. had 0.913). Consequently, Japanese firms are more productive with the same level of resources. Therefore, Hypothesis One is supported.

The estimate for differences in the propensity to add workers and the t values (in parenthesis) for equation (3) are:

$$TOTAL = -48.375 + 2.692\ FACTORY + 74.096\ JAPAN -1.648\ JAPAN(FACTORY)$$
$$(20.350) \qquad\qquad (1.222) \qquad\quad (-2.649)$$

$R^2=0.922$ n=40
Student "t" values in parentheses

As one may expect, the number of factory workers is the best predictor of the total employment in the firm. Although Japan has a larger number of non-factory workers than the U.S. for smaller factories (the coefficient of the Japan variable is positive), it is not significant until ($p>0.250$). The coefficient on the *JAPAN(FACTORY)* variable indicates that Japanese managers are less likely to add non-factory workers than their U.S. counterparts. Therefore, Hypothesis Two is supported.

IMPLICATIONS AND CONCLUSIONS

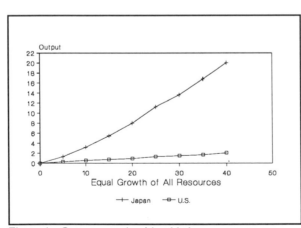

Figure 1. Output growth with added resources

To better visualize the productivity results, Figure 1 utilizes the above estimates to show a comparison between Japan and the U.S.. Both axes were drawn for relative scales with identical resources rather than using an actual scale to show comparisons based on the productivity estimates. The higher productivity of Japan suggests Japanese are more effective in utilizing their resources than their U.S. counterparts. However, there is a difference in the actual amount of resources each country has. In this study, Japan's average number of factory workers is 81.5 compared with 90.7 for the U.S. firms. The Japanese firms average 2.410 million spent on capital equipment compared to 6.245 for the U.S.. However, the largest difference is with non-factory workers, where Japan averaged 29.3 compared with 105.1 for the U.S.. Apparently the Japanese are more effective in utilizing their non-factory workers. However, it could be hypothesized that the U.S. has more government requirements than Japan, requiring U.S. firms to hire more less productive workers. This study will leave it for future studies to ascertain the impact of government reporting requirements on overall productivity.

Some of the previously mentioned limitations of this study resemble those of other studies of productivity. They are: data incompatibility problems, small sample size, and limited number of countries. However, some of the basic conclusions are important since they do indicate sources of differences in productivity:

1. Japanese machine tool firms are more productive than their U.S. counterparts, primarily because of their more effective use of non-factory workers.

2. Japanese firms are considerably less likely to hire non-factory workers than to hire factory workers. U.S. firms tend to hire 1.7 non-factory workers for every factory worker they hire, while Japanese firms tend to hire about 0.04 non-factory workers for every factory worker they hire.

3. Japanese firms make more effective use of their non-factory workers, as indicated by their higher productivity compared with their U.S. counterparts.

In short, it is highly likely that Japanese managers are more aware of productivity concerns when they hire non-factory workers than are U.S. managers. It is for this reason that they have better productivity from the same set of resources.

Future studies should investigate more industries with different countries to understand how pervasive are the differences in productive capabilities.

REFERENCES

1. G. Redding, "Some Perceptions of Psychological Needs Among Managers in South-East Asia," in: Y.M. Poortinga (ed.), *Basic Problems in Cross-Cultural Psychology*, Amsterdam, Swets and Zeitlinger, 1977.

2. G. Holstede, *Cultures Consequences: International Differences in Work-Related Values*, Beverly Hills, Sage Publications, 1980.

3. R. Hayes and S.C. Wheelwright, *Competing Through Manufacturing: Restoring Our Competitive Edge*, New York, John Wiley Co., 1984.

4.   R.W. Hall, *Zero Inventories*, Homewood, IL, Dow-Jones-Irwin Inc., 1983.

5.   J. Miller and T. Vollman, "The Hidden Factory," *Harvard Business Review*, September-October, 1985, 142-150.

6.   R. Coombs, P. Saviotti and V. Walsh, *Economics of Technological Change*, London, MacMillan, 1987.

7.* D.C. Whybark and B.H. Rho, "A Worldwide Survey of Manufacturing Practices," in: D.C. Whybark and G. Vastag (eds.), *Global Manufacturing Research Group, Collected Papers, Vol. 1*, Chapel Hill, NC, Kenan Institute, University of North Carolina, 1992.

8.   W. Krelle, "The Aggregate Production Function and the Representive Firm," in: F.G. Adams and B.G. Hickman (eds.) *Global Econometrics*, Cambridge, MIT Press, 1983.

9.   A.N. Link, *Technological Change and Productivity Growth*, Chur, Switzerland, Harwood Academic Publishers GmbH., 1987.

10. J. Neter and W. Wasserman, *Linear Statistical Models*, Homewood, IL, Irwin, 1982.

11. D.W. Jorgenson, "Econometric Methods for Modeling Producer Behavior" in: Z. Grilches and M.D. Intrilligator (eds.), *Handbook of Econometrics: Volume 3*, Amsterdam, North- Holland Publishing Co., 1986.

* This article is reproduced in this volume.

Are Customers Linked to Manufacturing?

D. Clay Whybark, University of North Carolina-Chapel Hill, USA

ABSTRACT

This study evaluates data compiled by the Global Manufacturing Research Group that would tend to disclose direct links between the customers and manufacturing. The results indicate that such linkages are rare and that they are associated mostly with forecasting inputs and shop floor activities. Much remains to be done by firms that wish to be "customer driven." The evidence is that Korean firms are a little more customer connected than those in North America and western Europe. The Chinese, on the other hand, have a long way to go to reach even the currently poor western standards.

INTRODUCTION

The motivation for this research stems from the calls for manufacturers to get closer to their customers. These appeals have come from popular business writers like Tom Peters, for example, who suggests that excellent firms are those that are customer driven (see Peters and Waterman [1]). In addition, almost all of the recent books on world class manufacturing have something to say about this [2], and even academic authors are looking at the strategic aspects of getting close to the customer in the design process [3]. Seminars on Delighting Your Customers or Customer Driven Manufacturing have been popular for managers from companies trying to learn ways to get closer to their customers. A primary interest in this research is to determine to what extent customers are *in fact* linked to manufacturing practices.

It is not just United States firms that are interested in the issue, it is a global phenomena. Western companies often find themselves compared to Japanese and other Asian firms on their ability to get close to the customer. Although it is clear that there is a global interest in serving the customer, it is not clear that links to the customer are the same in all parts of the world. Another motivation for the study was to see if there are geographical differences in the way that customers influence manufacturing practices.

A data base compiled by the Global Manufacturing Research Group (GMRG) on manufacturing practices around the world is located at the University of North Carolina. Although it was not the original purpose of GMRG work, some of the questions used to compile the data base can be used to shed light on the links between manufacturing and the customer. The primary methodology of this study was to analyze the answers to those questions.

The basic argument in this study is a very simple one. If there is to be any chance of getting close to the customer, there must be links between the customer and manufacturing. These links should be reflected in specific methods by which customers can influence manufacturing practices. Examining the extent to which firms in the data base actually use these methods provides one indication of the degree to which they are customer driven. By comparing responses from different areas of the world, any differences between regions should be revealed.

SOME BACKGROUND LITERATURE

The manufacturing strategy literature, from the early works of Skinner [4] to more recent writings like Adam and Swamidass [5], have acknowledged the importance of linking the business and marketing strategy to that of manufacturing. Cleveland, Schroeder and Anderson [6], in their work assessing production competence, start with market strategies when developing their framework. There is a section on market or customer oriented manufacturing in popular books on competitiveness like Moody [7], Hayes, Wheelwright and Clark [8], Harmon and Peterson [9], and Schonberger [10]. Swamidass's manufacturing strategy bibliography [11] has a section on business strategy, "because manufacturing strategy cannot exist apart from it." All these authors argue the importance of agreement in market, business and manufacturing strategies and many call for a closer linking to the customer in executing the strategy. Few, however, offer much help in how the linkages are to be achieved.

Hill [12] is an exception. He has developed a very comprehensive process for tying the marketing and the manufacturing infrastructure together. As with most other approaches, he starts with the business and marketing strategy and derives the tasks to be performed by manufacturing to support the strategy. Using the concept of "order winners," he focuses on the competitive battle in the market place for the customers' business. Hill's approach provides explicit guidelines for tailoring the manufacturing capability to the needs in the market or, when needed, changing the marketing strategy to match manufacturing strengths. Several case studies provide empirical evidence that there can be many benefits in making such changes in manufacturing practice.

A great deal of work has been done on the process of developing manufacturing strategy, and Hill carries that on through to tactical design and operational issues. There is little work, however, on the connections between customers and the shop floor practices in a firm or how the shop floor communicates information to customers. With the exception of some discussion of the use of customers on design teams, there is very little knowledge of how the market might be linked to manufacturing or how that might affect production practices. One purpose of this research is to identify some of the customer-manufacturing linkages that are reflected in the manufacturing practices data base and to see how they are used in various parts of the world.

THE DATA BASE

This paper presents an analysis of responses to selected questions from a data base on manufacturing practices gathered from firms around the world. To compile the data base, formal surveys were conducted in the non-fashion textile and small machine tool industries. These industries were chosen because they are found in nearly every country, including those that are less developed. and they have different manufacturing processes.

The small machine tool industry produces manufacturing equipment such as lathes, grinding machines, and milling machines. The processes are generally batch oriented. The non-fashion textile industry, on the other hand, produces items like sheets, towels, and industrial cloth, using production techniques that are more process oriented.

Data have been gathered in many countries. Starting in South Korea and the People's Republic of China, the effort has spread to Eastern and western Europe, other parts of Asia, North America, Australia, South America and the former USSR. The data were gathered using a common questionnaire for each of the industries and countries. A complete description of the project, data base, questionnaire, and data coding can be found in Whybark and Rho [13].

Some analyses of parts of the data have been made already. Country to country comparisons have been made (see, for example, Rho and Whybark [14,15]). Some interpretive work on the differences between Hungary and other parts of the world was done by Chikán and Whybark [16], and a multivariate analysis of the data from Hungary, western Europe and North America was performed by Vastag and Whybark [17]. These analyses have clearly shown several important differences in practice between various parts of the world. So far, however, no specific analysis of the linkages between customers and manufacturing practices has been performed.

SELECTING AND ANALYZING THE DATA

This study uses data from North America (the United States and Canada), western Europe (including several Northern European countries), South Korea, and the People's Republic of China. Since the movement for customer driven manufacturing arose in the west, the choice of North America and western Europe was obvious. In addition, the two regions have been found to be very similar [18]. South Korea has been compared to China and western Europe and has been found to be quite responsive to the market compared to the other regions [14,15]. China is the least market driven of the regions, so should provide a strong contrast with the others. Another factor that makes these attractive areas to study is the increasing amount of joint venture and other forms of strategic alliance taking place among them.

The sampling in North America and western Europe was random in the areas included. There is an underrepresentation of the western United States and Canada in the North American data and only Northern European firms are included in the western European data. In China, the sample is from the Shanghai area and areas of Northern and Southern China. While the Korean data is geographically dispersed, the firms were contacted through the Korea Productivity Center, so there may be a bias towards the more advanced firms. The sample size for each of these regions for each industry is shown in Table 1.

The data base has general information on the companies and their manufacturing practices. The manufacturing practices cover the areas of forecasting, production planning, shop

Table 1. Sample sizes by region and industry

	Industry	
Region	Machine tool	Textile
North America	45	50
Western Europe	34	24
South Korea	89	33
China	44	56

Table 2. Questions and responses for forecasting

What are the most important subjective factors in your forecast?

	NA*	WE	KO	CH
General economic and political situation	10	10	14	12
General company and industry situation	40	40	20	21
Trends in economic and political situation	9	9	13	31
Market research and customer/vendor information	39	39	53	32
Other	2	2	0	4

What are the most important uses of the forecast?

	NA	WE	KO	CH
Budget preparation	22	30	13	22
Material/inventory planning	22	23	21	41
Manpower planning	12	13	4	1
Facilities planning	3	17	7	26
Production planning	23	10	33	0
Sales planning	12	2	19	10
New product development	5	3	3	0
Other	1	2	0	0

What is the principal measure of your forecast error?

	NA	WE	KO	CH
No measure used	45	35	78	10
Percentage error	42	57	12	72
Average error	3	3	0	12
Average absolute error (MAD)	5	3	7	3
Other	5	2	2	3

* Percent of responses for North America (NA), Western Europe (WE), Korea (KO), China (CH)

floor control, and materials management and purchasing. The questionnaire had no direct questions on customer linkages or non-manufacturing related forms of tying to the customer. Consequently, the data used in this study comes from questions on manufacturing practices that can be impacted by customers. The contention is that without such an influence, claims of customer driven manufacturing must be questioned.

The questions chosen for analysis for this study are grouped into three broad categories: forecasting, planning, and execution on the shop floor. Those that were selected will be discussed under each of these categories. Some of the questions have to do with how manufacturing practices are affected by customer actions or other factors. Other questions are concerned with how the practices facilitate (or do not facilitate) responding to the customer.

Forecasting

The questions selected for the forecasting category are shown in Table 2 along with the summary of the responses. The first of these has to do with information used in forecasting. This question was chosen because one of the sources of information concerns market research and customer/vendor information. The degree to which this information is used is an indication of how the customer might influence the forecasts used for making manufacturing plans. The second question, relating to the uses of the forecast, specifically asks about sales planning. The responses to this question can be an indication as to how explicitly the customer is taken into account in planning. (Note that these two questions asked for two responses. Whenever this was done, both responses were counted in the statistics.)

The last question included in this section is on the evaluation of forecasts. What is of importance here is whether the forecasts are evaluated, not how. Closing the loop on forecasting by assessing their accuracy is an indication of concern for how well the customer needs are understood.

Planning

The production planning questions and responses shown in Table 3 cover direct customer input into planning and whether the plans can be easily changed to accommodate customer input. The first of these questions looks at whether customer plans are included in the input to the production planning process (as with the first two questions in the forecasting section, two responses were requested for this question and all responses are counted in the statistics).

The second question in this section looks at the customer due date impact on the sequence for releasing jobs to production. The final question asks about frozen production plans. Although providing stability to manufacturing, which might have advantages to customers, frozen plans are more difficult to change in response to customer requests.

Table 3. Questions and responses for production planning

What are the most important factors for your production plans?

	NA*	WE	KO	CH
Actual orders/backlog	45	28	47	30
Production capacity	13	5	1	3
Level of inventories	9	26	2	38
State plan	0	0	0	2
Previous sales	7	7	0	1
Customers' plans	10	10	4	5
The forecast	15	24	46	20
Other	1	0	0	1

What is the principal basis of the sequence for releasing jobs to production?

	NA	WE	KO	CH
Customer order due dates	64	55	63	48
Processing time required	17	24	6	11
Similarity of set-ups	4	4	3	2
Material availability	8	11	9	8
First-come first-served	1	4	7	0
Marketing preferences	4	0	5	12
Selling price of item	0	0	7	1
Management directive	2	0	0	9
Other basis	0	2	0	9

Is part of the production plan (or other plan or schedule) frozen or otherwise made very difficult to change in the near future?

	NA	WE	KO	CH
Yes	25	25	0	70
No	75	75	100	30

* Percent of responses for North America (NA), Western Europe (WE), Korea (KO), China (CH)

Execution on the shop floor

The questions chosen for this section and a tabulation of the responses are shown in Table 4. The first of these questions has to do with whether the products are made to order. An important observation from this data is that many products in China are still made to state plan. This question was chosen because make-to-order connotes a response to a specific customer's request and is an indication of being customer driven.

The other two questions on shop-floor linkages are concerned with relating to the customer before and during production. One of them is concerned with the process for setting the due date and the degree to which the customer influences that decision. The other looks at the ability of the customer to change priorities once production has started.

Table 4. Questions and responses for shop-floor execution

What is the percent of make-to-order and make-to-stock products?

	NA*	WE	KO	CH
Make-to-order	75	67	80	39
Make-to-State Plan	0	0	0	47
Make-to-Stock	25	33	20	14

How is the delivery date for customer orders determined?

	NA*	WE	KO	CH
Mostly by customer	*	26	68	26
Negotiation	35	52	21	63
Mostly by us	22	22	11	11
	43			

Which of the following factors has a heavy influence on the priorities once production has started on an order.

	NA	WE	KO	CH
Pressure from marketing	9	10	26	16
Pressure from customers	25	28	0	8
Orders from management	16	13	0	13
Changes in delivery dates	12	11	26	11
Sudden surges in demand	9	8	12	0
Manufacturing problems	8	7	12	7
Material shortages	11	11	18	23
Changes in sales plan	2	6	3	10
Engineering problems	7	5	3	9
Other	1	1	0	3

*Average percent reported for each region.
**Percent of responses for North America (NA), Western Europe (WE), Korea (KO), China (CH)

ANALYSIS AND DISCUSSION

The primary interest in this study is the extent to which the various possible linkages between the customer and manufacturing are actually used. This is determined by assessing the degree to which customer oriented responses are used in comparison to the alternative responses. The determination of any regional differences is based on comparing the responses among the four regions. Any differences in industry responses are also assessed since they could mask important regional differences. After some general remarks on the techniques used to analyze the data, the analysis will be discussed under the three sections used previously: forecasting, production planning and shop-floor execution.

Most of the questions selected for this study required the respondent to choose among several categorical responses (e.g., what is the principal measure of your forecast error, or how is the delivery date for customer orders determined). To determine how extensively customer oriented responses were used, they were compared to the alternative responses. Regional or industry differences in the answers were based on comparisons of the overall response patterns to the questions. For these categorical variables, statistical significance was determined using the chi-square non-parametric test.

There was one question for which there were numerical responses (what is the percent of make-to-order and make-to-stock products). The test applied to the responses to that question was the Kruskal-Wallis non-parametric test. The choice of these conservative non-parametric tests was based on preliminary evaluations of the data which indicated violations of the requirements for parametric tests (e.g., non-normality and non-equality of variances). Tests for which significance values were greater than five percent were declared not significant for this study.

In almost all circumstances, the evaluations of differences between the industries was not significant and the data was pooled for the geographical comparisons. In instances where there was a significant difference between industries, a check of the interaction effects indicated that the overall conclusions for the geographical differences remained unchanged. Therefore, all the results are reported as though the industry data were pooled. Where there were industry differences, they will be described in the discussion that follows.

Forecasting

Forecasting is an area where there should be a close relationship between the customer and the company. There are regional differences, however. There is a significant difference in the responses to all three of the questions in Table 2, but the data from the question on error measurement required some modification before the statistical test could be considered valid.

The first question, on forecast input, should have responses closely connected to the customer. Not surprisingly, then, market research and customer/vender information is a very frequently chosen response. The regional differences in response patterns is due partly to the other choices; notably for the general company and industry situation, and trends in economic and political situation. There is a difference in the regions' choices of customer oriented information as seen in Table 5 where the non-customer information responses are combined into "other". The differences in Table 5 are significant at the 0.003 level.

Table 5. Summary of choices of subjective factors in forecasting

	NA*	WE	KO	CH
Market research/customer information	39	41	53	31
Other	61	59	47	69

* Percent of responses for North America (NA), Western Europe (WE), Korea (KO), China (CH)

One potential use of the forecast, as reported for the second question of Table 2, is sales planning. But this is not a widely chosen response. The significant differences in responses on this question, however, do reflect differences in the choice of sales planning between regions. Most notably it is much less frequently chosen in western Europe. When geographical differences are summarized as responses to the sales planning choice versus all others, they are significant at least at the 0.001 level.

The data in Table 2 show that no measurement of forecast error is the most frequent choice in Korea and North America. The substantial use of percentage error by the Chinese firms may be related more to planning requirements than to a concern for customers. The data, however, also suggest that there are several very sparse cells. This suggests that the chi-square test may not give valid results.

In order to get around this problem, the data were summarized as shown in Table 6. For this summary, the differences between the regions are highly significant (at least at the 0.001 level). However, when the choices are summarized in this fashion (any measure versus no measure used), the industry differences are not significant.

Table 6. Summary of error measure use

	NA*	WE	KO	CH
No measure used	44	35	78	11
Other	56	65	22	89

* Percent of responses for North America (NA), Western Europe (WE), Korea (KO), China (CH)

The analysis of the forecasting questions provides little evidence of a close connection to the customer except as an input to the forecasts. Such input, however, is not more frequently chosen than the other inputs, except in Korea. The little use of forecasts for sales planning and the low use of forecast error measures suggest that there is not overwhelming evidence of strong linkages to the customer in this section.

Production planning

The data in Table 3 show the responses to questions in the production planning part of the questionnaire. For all three questions, there is a possible highly significant geographical difference indicated. For two of these questions (the factors used for production planning and the basis for releasing jobs to production), however, there are several sparse cells. This means that the chi-square test may give false conclusions. To get around that problem for the analysis, the data were summarized into customer oriented factors and others, as was done in the forecasting section. In no case was there a significant difference between the two industries.

The customers' plans are not very important in developing the production plans in any of the regions. The backlog of actual orders, on the other hand, is very important, particularly in North America and South Korea. To check the regional differences, the responses on the other factors were combined and compared to the responses on the customer plans and actual customer backlog. The data are shown in Table 7. South Korea makes the most use of customers' plans and China's use is probably related to government plans. The backlog seems to be a summary of customer plans in North America. The regional differences seen in Table 7 are significant at the 0.001 level.

Table 7. Summary of choices of factors for production planning

	NA*	WE	KO	CH
Customers' plans	10	28	48	30
Actual orders/backlog	45	10	4	5
Other	45	62	48	65

* Percent of responses for North America (NA), Western Europe (WE), Korea (KO), China (CH)

Customers do influence the release of orders to production. The customer order due dates are the single largest factor in all regions. Marketing preferences are another potential linkage between customers and manufacturing, although it is not highly used except in China. These data are summarized in Table 8. The regional differences, mostly the result of China's use of marketing preferences, are significant at least at the 0.01 level.

Table 8. Summary of choices of the basis for releasing jobs

	NA*	WE	KO	CH
Customer order due dates	64	55	63	48
Marketing preferences	4	0	5	12
Other	32	45	32	40

* Percent of responses for North America (NA), Western Europe (WE), Korea (KO), China (CH)

The use of a frozen plan is significantly (at least at the 0.001 level) different between regions. The Chinese firms, at one extreme, use it a great deal and the Koreans firms, at the other extreme, do not. It is moderately used in western Europe and North America. Thus, except for China, there seems to be some flexibility in the plan for responding to customer input. Again, however, with the exception of this flexibility and the use of actual customer orders and their due dates in production planning, there is little evidence of customer linkages in this section.

Execution on the shop floor

All the regional differences in shop floor execution seen in Table 4 are significant at least at the 0.001 level. There are no significant industry differences. Therefore, the industry data are pooled for the analysis of the shop floor questions. Only the first question (on make-to-order versus make-to-stock) required an analysis technique different from the chi-square test. A Kruskal-Wallis test was used for that question.

The firms in the western nations and Korea have a much higher percentage of make-to-order production than the Chinese firms, unless the make-to-state plan in China is considered as make-to-order. When the percent of production made to stock is tested, there are highly significant differences among the regions. China has lowest percentage. The Korean firms have the highest percentage of pure make-to-order, giving them the edge in responding to their customers.

When the data on determining delivery dates is considered, the Korean firms again are the most directly influenced by their customers. More than half of the firms say that the due dates are determined mostly by their customers.

The Korean firms, on the other hand, buffer the factory floor from direct pressure by customers. Unlike North America and western Europe, where the customer pressure on production priorities is direct, it is felt in Korea through marketing. Again, there is very little impact from the sales plan, except in China. This is probably related to the government plans for their firms as opposed to changes in a sales plan produced by the firm as indicated by the Chinese firms' relative response to sales planning in Table 2.

A summary of the customer related factors that can influence production priorities is provided in Table 9. Western European firms are influenced the most by customer oriented factors. Surprisingly, Korean firms are the lowest. Interestingly, the Chinese don't have demand surges and the Koreans don't have orders from management.

Table 9. Summary of the factors that influence production priorities

	NA*	WE	KO	CH
Pressure from marketing	9	10	26	16
Pressure from Customers	25	28	0	8
Changes in sales plan	2	6	3	10
Other	64	56	71	66

* Percent of responses for North America (NA), Western Europe (WE), Korea (KO), China (CH)

There is a little more overall evidence of linkages between the customers and manufacturing on the shop floor than in the other areas. The level of make-to-order production and the customer impact on production priorities are examples. The customer involvement in setting due dates in Korea is another. The differences between the regions is not consistent and clouded by the Chinese levels of make-to-state plan production.

SUMMARY

The analysis performed in this study of selected questions from the Global Manufacturing Research Group questionnaire does not provide much evidence of a close link between customers and manufacturing. There are differences around the world, however. The Chinese, as expected, seem to be the farthest removed from their customers, unless the State is considered a customer. The Korean firms had the closest linkage for many of the questions considered.

Overall, however, the picture is not encouraging. In the forecasting area, an area where there should be a particularly close relationship between the market and production, the only bright spot is the use of market research and customer/vender information in the forecast. The forecasts are not used for sales planning, however, and it is nearly as likely that no forecast error measurement is done as it is to have forecast accuracy formally measured.

In production planning, actual customer orders are important factors in both planning and sequencing production. Again, however, customer plans have little influence. The production plans, at least in the west and Korea, do have some flexibility built in as opposed to being frozen. In China, however, they are likely to be frozen.

It is in the shop-floor execution area that customer linkages seem to be most prominent. The picture is mixed, however. China has the lowest make-to-stock percentage when make-to-state plan production is considered the equivalent of make-to-order. Customers do have a substantial impact on setting due dates, especially in Korea. The least likely response, from any region except North America, is that the company sets the due dates on its own. Finally, customer pressure for priority changes is felt directly on the shop floor in North American and western European firms.

Without an overwhelming amount of evidence of customer impact in any of these regions, it seems that a marginal advantage in customer driven manufacturing goes to Korea. The western European and North American firms are quite similar for most of the questions, but are less linked to their customers than Korean firms. The Chinese firms, as expected, are quite a bit less linked to the customer then the other regions, unless the customer is the government.

This analysis provides some clear general conclusions for firms considering international production activities. There are differences between the regions and these differences are greater than those between the industries. For firms contemplating international sourcing, production, market expansion or product expansion partnerships, the implications are quite clear. They need to be concerned about the regional differences as they establish their partnerships. Some of the differences in practices could cause real difficulties in international relationships if not addressed early in the undertaking. In establishing international sources the firm can become a customer that is not well linked to the partner.

The implications of this study for the effective implementation of customer driven strategies are also quite clear. Firms desiring customer driven approaches must develop means for linking customers to their production if such strategies are to be made functional. Interestingly, even for production driven strategies, there must be a means for carrying out that strategy in the market. At the moment, the existence of such mechanisms is minimal. The mission is clear. The development of direct linkages to and from customers is a necessary precursor to effective implementation of any customer driven manufacturing strategy. Certainly it is critical for firms that claim they want to be customer driven.

REFERENCES

1. T.J. Peters and W.H. Waterman, *In Search of Excellence*, New York, Harper and Row, 1982.

2. T.G. Gunn, *Manufacturing for Competitive Advantage: Becoming a World Class Manufacturer*, Cambridge, MA, Ballinger, 1987.

3. J.A. Fitzsimmons, P. Kouvelis, and D.N. Mallick, "The Design Strategy and Its Interface with Manufacturing and Marketing Strategies: A Conceptual Framework," Decision Science Institute Annual Proceedings, San Diego, Nov. 1990.

4. C.W. Skinner, "Manufacturing: Missing Link in Corporate Strategy," *Harvard Business Review*, May-June, 1969.

5. E.E. Adam Jr. and P.M. Swamidass, "Assessing Operations Management from a Strategic Perspective," *Journal of Management*, 13, No. 2 (1989).

6. G. Cleveland, R.G. Schroeder and J.C. Anderson, "A Theory of Production Competence," *Decision Sciences*, Fall 1989.

7. P. Moody, *Strategic Manufacturing: Dynamic New Directions for the 1990s*, Homewood, IL, Dow Jones Irwin, 1990.

8. R.R. Hayes, S.C. Wheelwright, and K.B. Clark, *Dynamic Manufacturing: Creating the Learning Organization*, New York, The Free Press, 1988.

9. R.L. Harmon and L.D. Peterson, *Reinventing the Factory*, New York, The Free Press, 1990.

10. J. Schonberger, *World Class Manufacturing: The Lessons of Simplicity Applied*, New York, The Free Press, 1986.

11. P.M. Swamidass, "Manufacturing Strategy: A Selected Bibliography," *Journal of Operations Management*, August 1989.

12. T. Hill, *Manufacturing Strategy: Text and Cases*, Homewood, IL, Irwin, 1989.

13.* D.C. Whybark and B.H. Rho, "A Worldwide Survey of Manufacturing Practices," Discussion Paper No. 2, Bloomington, IN, Indiana Center for Global Business, Indiana Business School, May 1988.

14.* B.H. Rho and D.C. Whybark, "Comparing Manufacturing Practices in the People's Republic of China and South Korea," Discussion Paper No. 4, Bloomington, IN, Indiana Center for Global Business, Indiana Business School, May 1988.

15.* B.H. Rho and D.C. Whybark, "Comparing Manufacturing Planning and Control Practices in Europe and Korea," *International Journal of Production Research*, 28, No. 12 (Fall 1990), 2393-2404.

16.* A. Chikán and D.C. Whybark, "Cross-National Comparison of Production-Inventory Management Practices," Discussion Paper No. 8, Bloomington, IN, Indiana Center for Global Business, Indiana Business School, April, 1989.

17.* G. Vastag and D.C. Whybark, "Manufacturing Practices: Differences that Matter," Discussion Paper No. 23, Bloomington, IN, Indiana Center for Global Business, Indiana Business School, May 1988.

18.* G. Vastag and D.C. Whybark, "Comparing Manufacturing Practices in North America and Western Europe: Are There Any Surprises?" in: R.H. Hollier, R.J. Doaden and S.J. New (eds.), *International Operations: Crossing Borders in Manufacturing and Service*, Amsterdam, Elsevier, 1992.

* This article is reproduced in this volume.

ACKNOWLEDGEMENTS

An early version of this paper was published by the Center for Manufacturing Excellence at the Kenan-Flagler Business School at the University of North Carolina-Chapel Hill under the title, "Are Markets Linked to Manufacturing?" November 1990. The Center is gratefully acknowledged for supporting this research.

Manufacturing Practices in a Transition Economy

Attila Chikán and Krisztina Demeter,
Budapest University of Economic Sciences, Hungary

ABSTRACT

This paper summarizes the results of two surveys conducted in 1986 and 1991 to explore manufacturing management practices in Hungary. The dramatic changes of the region have exposed companies to a completely new situation, which is reflected in the survey results. Hungarian companies have apparently turned in new directions and their current manufacturing practices are definitely closer to those in the market economy countries than to what used to be--however there is still a long way to go to meet world standards.

INTRODUCTION

The fast and fundamental changes rocking the economies of Eastern Europe these days certainly have a great influence on the sphere of manufacturing. There are interconnected effects on both the macro and micro level, which finally will lead to a radically changed production structure and behind it, a not less radically changed production behavior of companies. This paper provides some points about the nature of the changes going on, using the results of a survey.

THE SURVEY

The survey is actually part of an international effort to discover the general nature and the cross-country differences of manufacturing practices world-wide. The idea came from Boo-Ho Rho of Sogang University in Korea and D. Clay Whybark of the University of North Carolina-Chapel Hill, USA, who, together with other colleagues, constructed a questionnaire in 1986 to conduct the first surveys in South Korea and China [1]. Later, other regions were also examined, and two years ago an international research venture, the Global Manufacturing Research Group (GMRG) was formalized, within which, under the leadership of Professor Whybark, the research process is continuing.

Hungary was among the first countries joining the project. A survey of 78 companies was carried out and evaluated, and later the results were compared with the findings in other regions of the world [2,3,4]. It was a straightforward idea to repeat the survey in 1991, when Hungary had already started the transformation into a market economy: we expected that interesting results could be achieved about the reflection of the overall changes in a specific economic activity, manufacturing. (It is also probably not a surprise that we intend to repeat the survey again in a few years.) So a questionnaire comparable to the one covering 1986 was administered in 77 companies in 1991.

This paper summarizes the results of the comparison of the two surveys and concludes with our findings about the effect of economic transition.

The questionnaire used internationally consists of questions in the following categories:
1. Company profile
2. Sales Forecasting
3. Production Planning and Scheduling
4. Shop Floor Control
5. Purchasing and Materials Management

The questionnaire used in Hungary has two additional sections, one concentrating on inventory control and the other covering the use of computers.

The number of companies in the survey were similar in both cases. We did our best in the second survey to use a group comparable to that of the first survey, and we believe that the differences are within a reasonable limit. The breakdown of the sample data by industry is indicated in Table 1.

Table 1. Distribution of the companies by industry

Industry	1986	1991
Machine	38.5%	49.4%
Textile	21.8%	35.1%
Other	39.7%	15.5%

The target industries of the international research are small machine tools and non-fashion textiles. We tried to have a comparable sample, but the small number of companies in these industries in Hungary does not make it possible to have a sample of appropriate size.

THE RESULTS

In the following we organize the results around four main topics that, in our view, reflect the profound changes taking place in Hungary:
1. Market reorientation and delivery performance
2. Changing means of adjustment
3. Outside pressures
4. Organization and methodology

Market reorientation and delivery performance

In the 1980s, there was a situation of generally high (excess) demand and the related sellers' market; a comfortable selling position. Consequently, we found that companies were not really keen on satisfying customers' needs. They were very much interested in obtaining hard currency; because of that they made a clear distinction between the various markets, preferring those where they could sell their goods for hard currency. This preference was also a very important influencing factor of manufacturing behavior, the effects of which were identified in our research as well.

The environment has changed in the time between the two surveys and so did the behavior of the companies. Moreover, these changes are not simple and painless. Let's see some of the effects.

Companies must be much more consumer oriented than previously, for two main reasons. First the collapse of the COMECON market and the continuing need to obtain hard currency forced the companies to turn more and more to markets in Western Europe and other regions with market economies. Second, the general recession and decreasing production in Hungary also make the internal market customers more demanding. The changing trade orientation can be seen in Figure 1.

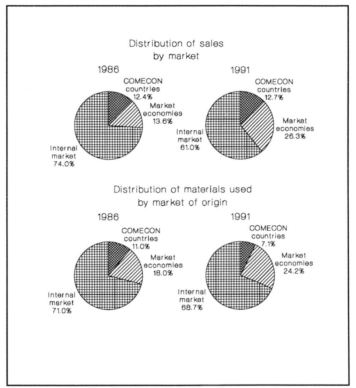

Figure 1. Trade orientation in both the sales and purchase markets

The changes shown in Figure 1 are really substantial when one considers that 1990 was the year in which trade reorientation was started.

The different requirements of the markets and the adaptation of companies to them can be seen clearly in the delivery time promises reflected in the data in Table 2. The results in the table are striking. They show two major changes: a radical improvement in overall delivery promise conditions; and a closing of the gap between delivery promises made to different markets. It has always been an issue in Hungary that companies' performance will not really be adequate until they can treat the various markets uniformly. No compa-

Table 2. Usual delivery time promised in various markets
(in % of companies in the survey)

Market		Delivery time in days		
		<60	60-80	>80
Internal	1986	29.8	54.3	15.8
	1991	51.3	43.2	5.4
Market economies	1986	40.8	44.9	14.3
	1991	52.7	41.8	5.4
COMECON countries	1986	20.0	44.4	35.6
	1991	32.2	50.8	17.0

ny can handle different requirements in discipline, quality, etc., efficiently in the same time.

It must be added that the survey shows a little larger proportion of deliveries are late in 1991 as compared with 1986, but one should not forget the much shorter delivery time promise. On the other hand, the average time late is shorter than in 1986 (see Table 3).

Table 3. Delivery performance

Market	Late deliveries as a percent of total deliveries		Average days late	
	1986	1991	1986	1991
Internal	10.2	12.6	33.0	21.4
Market economies	4.4	8.6	14.0	13.1
COMECON countries	6.8	6.5	32.0	17.7

These three data sets (delivery times promised, number of deliveries late and average lateness) together show a substantial improvement in delivery performance.

As for the causes of late deliveries (see Table 4), material shortage remained the most important cause (though with a smaller weight), while labor shortage plays a smaller role than before. The general decrease of production in the economy is a natural reason for the smaller weight of capacity problems. This is supported by the observation that the average capacity utilization rate of the companies in the 1986 sample was 75.0%, while those in the 1991 sample averaged 58.9%. The 1991 companies suffered an average 14% production decrease compared to 1986. There is a slight increase in the weight of quality problems and due date changes. However, analysis of this table leads directly into the second part of the results, namely how companies' means of adjustment are changing.

Table 4. Causes of late delivery

Characteristic Reason	Average weight (1-5)	
	1986	1991
Insufficient overall capacity	2.30	1.97
Labor shortage	2.96	2.36
Production bottlenecks	3.13	3.07
Materials shortage	4.37	3.92
Quality problems	2.77	2.94
Due date changes	2.51	2.76

Changing means of adjustment

Results of the 1986 survey support general knowledge about the production policies of companies in planned economies. Companies organized and managed their production according to their internal interest--provided that it was not prevented by resource constraints (notably, the availability of materials and manpower). Scarce resources were the most important determinants of production policy. In 1991, they were forced to react much faster to the changes in market demand and needed higher flexibility from the beginning to the end of their production process. Changes in the planning (beginning) part of the process can be seen very clearly in Table 5, which shows the average number of plan revisions per year in 1986 and 1991.

Table 5. Average number of revisions per year

Area	1986	1991
Sales forecast	3.3	8.9
Production plan	4.2	7.2
Inventory plan	*	4.6

* Data not available

Sales forecast revision frequency more than doubled, and production plan revisions also highly increased. The changing role of sales and production (their importance exchanged) support the hypothesis of market orientation instead of the production orientation.

The change in company policy toward a stronger market-orientation can also be seen if we take a look at the management of the production process itself. As an example, scheduling priorities were previously changed mainly on the basis of resource availability, while in 1991 they are based mostly on consumer-related aspects (see Table 6).

Table 6. Causes of priority changes

Causes	Average weight (1-5)	
	1986	1991
Pressure from marketing	3.4	3.9
Pressure from consumers	3.3	3.9
Orders from management	2.9	3.4
Changes in delivery dates	3.7	3.8
Sudden surges in demand	2.8	3.6
Manufacturing problems	3.5	3.5
Material shortages	4.3	3.8
Changes in sales plan	2.8	3.4
Engineering problems	2.7	2.9

Not only have the causes for priority adjustment changed, but companies were adjusting capacity differently in 1991 compared with 1986. This can be seen in Table 7.

Table 7. Alternative ways to change capacity

Increases	Average weight (1-5)	
	1986	1991
Hire additional workers	2.8	2.5
Use overtime	3.0	3.8
Subcontract production	2.9	3.0
Backorder the production	2.7	2.1
Lease temporary capacity	1.9	2.2
Decreases		
Lay off extra workers	1.7	2.2
Use undertime (idleness)	1.6	2,1
Reduce the work time	1.6	2.5
Build inventory	2.0	2.4
Lease capacity to others	2.0	2.5

Before the changes in manufacturing practices brought on by the economic reform it was much less characteristic to use manpower resources for adjusting capacity. By 1991, the main means for increasing and decreasing capacity are overtime and reduced working time, respectively. It is worth mentioning that the average weights have increased substantially overall, especially for the ways of decreasing capacity. This shows the concern for adjusting to the general decrease in production. In correspondence with other findings in this study, the decreased weight of backordering in cases of capacity shortage shows the increased importance of reliable delivery.

Under the new conditions companies are much more cost-sensitive than before. This is reflected in the make or buy decisions shown in Table 8. Companies had much different focuses when deciding on subcontracting in 1991 than five years earlier. Subcontracting to improve cost, quality and speed became more important, while problems with producing are less important reasons.

Table 8. Reasons for subcontracting

Reasons	Average weight (1-5)	
	1986	1991
Production load	3.3	2.8
Production difficulty	3.2	3.1
Company policy	2.6	2.6
Lower costs	2.9	3.7
Higher quality	2.5	3.3
Faster production	2.5	2.9

The inventory policy in 1991 also reflected customer-orientation and changing market conditions. When asked about the main factors influencing planning the inventory levels, the following weights were given by the managers (Table 9).

Table 9. Factors in planning inventory levels

Factors	Average weight (1-5)	
	1986	1991
Working capital considerations	4.5	3.8
Purchasing difficulties	3.4	3.3
Production smoothing	4.0	3.9
Consumer service	3.2	3.8
Production uninterruptedness	4.3	4.3
Sales forecast	3.2	3.5

The first priority is still to have uninterrupted production, but customer service and adaptation to sales forecast have increased in importance. It must be added that the "old" system put very severe financial restrictions on inventory increase. Lifting these restrictions caused the decrease of the weight of working capital considerations.

Outside pressures

Companies must adjust their policies and manufacturing practices to the changing environmental factors. Fortunately, the changing environment has provided more new means to adjust their actions than earlier. An aspect of this can be seen in Table 10.

Table 10. Which type of inventory can you influence most?

Element	1986	1991
Raw material	26.0%	50.0%
Work-in-process (WIP)	46.8	35.1
Finished goods	54.5	35.1

While in 1986 WIP and, especially, finished goods were the easily adjusted inventories, by 1991 it had shifted to raw material. That is, companies now feel they have more power to influence the raw material side. Moreover raw materials account for almost 70% of total inventories.

An important tenet of the socialist system was long term contracts to maintain a supply of material. These are being eliminated and the role of penalties for late delivery has been increased according to market economy standards, as can be seen in Table 11. Despite the changes, the highest score still goes to multiple sourcing.

Table 11. How is availability of materials and parts assured?

Method	Average weight (1-5)	
	1986	1991
Long term contract	3.1	2.7
Substantial penalty	1.6	2.1
Large volume	2.8	2.5
Multiple sourcing	3.7	3.7

External conditions sometimes may cause campaigns of building up or reducing inventory. The reasons for this are shown in Table 12. The transition from a sellers' market to a buyers' one is obvious. While in 1986 the main causes of inventory changes were supply difficulties or other shortages, in 1991 the role of financial considerations has been strengthened.

Table 12. Reasons for increasing or decreasing inventories
(in % of companies in the survey)

Increase	1986	1991	Decrease	1986	1991
Price increase foreseen	9.1	18.9	Lack of financing	37.2	51.0
Supply difficulties	40.0	34.0	Supply difficulties	25.0	10.4
Restrictions in financing	4.6	11.3	Improved supply conditions	9.5	7.3
Production increase	21.7	15.1	Lack of available credit	12.0	20.8
Shortage of complementary materials	24.6	20.8	Production decrease	15.5	10.4

The fast and fundamental changes in the environment increase uncertainty in company operations. This leads to a situation were objective factors play a less important role and expert judgements are more important for many decisions. It is reflected in areas like forecasting (Table 13) and inventory management. Overall, the weights of the subjective factors in sales forecasting have increased more than 10%, while the weights for the objective factors decreased more than 5%.

Table 13. Weight of factors in sales forecasting

	Average weight (1-5)	
Subjective	1986	1991
General economic and political situation	2.5	3.5
General company and industry situation	3.3	3.4
Trends in economic and political situation	2.0	3.5
Market research and customer information	4.8	3.7
Objective		
Past sales (demand) history	3.6	3.5
Current order backlog	4.8	4.3
Economic indices (GNP, prices, etc.)	3.2	2.7
Industry statistics (market shares, inventories)	2.4	2.7

Organization and methodology

We start with saying that organizational and methodological issues have not yet received as much attention as one would think seeing the economic policy changes. The main reason is that companies are very much involved in restructuring the main framework (legal form and organization of their operations) and do not have the time or resources for changing their internal operations. However, we will not have to wait long for the changes to come and they will be very substantial. Some signs of them can already be discovered.

There is a shift in emphasis on the different activities within the company. Production planning became more comprehensive and integrated, while the lifting of financial restrictions means that inventory planning is not as crucial as it used to be. This is reflected in the organizational level of these planning activities (Table 14). Production planning went up in hierarchy, while inventory planning went down from 1986 to 1991.

Table 14. Organization levels for production and inventory planning

Responsibility level for planning	Production		Inventory	
	1986	1991	1986	1991
CEO	5.0	5.0	3.9	1.3
Vice President/Director	25.0	35.9	48.7	23.0
Department/Division head	70.0	59.1	47.4	70.3
Group/Section manager	0.0	0.0	0.0	5.4

The average methodological level is still rather low--however there are some positive signs. Both the absolute level and trend can be observed in Table 15. As for the use of integrated production-inventory systems (MRP, JIT), there was in fact no possibility of using them in 1986, even though the planned economy was being reformed. These methodologies still have not penetrated Hungarian management. However, knowledge of these systems is increasing and in a short time we will see a dramatic change in their use.

Table 15. Exposure to MRP and JIT

Statement	MRP		JIT	
	1986	1991	1986	1991
Never heard of it	39.7	24.1	56.4	17.1
Using it and benefitting from it	10.3	6.3	0.0	5.2
Using it but not benefitting from it	2.6	2.5	0.0	1.3
Understanding it but feeling no necessity to introduce it	20.5	10.0	19.2	5.2
Just starting to introduce it	2.6	3.8	1.3	2.6
Trying to introduce it, but having difficulty doing so	3.8	4.9	0.0	3.9
Considering its introduction	17.9	15.2	0.0	19.7
It cannot be introduced under present conditions in Hungary	*	26.6	*	43.4
Other	2.6	6.3	6.4	1.3

* Data not available

A substantial change in the computerization of various activities has taken place, as shown in Table 16. The scale used for Table 16 was that 1 means the computer is not used at all and 5 means the activity is fully computerized.

We believe that the development of computer usage is remarkable, especially if we consider the difficult situation of companies in the period examined. Nevertheless, it has to be added that the current situation is still far from the standards of the developed world.

Table 16. Degree of computerization

Activity	Average weight (1-5)	
	1986	1991
Sales forecasting	1.4	1.8
Production planning	2.2	3.0
Capacity planning	2.0	2.6
Material usage planning	2.4	3.1
Shop order recording	2.3	2.4
Production scheduling	2.2	2.4
Quality control	1.6	1.6
Production design	1.6	2.1
Sales order recording	2.5	3.7
Purchase order recording	1.9	2.9
Inventory recording	3.1	3.5
Material accounting	2.8	3.1

CONCLUSIONS

This paper has shown that manufacturing practices, policies and methodology in Hungary have gone through fundamental changes from 1986 to 1991. These changes are closely connected to or as a consequence of the general economic changes stemming from the reforms. The main characteristics of the directions of change in Hungarian companies are summarized next.

Customer Orientation

The direction of change has clearly been toward a customer orientation away from a resource orientation. While materials, labor and other resources accounted for the majority of management attention some years ago, now companies have turned towards the output side of the business. They must be more sensitive to customer demands since they are forced to deliver earlier and react faster to markets.

Production Focus

Much more intensive efforts than before for cost-sensitive, efficient production are clearly evident from the survey. Corporate and lower level decisions are based now on clearer economic market reasons. Political aspects or possibilities of enjoying government subsidies have been eliminated.

Methodology

A slow but steady move towards more developed and sophisticated management method-ology is also evident. The first steps toward a developed information technology are being made through computerization. Hungarian companies are already getting accus-tomed to using computers in their operations. However, the lack of education, appropriate corporate culture, and also a more clear strategic approach to computer investments are still missing.

These findings support the hypotheses that macroeconomic changes unavoidably result in changing the manufacturing practices of companies, and show that Hungarian compa-nies are adjusting to a market economy. Though conclusions based on a survey always raise questions of general validity, the observations here are in accordance both with other research results and our everyday experience.

REFERENCES

1.* D.C. Whybark and B.H. Rho, "A Worldwide Survey of Manufacturing Practices," Discussion Paper No. 2, Bloomington, IN, Indiana Center for Global Business, Indiana Business School, May 1988.

2. A. Chikán, "Characterization of Production-Inventory Systems in the Hungarian Industry," *Engineering Costs and Production Economics*, 18 (1988), 285–292.

3. A. Chikán and K. Demeter, "Production and Inventory Management in the Hungarian Industry--An Empirical Study," *AULA Society and Economy, Quarterly Journal of Budapest University of Economic Sciences* (1990/2), 123–130.

4.* A. Chikán and D.C. Whybark, "Cross-National Comparison of Production-Inventory Management Practices," in: A. Chikán (ed.), *Inventories: Theories and Applications*, Amsterdam, Elsevier, 1990, pp. 149–156.

* This article is reproduced in this volume.

ACKNOWLEDGMENTS

A version of this paper appeared in the Pre-prints of the Seventh International Working Seminar on Production Economics, Igls, Austria, February 17-21, 1992.

SECTION FIVE

TECHNICAL CONSIDERATIONS

The three papers in this section are all concerned with issues involving the analysis and quality of the GMRG data base. The first two present approaches to the analysis of the data. The first paper provides a comprehensive framework for bilateral comparisons and the second provides some additional suggestions for data preparation and analysis. The last paper documents some potential data problems by comparing the data base contents with national statistics for selected countries. All three papers are useful for the researcher intending to use the GMRG data.

Gyula Vastag and D. Clay Whybark
An Analytical Framework for Bilateral Comparisons in the GMRG Data Base

The first paper in this section provides a framework for bilateral analyses of the data in the GMRG data base. This framework can serve as a guideline for future evaluations of the current data and can be extended to new data as it is gathered. The approach suggests grouping the data into basic units for statistical comparisons and introduces the idea of a pure effect, a conservative criterion for determining significance. The procedures take into account the different scales for the variables, and use different statistical approaches depending on the assumptions required of the data and the type of variable involved.

Gyula Vastag
Technical Considerations in Analyzing the Manufacturing Practices Data

Many of the researchers who have worked on the GMRG data have expressed an interest in some of the techniques that might be applied to the data. This article presents several techniques that might be useful. They include multi-variate statistical techniques, Chernoff faces (for depicting results), a technique for grouping companies, and a method for computing a regional compatibility index. In addition, several suggestions for transforming and coding the variables for analysis are provided.

John G. Wacker and Ling Feng Li
An Empirical Evaluation of the Worldwide Survey of Manufacturing Practices: Implications and Limitations

This article presents a comparison of national and industrial statistics with statistics on the companies surveyed for the GMRG date base. It provides information on what percent of the firms in each country and industry were sampled, the relative size of the firms (compared to the average for the country and industry) and other statistical comparisons of importance to the generality of the survey findings. The countries evaluated are China, Japan, South Korea, North America, Hungary and western Europe. Overall and within country comparisons are provided and suggestions made for the future data gathering and research activities of the GMRG.

An Analytical Framework for Bilateral Comparisons in the GMRG Data Base

Gyula Vastag, Budapest University of Economic Sciences, Hungary
D. Clay Whybark, University of North Carolina-Chapel Hill, USA

ABSTRACT

This paper provides a framework for analyzing bilateral differences in data collected on manufacturing practices around the world by the Global Manufacturing Research Group (GMRG). The data are housed at the Kenan Institute at the University of North Carolina and have been used for a variety of studies. Many techniques have been used in these studies, several of which have as an objective comparing manufacturing practices between regions. This paper describes in some detail an analytical approach for the comparison of two regions. There is no intention to claim an exclusive, "correct" way of performing the analysis, but, simply, to document a method that is technically sound and provides useful results. The framework presented here groups the data into basic units for statistical comparisons, accounts for the measurement scales of the variables, and uses different statistical approaches depending on the assumptions required of the data and the type of variable.

INTRODUCTION

This paper provides a framework for analyzing bilateral differences in data collected on manufacturing practices around the world by the Global Manufacturing Research Group (GMRG). The data comes from the non-fashion textile and small machine tool industries in several regions of the world. The data was gathered using a common questionnaire format in order to be able to compare manufacturing practices between the regions. Several areas of practice are addressed in the study, from forecasting to materials management. A description of the project, regions and questionnaire is provided in [1].

A variety of studies have been performed on the data base, which is housed at the Kenan Institute at the University of North Carolina. They range from within country studies, comparing industries or looking at firm size differences, to between region and multiple region studies. In some cases specific issues (like the relationship of the market to manufacturing or the productivity of Japanese factory labor) have been investigated. A number of different statistical techniques have been applied in these studies and several analysis formats used. The techniques have ranged from parametric tests (like the Student's t-test) to non-parametric tests (like the Kolmogorov-Smirnov test). Multi-variate approaches involving principal components analysis and special data representation techniques have also been used.

With the variety of techniques that have been used, and the interest in making statements about regional differences, we felt it desirable to describe in some detail an analytical approach to bilateral comparisons. There is no intention to claim an exclusive, "correct" way of performing the analysis, but, simply, to document a method that is technically sound and provides useful results. The framework presented here groups the data into basic units for statistical comparisons, accounts for the different scales for the variables, and uses the appropriate statistical approach depending on the assumptions required of the data and the type of variable. In many cases, the approach differs from that used in other studies comparing two regions (see, for example, Rho and Whybark [2,3]).

We start our discussion of the analysis by describing the grouping of the data for analysis and introduce the concept of a pure regional effect. We next describe the types of variables from the questionnaire that must be accommodated by the analysis. The remainder of the paper describes the statistical approaches used for each type of variable considered. The analysis described in this paper is illustrated with data gathered in North America (the United States and Canada) and western Europe (including several countries). Although our examples are concerned with comparing the two regions, the approach is quite general for bilateral comparisons.

GROUPING THE DATA

The four cells shown in Table 1 represent a natural way to group the GMRG data for bilateral comparisons since it was gathered in the two industries in all regions. In Table 1, 'ab' and 'cd' represent two hypothetical regions, while 'abm' and 'abt' denote the groups of data for the machine tool (m) and textile (t) industries in region 'ab.' Similarly, the machine tool and textile groups from region 'cd' are denoted by 'cdm' and 'cdt,' respectively. It is not the only way to group the data, however. Based on the responses to the questionnaire, other detailed subsamples could be developed by size of firm, by type of ownership, by export orientation, or by portion of in-house production.

Table 1. Grouping the data

Country/industry	Machine tool (m)	Textile (t)
ab	abm	abt
cd	cdm	cdt

The groupings shown in Table 1, however, permit us to make five different types of comparisons:
  (A) We can compare the manufacturing practices between different regions, namely 'ab' and 'cd,' to determine regional effects.
  (B) We can compare practices between the industries, namely machine tool and textile.
  (C) We can go into details and:
    (Ca) compare regions within the industry (e.g., compare 'abm' to 'cdm' or 'abt' to 'cdt') or
    (Cb) compare the industries within the country (e.g., compare 'abm' to 'abt' or 'cdm' to 'cdt') or
    (Cc & Cd) make cross industry-region comparisons (e.g., compare 'abm' to 'cdt' or 'cdm' to 'abt').

Each of these comparisons has a certain rationale and at the same time focuses on a different issue. The analysis in (A) emphasizes the regional differences and eliminates the industrial ones by combining the two types of industries (this analysis is stronger if it has been determined that the industries can be pooled for the statistical analysis). The analysis for (B) combines different regions in order to focus on the industrial characteristics (again, stronger if the regions can be pooled). The types of comparisons in (C) are the most elemental. Any industrial differences are not considered in (Ca) and regional ones are not considered in (Cb). In (Cc) and (Cd), the regional and industry differences are mixed.

THE CONCEPT OF A PURE EFFECT

In this paper we use the groupings of Table 1 to show how bilateral comparisons can be made using the (Ca) type of analysis. This does not require any prior determination of whether the industry data can be pooled. To make the comparisons we use the concept of a pure effect. A pure effect is illustrated in Figure 1 for a hypothetical example. In the example the groups represent the four cells of Table 1. The VAR(1) on the vertical axis is a hypothetical variable. The lengths of the bars represent the averages of the variable for each group.

Suppose that the differences between the regions in each industry is significant (compare 'abm' to 'cdm' and 'abt' to 'cdt'). Since the differences are in the same direction in both industries we define this as a pure regional effect. (Obviously, a pure industry effect can be defined by direct analog.) In the example in Figure 1, it appears that there is no statistically significant difference (for VAR(1)) between the industries within the same region (compare 'abm' to 'abt' and 'cdm' to 'cdt'). If the regional differences are the focus of the research, further analysis of the industrial effects may be eliminated. As long as the regional differences are significant and in the same direction, we have a pure regional effect regardless of what the direction or significance is for the industry values.

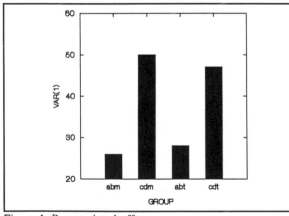

Figure 1. Pure regional effect

Using the pure effect to define differences is very conservative since it may overlook some interesting interactions between industry and region. On the other hand, since the effect is "pure," the description and interpretation of the differences are straightforward.

As a method of identifying regional differences that warrant discussion, it works quite well.

The use of the sub groupings like those of Table 1 and the concept of pure effects is not limited to specific types of variables, measures of central tendency (location) or statistical tests. Thus, it is a useful extension of the approaches used by others to analyze differences between groups of the GMRG data. The groups are quite useful for other purposes as well. For example, a firm considering a joint venture might be very interested in some of the other combinations, particularly those that involve regional differences for their industry--regardless of the other effects. The use of selected sub groups allows such comparisons and the exploration of other combinations of interest.

If we want to analyze the effects of two factors simultaneously we can extend our analysis to include the effect of the other grouping factor as well. Using hypothetical examples, Figure 2 gives illustrations for pure regional and industrial effects; and their combinations. This analysis can be extended to three or more factors, although it increases the complexity of the statistical analysis significantly.

THE GMRG DATA

The GMRG data are coded according to a standard set of conventions and units. A great effort has gone into assuring that the data is clean (e.g., the non-responses are correctly coded, that the units are correct on the variables, etc.) but a quick check of this is a key step in preparing to work with the individual variables. Once the data are clean, any variables derived from the data to be used in the comparisons can be produced.

The basic coding of the questionnaire has both numeric and character data. These distinctions are definitional, but have operational significance. Character data can be a number (the number "1" signifying a yes, for example) but the fact that it is a "number" does not mean that numeric operations can be performed on it sensibly. Numeric data can be manipulated numerically, while character data can not. In some cases, it is useful to change the definition of a character variable and/or transform it so that numeric operations are sensible and more analysis options are available.

The combination of converting variables and deriving new ones can add substantially to the data base. The questionnaire contains 65 original questions representing 95 variables divided into five sections (see [1] for details). Some of the studies that have been done used derived performance measures (for example, sales per employee or inventory turnover). Other variables result from individually analyzing each possible response to a multiple-response question. As an example, one of the questions asks the respondent to indicate how important each of several alternatives for changing capacity is for the company. The variable for that question is a string of characters indicating whether each particular alternative (e.g. hire workers, use overtime, etc.) was of high, moderate or little importance. To work with that variable, the string was broken down into individual variables for each possible response indicating the importance of that alternative.

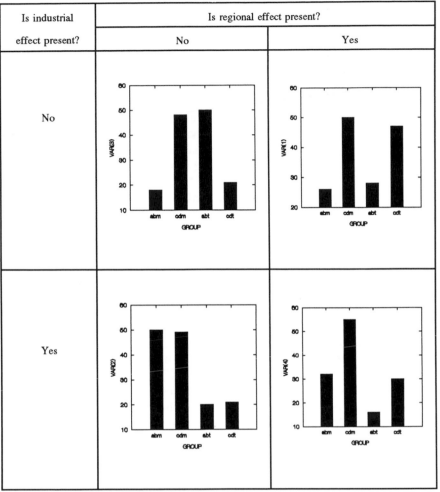

Figure 2. Pure and mixed regional and industrial effects

THE VARIABLE TYPES

The number of variables alone is not the only concern in the analysis of the GMRG data. The types of variables that are present are important as well. The variables associated with the questions asked in the GMRG questionnaire fall into three measurement categories: ratio scaled, ordinal and nominal. Each of these three measurement scales will be described below, since the analysis techniques used depends on the measurement of the variable involved (for a more complete description of measurement scales see [4]).

Many of the questions in the questionnaire have responses that are ratio scaled. Questions like; how many employees work at the company, what are the annual sales, or how long is the production planning horizon all result in a ratio scaled variable. The ratio between responses (e.g., 100 employees work in one company and 1000 in another) has meaning (the second response is 10 times greater than the first). Ratio scaled variables are the ones with which the operations management research community is most familiar. The production output of a factory, the values of a linear programming model and many of the outputs of simulation studies are measured on a ratio scale. Thus, the analysis of these kinds of variables is most common in the operations management literature.

Several of the questions of interest in the questionnaire have responses that are ordinal. Examples are questions specifying a position in the company (from president to supervisor), level of education (from PhD to high school) or whether the use of the computer is "extensive" to "a little". These ordinal variables are directional, but the distance between responses is not meaningful in the same way that it is for ratio scaled variables. One can say that the president of a company is "higher" than the vice president, but to try to say how much higher is not meaningful. Ordinal variables require different analysis approaches than those that are ratio scaled.

The final type of variable that is used to capture responses to the questionnaire is nominal. Questions that specify several alternatives and ask for a choice(s) like; what is the functional area of the person who makes the forecast, how are priorities communicated to the shop, or what are the reasons for changing priorities, all give rise to nominal variables. The values of the nominal variables can be said to be different from one another. They can't be said to be higher, bigger or better, however. For example, the functional area of the forecaster in one firm might be marketing and in another, production, but one can't say that one response is "better" than another, only that it is different.

The distribution of the measurement scales of variables used for each of the five sections of the questionnaire is shown in Table 2 for the comparison of North America with western Europe. The company profile section contains only ratio scaled variables. The other sections contain a mixture of all types.

Table 2. Distribution of measurement scales
in the questionnaire sections

Questionnaire section	Ratio	Ordinal	Nominal	Total
Company profile	16			16
Sales forecasting	4	3	8	15
Production planning	10	8	9	27
Shop floor control	6	30	11	47
Purchasing	6	2	6	14
Total	42	43	34	119

The different scales require different methods for determining statistical differences between the responses. Since we have indicated that we will only describe the differences that result from a pure effect, we need to determine when both regional comparisons are significant. In the next sections, we will discuss a method for making these comparisons for each type of measurement scale.

TESTING RATIO SCALED VARIABLES

A common method of testing for statistically significant differences between the mean responses of two groups of ratio scaled data is to use the Student's t-test. For the t-test to be valid when comparing two samples, the response populations must be at least approximately normally distributed, and have approximately equal variances. A preliminary analysis of the ratio scaled variables in the data revealed that these assumptions are not met (obvious non-normality, and different variance levels) in many cases. Applying the t-test to the data when the assumptions are not met could lead to false conclusions.

To overcome the lack of normality and the dissimilar variances, we turn to nonparametric tests. Because nonparametric tests usually require fewer assumptions, use less information and generally have a narrower acceptance band, they have been termed conservative. It has also been said that nonparametric procedures are approximate solutions to an exact problem, whereas parametric procedures are an exact solution to an approximate problem.

To test for statistically significant differences in the location parameter (the nonparametric equivalent of mean or median) of a ratio scale variable we use the Kruskal-Wallis one way analysis of variance of ranks. This is a nonparametric counterpart of the t-test. It is a test of the differences in the distributions of the responses of two groups based on the ranks of the individual responses (see [4] or [5] for details).

To illustrate the analysis of ratio scaled variables we will use the variable for the amount of the export sales for each company (from the company profile section of the questionnaires for North America and western Europe). The question, mean values and standard deviations for the responses of each of the groups are given in Table 3. The designation 'nam' stands for the North American machine tool industry, 'nat' for the textile industry from the same region, 'wem' for the machine tool industry in western Europe and 'wet' for the western European textile industry. For the data in Table 3, non-normality is confirmed by the Lilliefors test. The standard deviations shown in Table 3 indicate that the variances are not equal.

The first step in applying the Kruskal-Wallis text is to rank all the responses from the two groups from highest to lowest. After ranking all responses, the rank for each response is assigned to the appropriate group, providing a distribution of the ranks for each group. If the two groups were completely different, all the top ranks would be with one group and all the bottom ranks would be with the other. The Kruskal-Wallis test evaluates the distributions of these ranks to determine significant differences. Applying the Kruskal-Wallis test to the question on export sales of Table 3 gives the results shown in Table 4. There is a significant effect for both industries.

Table 3. Example question, mean responses and standard
deviations for a ratio scaled variable

Question:
What were the export sales in the last year?

_____ (US$1,000,000)

Mean values and standard deviations of the responses by group:

	nam	nat	wem	wet
Mean export sales (US$1,000,000)	2.81	2.12	24.10	9.55
Standard deviation (US$1,000,000)	3.85	3.43	48.20	20.06

Table 4. Kruskal-Wallis test results for export sales

Test comparison	nam-wem	nat-wet
Significance level	0.013	0.021

DEPICTING RATIO SCALED DATA

Discovering that responses from different groups are significantly different is only the first part of the question. The second is determining whether they are different in the same direction, the other necessary condition to have a pure regional effect. This is fairly easily determined for ratio variables, since the mean or median values can be consulted directly.

Another way of determining the direction of differences is to use a box plot to visualize them. The box plot is a way of depicting the data that provides a great deal of information and opens up another nonparametric test for evaluating the differences between variables. It is based on the median and dispersion of the values of the observations for the group. We use the hypothetical example in Figure 3 to describe the box plot. Table 5 shows the data, the values of VAR(1), used for preparing Figure 3. It shows frequencies of the values of VAR(1) ranging from 1 to 9 in groups A and B.

In order to draw a box plot first we have to determine the median, and the upper and lower quartiles. By definition the median is the value which splits an ordered (ranked by values) list in half. Half the observations will have values lower than or equal to the median and half the observations will have values greater than the median. The quartiles cut an ordered list into quarters, with the first or lower quartile dividing the lower one-fourth of the data from the upper three-fourths. The second quartile is the median: it divides the lower half of the data from the upper half. The third or upper quartile separates the lower three-fourths of the data from the upper one-fourth.

Table 5. Data for Figure 3 showing number of
responses per value of VAR(1) by group

Group	1	2	3	4	5	6	7	8	9	Total
A	2	4	3	2	1	0	1	1	1	15
B	1	1	0	0	1	4	4	1	3	15

Figure 3 depicts the quartiles and other information on the distribution of the values of each group. The middle horizontal line in the "box" for each group shows the median of the data for that group (for group A the median is 3, for group B it is 7). The upper and lower horizontal lines of each box show the upper and lower quartiles. It follows from the definition of the quartiles that fifty percent of the observations for each group lie between the values of the upper and lower quartiles (for group A the upper and lower quartile values are 4.5 and 2, while for group B they are 7.5 and 6).

The distance between the upper and lower quartile is defined as the interquartile range, a value which is used to provide additional information about the observations. The interquartile range is used to define two pairs of points called fences (upper and lower, inner and outer fences). The lower inner fence is one and one-half interquartile ranges below the lower quartile, while the upper inner fence is one and one-half interquartile ranges above the upper quartile. The outer fences are three interquartile ranges from the lower and upper quartiles.

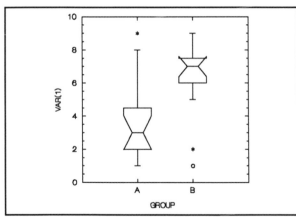

Figure 3. Box plot for two hypothetical groups

The inner and outer fences are not shown on the box plot but are used to highlight some of the actual data. From the data in Table 5, the lower inner fence for group A is: 2-(1.5)(4-.5-2) = -1.75. For group B it is: 6-(1.5)(7.5-6) = 3.75. The upper inner fences for groups A and B are 8.25 and 9.25. The outer fences for group A are -5.5 and 12, and for group B they are 1.5 and 12.

Actual observations that are inside and outside these fences are shown on the box plot. For example, vertical lines are drawn from the upper and lower quartiles to the observation closest to the inside of the inner fences. The ends of the vertical line at 8 and 1 indicate those observations for group A. The observations between the inner and outer fences are denoted by asterisks (for example, the observation with the value of 9 in group A). Those observations which are out of the outer fences are denoted by circles (for example, the observation with the value of 1 in group B). Using these indicated points, one can see that the values for group A are skewed to the higher values of VAR(1), while group B values exhibit the opposite skew.

In addition to the basic information on the distribution of the values, the box plot can be used for testing the differences between the medians. The boxes are notched at the median and return to full width at the upper and lower 95% confidence interval values of the median. The confidence limits can extend beyond the quartile limits (for example, in group B at 7.5) depending on the dispersion of the values. It is these confidence limits that are used to perform the nonparametric test of differences in the medians of the groups.

Wilkinson [6,7] describes how McGill, Tukey and Larsen use the confidence intervals around the median to determine if there is a significant difference between them. They showed that when the confidence intervals around two medians do not overlap, you can be confident at about the 95% level that the two population medians are different. Although the median test is not as powerful as the Kruskal-Wallis test it can be used with the box plot to determine significant differences. In Figure 3, for example, it is clear that the medians differ, since the confidence intervals are not overlapping.

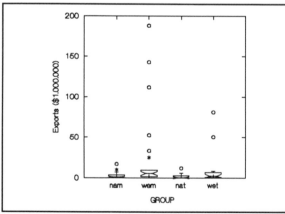

Figure 4. Export sales

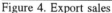

The actual data for export sales is shown in Figures 4 and 5 for North America and western Europe. In Figure 4, the scale is large enough to incorporate all of the values in the plot. The skew of the data, especially for western Europe, is very clear. The scale is too large to permit any significance testing, however.

Note that many of the points visible in Figure 4 are not shown in Figure 5. Figure 5 expands the scale so that the boxes and confidence intervals are clear. The median test shows that we can be confident at about 95% level that the medians of the North American and western European machine tool companies are different. The confidence intervals for the textile companies are slightly overlapping, however, which indicates that

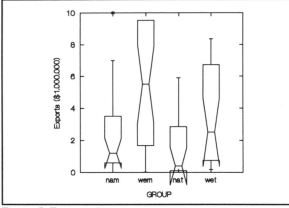

Figure 5. Export sales

the confidence level for a difference in the medians of these groups is less than 95%.

TESTING ORDINAL VARIABLES

The largest number of variables from the questionnaire are ordinal scaled. Since the responses for those questions are ordered (the differences between responses have a "direction"), a meaningful cumulative percentage can be developed. A test for determining significant differences between these cumulative distributions is the Kolmogorov-Smirnov test. The test is based on the maximum difference between the cumulative distributions of the two groups of interest.

The tests of ordinal variables will be illustrated using the example question and data in Table 6. The cumulative percentage for the responses, going from the lowest percentage of engineering changes to the highest, is shown in Table 7, along with the significance levels. The significance levels show that a pure regional effect does not exist; however, a pure industrial effect might.

Table 6. Example question and responses with ordinal values

Question:

On what portion of the orders do engineering or design changes occur after starting production on an order (check one)?

Less than 10% ___ 10%-30% ___ 30%-60% ___ 60%+ ___

Responses (percent of responses in each group)

Group	<10%	10-30%	30-60%	60%+
nam	31.11	22.22	22.22	24.44
nat	81.63	12.24	6.12	0.00
wem	52.94	17.65	23.53	5.88
wet	91.30	8.70	0.00	0.00

To determine if the directions are the same, one could plot the cumulative distributions, but it is not necessary. The differences in the directions of the data can be seen directly. If visualization of this were important, cumulative distribution graphs or stacked bar charts would make it clear.

Table 7. Cumulative distributions, differences and significance for Table 6 example
(Cumulative percent of responses in each group)

Group	<10%	10-30%	30-60%	60%+
nam	31.11	53.33	75.55	100.00
nat	81.63	93.87	100.00	100.00
wem	52.94	70.59	94.12	100.00
wet	91.30	100.00	100.00	100.00

Comparison	Maximum absolute difference	Significance level
nam-wem	21.83	0.282
nat-wet	9.67	0.996
nam-nat	50.52	0.001
wem-wet	38.36	0.031

It can be seen directly, from Table 6, that there are more responses at the lower end of the distribution for the textile industry than for the machine tool industry in both western Europe and North America. Therefore, there are fewer changes in the textile than the machine tool industry, a reasonable outcome. This is a pure industrial effect that can be seen clearly in a stacked bar chart showing the cumulative distributions. Such a chart is provided in Figure 6.

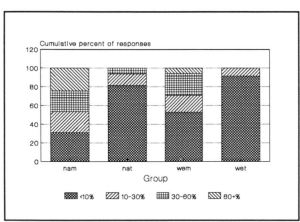

Figure 6. Frequency of engineering change occurrences.

TESTING NOMINAL VARIABLES

The question of interest in analyzing nominal variables is usually whether there is a significant difference between the response patterns of the groups. A natural way to think of performing this analysis is with the Kolmogorov-Smirnov test. The difficulty is with the order of the categories of response. The maximum difference in the cumulative distributions can change with a change in the order of the categories. Since there is no fixed order for the categories, a conservative approach would be to test all the order permutations and see if the minimum of the maximum differences was significant. There are substantial computational difficulties with carrying out this idea, however.

To illustrate the analysis of the nominal variables, we will use the educational field of the person responsible for production planning. There can be no order to the response, since it can't be said that one educational field is better, higher, bigger, etc., than any other. The question is shown in Table 8.

Table 8. Example question with nominal responses

Question:

What is the educational field of the person that is responsible for production planning (check one)?

Business ( ), Arts ( ), Sciences ( ), Engineer ( ), Other ( )

Table 9 shows the percentage responses for each group to the two categories most frequently chosen (Business and Engineer) and the total for Arts, Sciences and Other.

Table 9. Percentage distribution of the production planners by educational fields

Group	Business	Engineer	Other
nam	44.12	35.29	20.59
wem	28.57	57.14	14.29
nat	52.94	17.65	29.41
wet	50.00	35.71	14.29

The problem of the changing maximum differences is illustrated in Table 10, for the example of Table 9. By changing the order of the categories, the maximum difference between the resulting cumulative distributions changes from 15.55 to 21.85.

Table 10. The effect of changing the order
of categories for a nominal variable
(Cumulative percent of responses in each group number 1)

Group	Business	Engineer	Other
nam	44.12	79.41	100.00
wem	28.57	85.71	100.00
lnam-weml	15.55	6.30	0.00

Maximum difference=15.55

(continued)

Table 10, continued

(Cumulative percent of responses in each group number 2)

Group	Engineer	Business	Other
nam	35.29	79.41	100.00
wem	57.14	85.71	100.00
Inam-weml	21.85	6.30	0.00

Maximum difference=21.85

The data in Table 11 are the responses to the question in Table 8. These data will be used to illustrate the two approaches to analyzing nominal data, the chi-square and binomial tests. Sample size is a concern with the chi-square test, as it is with the tests for the other types of variables. For example, the rule of thumb is that no more than 20% of the cells should have fewer than five expected responses for the chi-square test to be valid.

Table 11. Number of responses for educational
field of the production planner

Group	Business	Engineer	Other	Total
nam	15	12	7	34
wem	6	12	3	21
nat	18	6	10	34
wet	7	5	2	14

One way to attempt to overcome the sample size difficulty is to combine several categories into one to create cells of sufficient sample size--as we did by combining the categories of Arts, Sciences and Other into the new "Other" category. Although some data is lost in doing this, the differences in response patterns can still be tested.

Unfortunately, even with this combination, the problem is not completely overcome. While there is no problem in the machine tool comparison, in two cases the expected number of responses is less than five in the textile industry. The western European textile industry has less than five expected responses in the "Engineer" and "Other" categories. This is calculated by recognizing that the proportion of production planners in the textile industry that are "Engineers" is 11/48, while the proportion of western European textile firms is 14/48. The expected frequency of both is the product of the proportions times the number of observations or about 3.2 (($11/48)\times(14/48)\times48$). Similarly, the expected frequency in the "Other" west European textile category is about 3.5. In testing the differences in the textile industry it means that two out of six cells have expected responses less than five so the use of the chi-square test is questionable. The chi-square test can be used for the machine tool comparisons, however, and the results are shown in Table 12.

Table 12. Significance levels for tests of differences
in educational levels of production planners

Comparison	Chi-square test
nam-wem	0.283
nat-wet	0.311*

* May not be valid due to small expected frequency
in two cells of six.

We may be able to overcome the problem of sparse cells by using the binomial test. We illustrate the procedure using the textile industry. The data collected from 34 companies in the North American textile industry revealed that the proportion of the production planners having a "Business" educational background was 52.94% (18 of 34). In the western European textile industry this proportion is 50.00% (7 of 14).

We can ask then: how likely are we to see 7 production planners with a "Business" background if a sample of 14 was taken randomly from a population whose overall proportion is 52.94%? In other words, what is the probability that the proportions between the two regions (nat and wet) are the same? If the probability is low then we can conclude that the percentage of production planners with "Business" background is different in the western European textile industry than in the North American one. A high probability, on the other hand, would indicate that the differences in percentages could easily be due to chance.

The probability of having 7 occurrences out of 14 from a population with a percentage of 52.94% is 20.40%, which is quite high, therefore we can conclude that there is no significant difference between the North American and western European textile industry in the percentage of production planners having a business education. For this data there is not a pure regional effect.

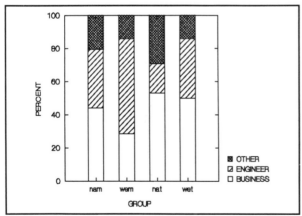

Figure 7. Education of production planner

Stacked bar charts can be used to illustrate the differences (or in this case the similarities) in the nominal and ordinal variables. Figure 7 shows the data from Table 9 as a stacked bar chart.

In this example of the analysis of nominal data, we combined categories to increase sample size. That means only some parts of the data were explicitly evaluated. When this is done it means that care must be taken in framing the question to be answered in order to guide the choice of what is included in the analysis. It also means that care must be taken in the interpretation of the results.

SUMMARY

In this paper we have outlined an approach to making bilateral comparisons using the Global Manufacturing Research Group data base. We illustrated the approach using regional comparisons of western Europe and North America (for a complete analysis of these regions, see Vastag and Whybark [8]). The analytical techniques varied with the type of data to be analyzed. In all instances we have tried to illustrate a conservative approach in the manner in which significant differences are declared. The concept of a pure

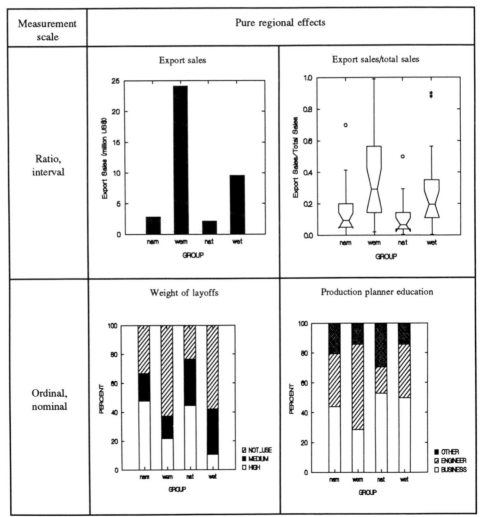

Figure 8. Pure regional effects on different measurement scales

bilateral effect is part of the approach. Figure 8 illustrates some pure regional effects for different measurement scales using different graphical representations of some of the North American and western European data.

The choice of the criterion of a pure effect is one aspect of conservativeness, the use of nonparametric tests another. The use of very stringent levels of significance would do the same. This conservativeness may mask some important differences, but these can be left for more detailed questions. When the intention is to paint a broad brush picture of differences in bilateral comparisons, it is only the clear, unconditional differences that one wants to find and describe. These techniques should help do that.

REFERENCES

1.[*] D.C. Whybark and B.H. Rho, "A Worldwide Survey of Manufacturing Practices," Discussion Paper No. 2, Bloomington, IN, Indiana Center for Global Business, Indiana Business School, May 1988.

2.[*] B.H. Rho and D.C. Whybark, "Comparing Manufacturing Planning and Control Practices in Europe and in Korea," *International Journal of Production Research*, 28, No. 12 (Fall 1990), 2393-2404.

3.[*] B.H. Rho and D.C. Whybark, "Comparing Manufacturing Practices in the People's Republic of China and South Korea," Discussion Paper No. 4, Bloomington, IN, Indiana Center for Global Business, Indiana Business School, May 1988.

4. W.W. Daniel, *Applied Nonparametric Statistics*, 2nd edition, Boston, PWS-Kent Publishing Company, 1990.

5. S. Siegel, *Nonparametric Statistics for the Behavioral Sciences*, New York, McGraw-Hill, 1956.

6. L. Wilkinson, *SYSTAT: The System for Statistics*, Evanston, IL, SYSTAT, Inc., 1990.

7. L. Wilkinson, *SYGRAPH: The System for Graphics*, Evanston, IL, SYSTAT, Inc., 1990.

8.[*] G. Vastag and D.C. Whybark, "Comparing Manufacturing Practices in North America and Western Europe: Are There Any Surprises?" in R.H. Hollier, R.J. Doaden and S.J. New (eds.), *International Operations: Crossing Borders in Manufacturing and Service*, Amsterdam, Elsevier, 1992.

[*] This article is reproduced in this volume.

ACKNOWLEDGEMENTS

The authors would like to acknowledge the Center for Manufacturing Excellence and the Kenan Institute, both at the University of North Carolina, Chapel Hill, for supporting this research. Also, Gyula Vastag was Citibank International Fellow at the Kenan Institute during the preparation of this paper.

Technical Considerations in Analyzing
the Manufacturing Practices Data

Gyula Vastag, Budapest University of Economic Sciences, Hungary

ABSTRACT

On one hand this paper suggests a standardized approach to the analysis of the manufacturing practices data and on the other it offers some new concepts and techniques borrowed from multivariate statistics and the theory of satisficing decisions. A part of the paper includes a description of Chernoff faces (a novel way of graphical clustering) and the aspiration level based approach to decision-making. The last part of the paper has a discussion of the compatibility index, a method for comparing regions.

INTRODUCTION

The worldwide survey of manufacturing practices has been going on for a few years now [1]. Several papers [2,3,4,5,6,7] have been published on the comparisons of different regions but there is still a lot to do, especially in getting at all the pieces of information hidden in the hundreds of questionnaires.

This paper has two objectives: on one hand it tries to give guidelines to the analysis of the manufacturing practices data and on the other hand it introduces some concepts and techniques which may be new and useful in this field of application.

Seemingly these two objectives are conflicting: following the first one to a great extent we would be likely to increase the level of standardization and sophistication of the approaches in order to be able to get more information out of the data than with the previously used approaches. The second implies building upon the standardization we introduced, but new concepts and techniques increase the variety of methods used.

THE FIRST STEPS OF THE ANALYSIS: DATA TRANSFORMATION AND NEW VARIABLES

The field data which are available to the researchers were collected and coded in dBASE III in both numeric form (e.g., numbers of employees or percentage of inventory in finished goods) and character form (e.g. what department develops the production planning data or does the company know about JIT). Obviously, data gathering is the most crucial step taking into account the different languages, different economies, different cultures (where some expressions common in one culture do not even exist in the other) and the coding errors which may come up. Therefore the first step of the analysis should be some kind of a validity check based on common sense or some sophisticated techniques can be used to filter out the outliers or the coding errors.

Supposing that there is no problem with our data, we have to decide how to begin our analysis. One of the options is to go on with our dBASE III files and carry out some parametric and/or non-parametric tests comparing two countries or two industries variable by variable. Or what is even simpler: we do not go beyond the graphical representation and descriptive statistics of the variables. However, this simple approach has some obvious limitations: (1) it uses information only from one variable; (2) it is difficult to build up the whole picture from small pieces--i.e., one can get lost among details.

The other option is to use a statistical software package and to carry out all investigations in a comprehensive framework. From the top three (SYSTAT, SPSSPC+ and STATGRAPHICS) I chose SYSTAT and SYGRAPH [8,9]. Technically, it is easy to "export" dBASE III files to the chosen package, but in doing so we have to clarify the difference between the numeric and character fields in the dBASE III files.

In this case the difference in form indicates a difference in the nature of the data. The character data consists almost exclusively of nominal scaled variables while the numeric data represents at least interval scaled variables. Different methods of analysis are required for the different scales but it is not necessary at all to use different data forms for the different scales. So the first step of the data manipulation is recoding. Here recoding means three things: (i) change the character form into numeric (it can easily be done in dBASE III); (ii) recode the old character variables into shorter numeric ones (it is easy to do in SYSTAT); and (iii) cut the long character strings into smaller "sub" variables.

The change in data form does not alter the nature of the data: the "new" numeric variables represent the same nominal scale properties. When doing (ii) the general idea was that a long character string with a 1 in the ith position should be replaced by number "i," while a string with two ones (in the ith and jth positions) should be recoded as number: "ij." The CHGPRIORTY and CAPACITY variables were recreated in the form of ten and eleven new variables using this method.

Moreover, we introduced new variables. On one hand these were grouping variables, such as group (country + industry together), region (country), industry (machine tool or textile), and the type of the economy (planned or market). On the other hand we introduced some derived variables, too. Table 1 shows these variables.

Table 1

New Variable	Method of Derivation
SALES (total sales)	DOMEST+XPORTS
RPW (ratio of physical workers)	FCTRY/EMPLS
IEL (investment in equipment per employee)	EQPMT/EMPLS
PLPROD (physical labor productivity)	SALES/FCTRY
LPROD (labor productivity)	SALES/EMPLS
XPR (export ratio)	XPORTS/SALES
REQ (return on equipment)	SALES/EQPMT
TURNS (inventory turns per year)	SALES/TOTAL

Using these derived variables we wanted to eliminate or at least to decrease the bias caused by different monetary systems and price structures in different regions.

MULTIVARIATE TECHNIQUES AVAILABLE FOR THE ANALYSIS

The multivariate techniques can be categorized into two groups [10]. If interest focuses on the association between two sets of variables, where one set is the realization of a dependent or criterion measure, then the appropriate class of techniques is designated as dependence methods. If interest centers on the mutual association across all variables with no distinction made among variable types, these are the interdependence methods. The dependence methods seek to explain or predict one or more criterion measures based upon the set of predictor variables. Interdependence methods, on the other hand, are less predictive in nature and attempt to provide insight into the underlying structure of the data by decreasing the complexity, primarily through data reduction.

From the available techniques I give an overview only of those which we used either in a multivariate study [7] or later and proved useful in this special field of application, too.

In order to remove the limitations of the classical descriptive methods and to get an insight into the nature of the problem it is logical to begin our analysis with some interdependence methods.

In [7] a principal components analysis was done first. Principal components analysis, which belongs to the factor analysis family, is concerned with the identification of structure within a set of variables. It establishes relationships within the data and can serve as a data reduction technique as well. We used this technique to see if it was possible to reduce the number of (at least) interval scaled variables, while keeping as much of the original information as possible. If reduction is successful, the reduced set of variables can account for most of the variance in the data and provide the advantages of reduced dimensionality. Moreover, the dimensions (principal components) calculated in the analysis (which are linear combinations of the variables) may help to classify the objects (companies). In this research the classifications would help in establishing the strength of the regional, industrial or other relationships.

The principal components analysis results in a certain number of principal components (factors) and in the coefficients for computing the factor scores for each observation (company). It still remains to be seen, however, which, if any, of these factors differ with respect to regions and/or industries. Individual comparisons of these factors could be performed, but if the number of combinations is relatively large one has to decide on the factors to compare. In order to (possibly) reduce the number of factors to be analyzed, a one-way-analysis of variance can be carried out on the factor scores using the regions and the industries as grouping variables.

Cluster analysis can be considered as another technique for data reduction. The objective of this technique is to identify a smaller number of groups, such that elements residing in a particular group are, in some sense, more similar to each other than to elements belonging to other groups. The construction of the homogeneous subgroups is generally based on some (dis)similarity measure between the objects.

There is a similarity between cluster and factor analysis: by transposing the data matrix, clustering can be viewed as a factoring technique and vice versa.

The primary purpose of multidimensional scaling (it is another data reduction technique) is to represent the proximities (similarities/dissimilarities) between objects (e.g. companies) viewed in space just as in a map. It maps the objects in a multidimensional space (whose dimensions are fewer than the original data space) in such a way that their relative position in the space reflect the degree of proximity between the objects.

The analysis of the character variables from the data base (these were transformed into numeric nominal/ordinal variables) can be done to some extent in the same fashion as in the case of numeric interval scaled variables. The set of potential analytical techniques for these variables is very limited, however, by the nominal/ordinal scale properties.

From the dependence methods discriminant analysis involves deriving linear combinations of the independent variables that will discriminate between the a priori defined groups in such a way that the misclassification error rates are minimized.

NEW APPROACHES: CHERNOFF FACES AND ASPIRATION LEVEL BASED OBJECTIVE FUNCTION

Chernoff [11] introduced a novel way of representing multivariate data. Since its introduction in 1973, this method has become a useful (but sometimes contentious) tool for representing multivariate data as faces in exploratory data analysis. The SYGRAPH program allows the use of up to 20 variables to represent a single face.

The key step in the creation of a face is the assignment of variables to facial features. This assignment may be done by some procedure or randomly. Much of the criticism of Chernoff faces has arisen over the subjectivity of the procedures or over interpretation when the assignment is random. As Chernoff wrote: "This approach is an amusing reversal of a common one in artificial intelligence. Instead of using machines to discriminate between human faces by reducing them to numbers, we discriminate between numbers by using the machine to do the brute labor of drawing faces and leaving the intelligence to the humans, who are still more flexible and clever."

This method may serve as a special technique of graphical clustering. After having drawn the faces representing companies/regions/industries one can ask a group of people to categorize the faces according to their similarity. The result of the classification can be compared with that of an "objective" clustering technique (e.g. k-means clustering, where the number of groups is given in advance). Moreover, Chernoff faces may serve as mnemonic device for remembering major conclusions and in this respect they are useful in education, too.

The idea of aspiration level based decision support was raised in [12]. I think that with some modifications this idea can be used for grouping the companies.

Suppose that we have chosen a set of at least interval scaled variables for which we can give "aspiration" and "reservation" levels. The aspiration level of a variable is such a value that it can be considered as an internationally accepted good level. Above this level the companies can be considered as good. For example, if we are speaking about labor productivity we can determine a value that means the beginning of an interval above which companies of high labor productivity are found. The reservation level is above the lower bound of the variable. Below this level are companies of unacceptably low labor productivity.

It is true that it may be difficult and/or subjective to determine these values. One of the possibilities to ease this process is to combine this approach with a sensitivity analysis. The lower bounds and upper bounds are given by the sample and the aspiration and reservation levels can be determined as percentiles.

The essence of the approach is to order a scalar of all companies using their values of the selected variables. The idea is that a company whose values of all variables are at the upper bound (here upper means better) should score greater than 1. A company whose values are at the aspiration levels in all variables should get 1 and a company with data at the reservation levels should get zero. Finally the company with values less than the reservation level down to the lower bound gets less than 1. There are several functions which can achieve this, but at the beginning we can use the simple piecewise linear function introduced in [12].

After having obtained the objective function values of the companies we group them based on these values using some non-parametric tests (e.g. Wilcoxon). Moreover, we can analyze the effect of changing the aspiration and reservation levels on the stability of the classification.

NEW APPROACHES: THE COMPATIBILITY INDEX

In this section we discuss an approach to capturing the differences between regions called the compatibility index. A compatibility index was applied to all regions in the data base by Vastag and Whybark [13]. After describing some of the technical issues with the index, the full data base will be used to illustrate the points and to provide a comparison with the previous work.

Figure 1 describes the company-variable facet of the basic data relation matrix (BDRM) [14]. The observations (i.e., the companies) are represented by the rows, the variables by the columns of the matrix. Depending on the focus of our analysis, we can carry out R (variable) or Q (company) type of analysis. If, for example, we calculated the correlation coefficient between variables v and w over the companies, it would be an R type of analysis. On the other hand, if we calculated the correlation between companies i and j over the variables, it would be a Q type analysis.

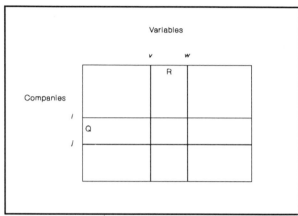

Figure 1. The company-variable facet of the basic data relation matrix

Variables

All regions have to have observations for the same set of variables ($v=1,2,...,V$) but there may be missing values. For company i from region r, the number of variables with non-missing values is l_{ri} (l_{ri} is less than or equal to V).

Companies

The assumption is that the companies are identified within a region, so each one has an individual identifier (i) and a regional identifier (r) ($i=1,2,..,n_r$).

Regions

We compare the regions from the point of view of region k, where k can be any region in the data base. In region r ($r=1,2,..,R$) there are n_r companies. Because of missing values, the actual number of observations for variable v in region r, n_{rv}, can be less than or equal to n_r.

Compatibility indicator

The compatibility indicator, $c_{rvi}^{(k)}$, is 1 if company i ($i=1,2,...,n_r$) from region r ($r=1,2,...,R$) has the modal value or modal range of region k ($k=1,2,...,V$) for variable v ($v=1,2,...,V$), and 0 otherwise.

THE COMPATIBILITY INDEX

Depending on the focus of our investigations, we can define two types of compatibility indexes, which are directly related to the above described R and Q types of analyses. The variable oriented index (R analysis) focuses on the variables, while the company oriented index (Q analysis) focuses on the observations for the companies in the data base.

Variable (R) type analysis

We define the variable compatibility of region r with region k (region k is the basis) as:

$$\bar{C}_{rv.}^{(k)} = \frac{\sum_i c_{rvi}^{(k)}}{n_{rv}}$$

It is straightforward that: $0 \leq \bar{C}_{rv.}^{(k)} \leq 1$.

Company (Q) type analysis

We define the company compatibility index of region r with region k for variables $v=1,2,...,V$ as:

$$\bar{C}_{r.i}^{(k)} = \frac{\sum_v c_{rvi}^{(k)}}{l_{ri}}$$

As before: $(0 \leq \bar{C}_{r.i}^{(k)} \leq 1)$.

The compatibility index

We define a compatibility index for region r from the point of view of region k. Using the variable and the company compatibility indexes we can compute two aggregate indexes: one by aggregating the variable compatibility indexes (derived from R analysis) over the variables and one by aggregating the company compatibility indexes (derived from analysis Q) over the companies from the same region. These indexes are also between 0 and 1.

R analysis:

$$\bar{C}_{r(R)}^{(k)} = \frac{\sum_v \bar{C}_{rv.}^{(k)}}{V}$$

Q analysis:

$$\bar{C}_{r(Q)}^{(k)} = \frac{\sum_i \bar{C}_{r.i}^{(k)}}{n_r}$$

These indexes have two interesting characteristics. First, they are not symmetric: the compatibility of region r with region k (k is the base) can be different than the compatibility of region k with region r (r is the base). Second, region r can be more compatible to region k than region k to itself.

In order to introduce some kind of standardization we divide the aggregate indexes by the base region values. In principle, these adjusted indexes can be greater than 1. To standardize we use:

R analysis:
$$\bar{C}_{r(R)}^{(k)*} = \frac{\bar{C}_{r(R)}^{(k)}}{\bar{C}_{k(R)}^{(k)}}$$

Q analysis:
$$\bar{C}_{r(Q)}^{(k)*} = \frac{\bar{C}_{r(Q)}^{(k)}}{\bar{C}_{k(Q)}^{(k)}}$$

APPLICATION

The index used by Vastag and Whybark was an overall index based on the sum of the compatibility indicators for all companies and variables [13]. There can be differences in the index values between these approaches. This can be seen in Table 3 which shows the two compatibility indexes (R and Q type) and the overall index for all twelve regions.

The R and Q measures of compatibility are highly correlated ($r=0.95$) and they both are highly correlated with the overall measure. Therefore, the differences between the two types of analysis may not seem significant at this level of aggregation. However, if we group the regions together using k-means clustering (for the number of clusters we chose four) and the three measures, we get different results. Table 4 shows these results. The higher the cluster number, the farther the regions are from the manufacturing practices of the United States and Canada, the base region.

Several factors should be taken into account in evaluating the results. First, we are analyzing sample company groups and not the regions themselves. Sampling and survey methodology differences certainly influence the conclusions. Second, the selected set of variables--although it is based on the results of previous studies--is arbitrary: the conclusions drawn could change if other variables were selected. Third, the technical details of the analysis (modal value determination, scoring and derivation of the compatibility indexes) certainly have an influence on our results.

However, despite of all these difficulties and problems, we feel that the compatibility analysis may become a useful new tool in the hands of the researchers. The advantage of the index as a measure of compatibility is the ease with which it is computed. This

Table 3. Compatibility indexes for the regions

Region	Overall	Q based $\overline{C}_{r(Q)}^{(k)*}$	R based $\overline{C}_{r(R)}^{(k)*}$
Australia	0.939	0.931	0.977
Bulgaria	0.892	0.907	0.884
Chile	0.740	0.759	0.791
China	0.690	0.735	0.746
Finland	0.856	0.919	0.935
Hungary	0.812	0.792	0.821
Japan	0.774	0.809	0.803
Mexico	0.692	0.730	0.763
South Korea	0.819	0.801	0.875
USA and Canada	1.000	1.000	1.000
USSR	0.798	0.828	0.877
Western Europe	0.827	0.835	0.865

Table 4. Clustering the regions based on the three indexes.

Cluster	Overall	Q analysis	R analysis
1	USA and Canada	USA and Canada	USA and Canada Australia Finland
2	Australia Bulgaria	Australia Bulgaria Finland	Bulgaria Western Europe South Korea USSR
3	Western Europe Finland Hungary Japan South Korea USSR	Western Europe Hungary Japan South Korea USSR	Chile Hungary Japan
4	Mexico Chile China	Mexico Chile China	Mexico China

means quick assessment of the relative similarity of manufacturing practices between countries, regions, or industries. This has value to researchers and practitioners alike and makes the extension to evaluating other relationships with compatibility much easier.

REFERENCES

1.* D.C. Whybark and B.H. Rho, "A Worldwide Survey of Manufacturing Practices," Discussion Paper No. 2, Bloomington, IN, Indiana Center for Global Business, Indiana Business School, May 1988.

2.* A. Chikán and D.C. Whybark, "Cross-National Comparison of Production-Inventory Management Practices," *Engineering Costs and Production Economics*, 19, 1-3 (1990), and in A. Chikán (ed.), *Inventories: Theories and Applications*, Amsterdam, Elsevier, 1990.

3. K.K. Kwong and C. Li, "Sales forecasting in China Europe, Japan, Korea, and the U.S.," Proceedings of the Decision Sciences Institute annual meeting, New Orleans, LA, November 1989.

4. K.K. Kwong, W. Fok and C. Hiranyavasit, "Comparing Manufacturing Practices in the People's Republic of China, South Korea and Japan," Proceedings of the Decision Sciences Institute annual meeting, New Orleans, LA, November 1989.

5.* B.H. Rho and D.C. Whybark, "Comparing Manufacturing Practices in the People's Republic of China and South Korea," Discussion Paper No. 4, Bloomington, IN, Indiana Center for Global Business, Indiana Business School, May 1988.

6.* B.H. Rho and D.C. Whybark, "Comparing Manufacturing Planning and Control Practices in Europe and Korea," *International Journal of Production Research*, 28, No. 12 (Fall 1990), 2393-2404.

7.* G. Vastag and D.C. Whybark, "Manufacturing Practices: Differences that Matter," Pre-prints of the Sixth International Working Seminar on Production Economics, February 19-23, 1990, Igls/Innsbruck Austria, pp. 1095-1108.

8. L. Wilkinson, *SYSTAT: The System for Statistics*, Evanston, IL, SYSTAT, Inc. 1988.

9. L. Wilkinson, *SYGRAPH: The System for Graphics*, Evanston, IL, SYSTAT, Inc., 1988.

10. W.R. Dillon and M. Goldstein, *Multivariate Analysis (Methods and Applications)*, New York, John Wiley and Sons, 1984.

11. H. Chernoff, "The Use of Faces to Represent Points in k-Dimensional Space Graphically," *Journal of the American Statistical Association*, 1973, 68 (342), pp. 361-368.

12. A. Lewandowski and A. Wierzbicki, "Theory Software and Testing Examples for Decision Support Systems," International Institute for Applied Systems Sciences WP-87-26.

13.* G. Vastag and D.C. Whybark, "Compatibility of Manufacturing Practices," Global Manufacturing Research Center Working paper, March, 1993.

14. R.B. Cattell. *Handbook of Multivariate Experimental Psychology*, Chicago, Rand McNally and Company, 1966.

* This article is reproduced in this volume.

An Empirical Evaluation of the Worldwide Survey of Manufacturing Practices: Implications and Limitations

John G. Wacker, Iowa State University, USA
Ling Feng Li, Hangzhou Institute of Electronic Engineering, China

INTRODUCTION

A wide variety of results from the World Wide Manufacturing Practices Survey (hereafter referred to as the Survey), conducted by the Global Manufacturing Research Group (GMRG), is published [1-13]. How generalizable the results from the Survey are is important, since any conclusions drawn are limited by the data's representativeness of the population. From the theoretical perspective, the key question is: What do the Survey data represent? If the data are representative of the country, then researchers are able to infer differences in manufacturing practices resulting from international, institutional and cultural differences. There are three ways of measuring the representativeness of a sample: One, by the proportion of the industry sampled; two, by the average size of the firm sampled; and three, by the relative productivity of the firms sampled. However, if the data are not representative of the country as "typical" or "average" firms, then the data represent the firms sampled and firms of comparable size inside each country.

Another important related question is: Are the size of the firms and measures of productivity more similar within industries (between countries) or are they more similar within countries (between industries). The between industries and between countries question will be addressed by measures of relative size of firms within industries (for between country comparisons) and between industries (for within country studies).

In short, the basic empirical questions are:
> Q1: How representative are the data derived from the Global Manufacturing Research Group of the general population of manufacturing for each country?
>> Q1.A: Are these data representative in terms of proportion of the entire industry size?
>>> Q1.A.i: Size as measured by percent of firms?
>>> Q1.A.ii: Size as measured by percent of sales?
>>> Q1.A.iii: Size as measured by percent of workers or percent of employees?
>> Q1.B: How representative are the data in terms of relative size of the individual firm compared to Yearbook data?
>> Q1.C: Are these data representative in terms of employee and worker productivity for the GMRG data?
> Q2: Are firms more similar within industries regardless of country? Or are firms more similar within countries between industries?
>> Q2.A: by average size of firm?
>> Q2.B: by average size of work force?
>> Q2.C: by average productivity?

In summary, this study investigates and compares basic statistics of firm size and productivity from the Survey with United Nations Yearbook data. Its conclusions are that the firms surveyed are generally larger and more productive than those of the general population. Another conclusion is that firms are more similar within countries between industries than between countries within industries.

METHOD

The basic method for analysis is to utilize data from the United Nations Yearbook and compare it with the same year's data gathered by the Global Manufacturing Research Group. Key aspects for the comparison are limited by the limited data available from the United Nations Statistical Yearbook. However, several key aspects can be derived from the comparisons. Unfortunately, the standard deviations of the data from the Yearbook are not available. However, the standard deviations computed from the Worldwide Manufacturing Practices Survey could be utilized for statistical comparison. The Survey had data that may be from larger firms so that it could restrict the size of the standard deviation. But because of the limited sample size and firms selected, all the results would be statistically significant. A conclusion will be drawn from the computations later in this paper.

In most countries, a priori, it would be suspected that larger firms would be more apt to cooperate with the survey since they would have the resources necessary to dedicate to the data collection effort. Further, it could be argued that the larger firms have a more rational planning process because they have the resources necessary to dedicate to planning. If this is the case, larger firms' planning procedures could be considered aspiration levels for smaller firms.

To compare the relative size of firms between countries, ratios of average sizes will serve to make comparisons. The average sizes will compare average sales per enterprise, average number of employees per enterprise, and average number of workers of the Survey data to the U.N. Yearbook data.

DATA COMPUTATIONS

The data from the Survey generally is more detailed than the data in the Yearbook. Yearbook monetary data was converted into U.S. dollars using the 1987 exchange rate. Because the Survey data for Europe are from more than one country, there was no easy method of converting the Yearbook data for comparisons of relative labor productivity.

COMPARISONS OF THE RELATIVE FIRM SIZES

From Table 1, most of the countries had only a small proportion of their enterprises surveyed. Eleven of the twelve surveys (two industries each for China, Japan, Korea, USA, Hungary and Europe) had less than three percent surveyed. Only Hungary's machine tool industry had more than five percent (5.37%) surveyed.

Table 1. Percentage of each industry sampled by number of enterprises,
sales, number of employees and number of workers

		China	Japan	Korea	USA	Hungary	Europe
Machine tool	Percent of enterprises	0.09	0.04	2.53	0.09	5.37	n/a[*]
	Percent of sales $ millions	2.83	1.32	19.89	0.43	66.05	0.07
	Percent of employees	0.94	0.95	28.41	0.55	34.71	0.96
	Percent of workers	n/a[*]	1.48	19.89	0.40	32.81	0.69
Textile	Percent of enterprises	0.22	0.09	0.48	0.78	19.32	0.37
	Percent of sales $ millions	5.66	0.91	2.31	1.83	77.40	12.24
	Percent of employees	1.51	0.41	4.03	3.04	47.76	1.74
	Percent of workers	1.06	0.68	3.91	2.22	47.61	1.56

* Not available

Unless the samples were entirely random, the conclusions may not be representative of the firms within each country. However, the representativeness of the sample using percentage of total sales varied from about (0.5%) for the USA to (66%) for Hungary. In most cases, the percentage of sales is higher than the percentage of the enterprises sampled. Additionally, the percentage of employees and percentage of workers vary widely between countries. Yet, generally, the percentage of sales, employees, and workers is larger than the percentage of enterprises surveyed.

The next question is: What is the relative size of firms represented in the Survey compared with the population as measured by the Yearbook? There are three measures utilized to indicate relative size: average sales, average employees, and average workers. In order to compare the surveyed firms with the U.N. Yearbook data, the ratio of these two will indicate relative size. From Table 2, the size of the surveyed firm is generally very large compared with the industry as a whole. For average sales per enterprise, the ratio ranges from 3.9 in the USA textile industry to over 30.8 for the Chinese machine tool industry. Using total employment as a measure of size, the range went from 2.47 for the Hungarian textile industry to 23.74 for the Japanese machine tool industry. Using workers (same ratios of total workers per enterprise), the range went from 2.46 for the Hungarian textile industry to 36.72 for the Japanese machine tool industry. In short, the size of the average firm in the Survey ranged from over twice as large as the average firm in the U.N. Yearbook up to 36 times as large. This leads to the conclusion that the firms in the Survey were generally very large compared with the typical firms in each country.

Table 2. Relative size ratios between the average firm in the survey
and the average size firm in the United Nations Yearbook

	China	Japan	Korea	USA	Hungary	Europe
Machine tool sales	30.18	7.72	13.68	5.86	13.75	n/a
Textile sales	27.78	10.52	8.36	3.09	4.01	n/a
Machine tool employees	10.04	23.74	12.08	6.39	6.47	n/a
Textile employees	6.87	4.71	8.37	3.90	2.47	n/a
Machine tool prod. workers	na*	36.72	10.96	4.73	6.11	n/a
Textile prod. workers	5.08	7.82	13.68	2.90	2.46	n/a

* Not available

There are interesting things to note. For example, in all countries (except Europe which has no comparable data), the machine tool industries were generally larger by all measures (sales, employees, and workers) relative to the textile industries. Thus, we can conclude that the textile industries are closer to the average firm size than the machine tool industries.

RELATIVE LABOR PRODUCTIVITY COMPARISONS

The term relative productivity may be misleading since the data on output are not given in either the U.N. Yearbook data or in the Survey. Ideally, the best measure of output is uniform physical units produced. However, since modern day production facilities are diverse, the next best measure should be value-added [14] (even though there is considerable difficulty with its measurement caused by different levels of profitability, wages, and managerial expertise). However, ideal measures of output are never possible in the modern world of diverse products with different levels of vertical integration. Consequently, for this study, sales productivity and other measures indicated will be used. Surveyed firms were less productive than the general population. These results mean no overall conclusion can be made regarding the relative productivity of firms in the Survey and those in the general population, as measured by sales per employee or worker.

Table 3a gives the between country comparison between country but within industry. The largest ratio (maximum productivity/minimum productivity) between countries for productivity ranged from (15.6771) to (27.8897), while the smallest ratios ranged from (1.3964) to (2.6283). Table 3b shows that within country differences are all on the order of the smallest between industry differences. These results indicate that the within industry variations are wider than between industry variations in productivity. These results lead further to the general conclusion that within an industry there are vast differences in productivity between countries, but within countries there are more manufacturing similarities than between countries. These differences are supported by the U.N. Yearbook data, which gave similar results. In short, the environment for each firm within an international industry is far different from those between countries. The reason for this result is not be entirely obvious, but considering differences in legal, cultural, economic, and sociological climates, exchange rates, and operating environments, these results may be easier to explain.

Table 3a. Comparisons of the relative sales
per employees and workers for the survey

	Between country comparison	Between country comparison excluding Europe
Machine tool workers	24.115 [2.20]*	24.115 [2.20]
Textile workers	40.104 [1.44]	18.18 [1.44]
Machine tool employees	27.89 [2.63]	27.89 [2.63]
Textile employees	35.718 [1.40]	15.68 [1.40]

* Numbers in brackets are the minimum ratios between all countries

Table 3b. The ranking of differences in ratios of machine tool productivity
to textile productivity within country between industry
(Except where noted by *)

	Machine to textile industry ratio for employees	Machine to textile industry ratio for workers
China	3.4695	2.5475*
Japan	1.0463	1.1661
Korea	1.5460	1.9371
USA	2.0863*	2.3453
Hungary	1.3365	1.6208
Europe	8.0808*	5.2059*

* Ratio is reversed for purposes of comparing relative productivity (textile to machine tool ratio)

Consequently, the result is that, *a priori*, differences in manufacturing techniques are more different between countries than between industries. To most production/operations management academics, this conclusion is contrary to traditional theory which propounds large differences between the manufacturing techniques in textile (continuous manufacturing) and machine tool (intermittent manufacturing). It could be hypothesized that these differences, although they may be very significant, are overwhelmed by between country manufacturing environments. Comments on individual countries follow.

THE SURVEY IN CHINA

The detailed original data for the individual countries are provided in the Appendix. The comparisons here are based on that data and Tables 3a and 3b. The portion of the industry sampled is 0.09 and 0.22 percent for the machine tool and textile industries, respectively. However, the firms represented 2.83% and 5.66% of the total manufacturing by sales, and between 6.87% and 10.04% of the total workers in the sample. In short, if the measure of proportion of the industry is best measured by sales or workers, then it could be concluded that a significant proportion of the market was sampled.

Additionally, the data indicate the Survey data are from firms in both industries that are larger in sales, employees, and production workers. The Survey's firms are five to ten times larger by employment and approximately thirty times larger by sales than the industry's average. Additionally, the survey's firms are three to four times as productive using sales per employee as firms in the industry as indicated by the Yearbook.

Conclusion on China's Survey Data

The Survey data represent larger and more productive firms in the People's Republic of China. Yet these firms are the least productive internationally.

THE SURVEY IN JAPAN

The data gathered from the Japanese industries have several notable anomalies. First, the proportion of firms represented in the sample is very small, 0.04% for the machine tool and 0.09% for the textile industry. Also, by using percentage of sales, employees, or workers the proportion sampled is still quite small, approximately one percent or less. Second, machine tool Survey firms are nearly eight (7.72) to thirty seven (36.82) times larger by sales, workers, and employees than the population as indicated by the Yearbook. In the textile industry, the results indicate the size as measured by percentage of sales, employees, or workers is 4.71 to 10.52 larger. In short, Japanese Survey firms are significantly larger.

However, the Survey's machine tool industry firms are 11% less productive using the average sales per worker, but 38% more productive using total employees, than firms in the Yearbook. From these results, it could be hypothesized that the non-factory workers add significantly to the productivity of the firms. In the textile industry, the Survey's firms were more productive than Yearbook firms as indicated by the average sales per worker or average sales per employee.

Conclusion on Japan's Survey Data

Larger firms were represented in the Survey. Yet the data on relative productivity leads to the conclusion that the larger firms in the textile industry are relatively more productive than the smaller firms, while just the reverse is true of the machine tool industry.

Japanese firms are, in general, the most productive firms of the countries in the Survey. They had over $250,000 annual sales per employee in the textile industry and over $109,000 per employee in the textile industry. In short, the evidence presented in this study supports the conventional belief that Japanese firms are more productive than firms in comparable industries in other countries.

THE SURVEY IN SOUTH KOREA

The data gathered indicated several basic tendencies. The proportion of firms surveyed in the machine tool and textile industries is 2.35% and 0.48% respectively. However, a greater percentage of sales, employees, and workers was sampled for each industry (around 20% for the machine tool industry and about 4% for the textile industry).

In the machine tool industry, the firms are nearly eleven (10.96) to over thirteen (13.68) times larger than those in the Yearbook as indicated by sales, workers and employees. In the textile industry, the Survey's firms are about eight times larger.

The Yearbook's machine tool firms are less productive than the Survey's as indicated by the ratios of the average sales per employee (0.70) or worker (0.77). Similarly, the textile firms are less productive, as indicated by the sales per employee (0.57) or sales per worker (0.59) ratios. These conclusions are very different from other countries participating in the Survey.

Conclusion on Korea's Survey

Generally, the machine tool industry is better represented in the Survey than the textile industry with a larger percentage of firms, sales, employees, and workers included. However, textile firms participating in the Survey are less productive than the population.

THE SURVEY IN THE UNITED STATES

The percentage of firms participating is quite small, 0.09% for the machine tool and 0.78% of the textile industry. The proportion of firms surveyed as indicated by sales, employees, and workers is quite small for the machine tool industry, about one half percent. The textile industry has a larger percentage participating than the machine tool industry as indicated by the sales, employees, and workers.

The Survey indicated that the machine tool industry's firms are three to six times larger than the Yearbook's average indicated for both the machine tool and textile industries.

However, the productivity in the United States machine tool industry is what the Yearbook's indicated (within ten percent) as indicated by output per employee or worker. The U.S. textile industry results are similar to the Korean's, namely the firms' productivity from the Survey was less than that of the Yearbook's.

Conclusion on USA's Survey Data

Only a small percentage of USA's machine tool industry is represented in the Survey. The textile industry is better represented in the Survey. The productivity of the workers is lower for textile firms than the industry as a whole and about the industry average for the machine tool industry.

THE SURVEY IN HUNGARY

The data gathered from Hungarian industry have several aspects worthy of mention. The proportion of firms surveyed is quite large, 5.37% for the machine tool industry and 19.32% for the textile industry. If percentage of sales, employees, and workers is used as an indicator, a very significant proportion of the industry was sampled--over thirty percent of the machine tool industry and almost fifty percent of the textile industry. The proportion of firms sampled is larger in the textile industry than in the machine tool industry. The machine tool firms in the Survey have seven to ten times more sales, workers or employees than the those in the Yearbook. Survey firms are approximately twice as productive as the firms in the industry. For the textile industry, the Survey firms are over twice as large as those in the Yearbook, as indicated by the sales, workers, or employees. Productivity in the Survey's textile firms is almost twice as high as those in the Yearbook as indicated by sales per worker or sales per employee.

Conclusion on Hungary's Survey Data

The Survey captures a significant proportion of both Hungary's machine tool industry and textile industry. The Survey data represent larger firms that were more productive than smaller firms.

THE SURVEY IN EUROPE

Some of the most interesting research questions could be addressed with proper data from some of the major manufacturing countries, such as Germany, United Kingdom, France, Denmark, Switzerland, etc. Unfortunately, current data is much too sparse for individual countries to make any international comparisons.

The Survey data from Europe indicated a wide dispersion of data. From a statistical perspective, it seems apparent that the comparisons of the data are not representative of Europe as a whole.

Conclusion on Europe's Survey Data

The European Survey data should only be used with extreme care since the data are from diverse countries.

DIRECTIONS FOR FUTURE SURVEY DATA GATHERING

Since productivity is a key concern for manufacturing, not having adequate performance data available limits the importance of the empirical questions because of differences in manufacturing practices. In short, without having productivity data, there is no way of making inferences about which manufacturing practice is superior to another practice. One step in this direction is to add questions on material costs from vendors which would facilitate the computation of value-added.

A similar argument could be made for quality data. Without having data available on quality, there is no way to compare the differences in manufacturing practices and their relative importance to improve quality.

Data on competitive strategic advantage such as cost, product variety, delivery speed to customer, and comparative quality would prove useful since it would indicate what the firm considered its primary competitive advantage. It would indicate "how" these factors affect manufacturing practice procedures to lead to competitive advantage.

DIRECTIONS FOR FUTURE RESEARCH

There are several very important questions that need to be addressed in future research. Probably the most important is the development of a structural equation model for relating manufacturing practices to productivity and quality of output. Currently, most empirical research is flying blind, using data mining techniques to discern some identifiable pattern to draw conclusions. The key question is not: What are the statistical relationships between productivity and quality with certain manufacturing practices? Rather, the key question is: What are the underlying theoretical (mathematical) relationships between productivity and quality with all manufacturing practices?

Other important empirical questions relate to the specific tying of a manufacturing strategy with the manufacturing practices [15]. In this context, the key questions are related to specific relationships of each level of strategy to the infrastructure decisions of manufacturing practices.

REFERENCES

1.* A. Chikán and D.C. Whybark, "Cross-National Comparison of Production-Inventory Management Practices," *Engineering Costs and Production Economics*, 19, 1-3 (1990), and in A. Chikán (ed.), *Inventories: Theories and Applications*, Amsterdam, Elsevier, 1990.

2.* D.C. Whybark, "An Analysis of Global Data on the Impact of the Market on Manufacturing Practices," Discussion Paper No. 38, Bloomington, IN, Indiana Center for Global Business, Indiana Business School, April 1990.

3. C.J. Guo, D.C. Whybark, M.Z. Dai, and C.Z. Xiao, "Survey on the Situation of Operations Management in the Chinese Textile Industry," presented at the Global Manufacturing Research Group Workshop, Shanghai, China, May 1990.

4. Y.X. Jiang, D.C. Whybark, M.Z. Dai, and C.Z. Xiao, "Operations Management in Chinese Machine Tool Firms," presented at the Global Manufacturing Research Group Workshop, Shanghai, China, May 1990.

5.* A. E. Kovacevic, J.C. Lopez, and D.C. Whybark, "Manufacturing Practices in Chile," Discussion Paper No. 40, Bloomington, IN, Indiana Center for Global Business, Indiana Business School, April 1990.

6. Z.S. Ren and S.Y. Lu, "A Survey of the Shanghai Textile Industry," presented at the Global Manufacturing Research Group Workshop, Shanghai, China, May 1990.

7.* B.H. Rho and D.C. Whybark, "Comparing Manufacturing Practices in Europe and South Korea," Discussion Paper No. 3, Bloomington, IN, Indiana Center for Global Business, Indiana Business School, April 1988.

8.* B.H. Rho and D.C. Whybark, "Manufacturing Practices in Korea," Discussion Paper No. 43, Bloomington, IN, Indiana Center for Global Business, Indiana Business School, May 1990.

9.* B.H. Rho and D.C. Whybark, "Comparing Manufacturing Practices in the People's Republic of China and South Korea," Discussion Paper No. 4, Bloomington, IN, Indiana Center for Global Business, Indiana Business School, April 1988.

10.* G. Vastag and D.C. Whybark, "Manufacturing Practices: Differences that Matter," Discussion Paper No. 23, Bloomington, IN, Indiana Center for Global Business, Indiana Business School, February 1989.

11. D.C. Whybark and C.Z. Xiao, "The Impact of the Economic Reforms on China's Enterprise Managers," Discussion Paper No. 1, Bloomington, IN, Indiana Center for Global Business, Indiana Business School, Feb. 1988.

12. L.F. Wu and C.Z. Xiao, "Comparing Manufacturing Practices Between The People's Republic of China and Countries in Western Europe," presented at the Global Manufacturing Research Group Workshop, Shanghai, China. May 1990.

13. C.Z. Xiao, W. Xu, "Comparative Research on Manufacturing Practices with The Data Envelopment Analysis (DEA) Method," presented at the Global Manufacturing Research Group Workshop, Shanghai, China, May 1990.

14. W. Krelle, "The Aggregate Production Function and the Representative Firm," in: F.G. Adams and B.G. Hickman (eds.), *Global Econometrics*, Cambridge, MA, The MIT Press, 1983.

15. R. Hayes and S.C. Wheelwright, *Competing Through Manufacturing: Restoring Our Competitive Edge*, New York, John Wiley Co., 1984.

* This article is reproduced in this volume.

ACKNOWLEDGMENTS

An earlier version of this paper appeared in the Proceedings of the First International Conference of the Decision Sciences Institute, Brussels, June 1991.

APPENDIX
Tables of original data

Table A1. A comparison of the Global Manufacturing Research Group Survey data
and the United Nations Yearbook data: China and Japan.

	China Machine tool		China Textile		Japan Machine tool		Japan Textile	
	Survey	Yrbk.	Survey	Yrbk.	Survey	Yrbk.	Survey	Yrbk.
Number of enterprises	44	46597 [0.09]*	56	25505 [0.22]	18	44771 [0.04]	36	41665 [0.09]
Employees	94839	10078000 [0.94]	64899	4300000 [1.51]	11884	1245000 [0.95]	2598	639000 [0.41]
Workers	53796	n/a	50129	4745000 [1.06]	7795	528000 [1.48]	1899	281000 [0.68]
Sales $ millions	872	30850 [2.83]	2071	36578 [2.83]	3049	230956 [1.32]	637	70081 [0.91]

* Numbers in the brackets below the U.N. Yearbook data are the Survey data percentage of the Yearbook data.

Table A2. A comparison of the Global Manufacturing Research Group data
and the United Nations Yearbook data: Korea and the United States

	Korea Machine tool		Korea Textile		USA Machine tool		USA Textile	
	Survey	Yrbk.	Survey	Yrbk.	Survey	Yrbk.	Survey	Yrbk.
Number of enterprises	89	3783 [2.35]*	33	6860 [0.48]	45	52135 [0.09]	50	6412 [0.78]
Employees	41652	146600 [28.41]	15959	396300 [4.03]	10812	1961000 [0.55]	25248	831000 [3.04]
Workers	28178	109300 [25.78]	13525	345600 [3.91]	4851	1214000 [0.40]	15684	707000 [2.22]
Sales $ millions	1007	5063 [19.89]	250	10780 [2.31]	931	218100 [0.43]	1284	70100 [1.83]

* Numbers in the brackets below the U.N. Yearbook data are the Survey data percentage of the Yearbook data.

Table A3. A comparison of the Global Manufacturing Research Group Survey data and the United Nations Yearbook data: Hungary and Europe

	Hungary Machine tool		Hungary Textile		Europe Machine tool		Europe Textile	
	Survey	Yrbk.	Survey	Yrbk.	Survey	Yrbk.	Survey	Yrbk.
Number of enterprises	19	354 [5.37]*	17	88 [19.32]	34	n/a	24	6412 [0.37]
Employees	42000	121000 [34.71]	46800	98000 [47.76]	20258	2104000 [0.96]	12738	732000 [1.74]
Workers	28544	87000 [32.81]	38563	81000 [47.61]	9853	1426000 [0.69]	9617	616000 [1.56]
Sales $ millions	1226	1856 [66.05]	1022	1319 [77.40]	1400	2091159 [0.07]	7116	58129 [12.24]

* Numbers in the brackets below the U.N. Yearbook data are the Survey data percentage of the Yearbook data.

SECTION SIX

COMPREHENSIVE COMPARISONS

This section of the book presents two papers that look at the GMRG data comprehensively. The first paper is an application of multi-variate statistics to three of the regions in the data base. As such it considers both common and differentiating factors among the regions. The second paper uses a simple method of determining compatibility between regions to evaluate all twelve of the regions in the data base. Both papers present methods of broadening the basis of comparison to a large number of the regions at the same time.

Gyula Vastag and D. Clay Whybark
Manufacturing Practices: Differences that Matter

In this paper models for factor analysis (based on principal components) and clustering were used to develop natural groupings in the data. The primary focus was on the numeric data, but some analysis was made of the character data that was at least ordinally scaled. In addition to the analytical techniques, display techniques like Chernoff faces were used to portray the data and illustrate the findings. The paper opens the way to applying a very rich set of techniques to the data to improve overall understanding of the differences in manufacturing practices around the globe.

Gyula Vastag and D. Clay Whybark
Compatibility of Manufacturing Practices

This paper describes a simple method for scoring the practices of a company against the modal practices used in North America. The scores of all firms in a region are then combined into an index of compatibility of the practices in that region with those in North America. This simple measure appears to be a useful approach to determining compatibility for purposes of making comparisons and to associate with performance measures. The paper points out that much research remains to be done, however, both on the index itself and its use in other studies.

Manufacturing Practices:
Differences that Matter

Gyula Vastag, Budapest University of Economic Sciences, Hungary
D. Clay Whybark, University of North Carolina-Chapel Hill, USA

ABSTRACT

This paper presents a comparison of manufacturing practices between Hungary, western Europe and North America. The supposition is that differences in operational practice may matter in the success of joint ventures or other strategic alliances. The comparison is based on a survey of firms in the small machine tool and non-fashion textile industries. The survey covered practices ranging from forecasting and planning procedures to shop floor decision making. Multivariate analyses were performed to find those areas of practice for which there were differences between the regions and industries. The differences were grouped into three broad categories: "metabolism" (the frequency, horizon, and increment for planning, forecasting and reacting to change), external orientation (the closeness to the market and degree of export sales), and managerial practices in several areas. The differences between the industries were judged less important than those between regions.

INTRODUCTION

There is currently a great deal of interest in initiating joint ventures and other collaborative schemes between North America, the Common Market, and the eastern European countries. The attention of managers in each of these areas is now focused on gaining access to expanding markets, securing low cost sources of components and parts, and/or establishing new manufacturing locations. As the reforms in eastern Europe continue and the 1992 date approaches, this interest will heighten. This paper looks at one aspect of these potential alliances; how closely matched are the actual manufacturing practices in these parts of the world. The intent is to identify those practices that might be impediments to making these partnerships work effectively.

Behind this study runs the theory of the "Martini Merger." Such an alliance is conceived in general terms, signed by senior managers (often in a remote spot like Bermuda), and is applauded by important political figures. The deal is then left to operational people to "work out the details." If there are differences between the two firms' operating practices, the theory goes, it may be very difficult (or impossible) to realize the benefits initially contemplated from the merger. A first step in evaluating this theory is to determine if there are significant differences in practice between firms in different parts of the world. That is what this paper is about.

This paper looks specifically at how the manufacturing practices are conducted in firms in the non-fashion textile and the small machine tool industries in Hungary (as a representative of the eastern European countries), North America (the United States and Canada) and western Europe (in which several countries are represented). The practices

that are evaluated range from forecasting and planning activities, to shop floor information and control capabilities. Multivariate analysis techniques are used to help identify
general factors (combinations of practices) that define differences, to highlight those
factors where differences are statistically significant, and to help portray the results of the
analysis.

BACKGROUND

This study is a part of a major worldwide effort to gather and analyze data on manufacturing practices. Initially conducted in South Korea (by Boo-Ho Rho of Sogang
University) and in China (by D. Clay Whybark of Indiana University), the study has been
expanded to many other countries and researchers. Information on the details of the
study, copies of the survey form, and the structure of the data base are available in [1].
Examples of some of the early analyses of the manufacturing practices data can be found
in [2,3,4,5,6].

Two industries were surveyed, the non-fashion textile industry and the small machine tool
industry. In the textile industry, the firms run the gamut from some that are integrated
(spinning and weaving through finishing) to some that finish only. Small tool manufacturers include firms that produce small drill presses, milling machines and other forming
equipment, some of which is computer-controlled. These two industries are found
throughout the world in both advanced and developing nations. Also, the two industries
represent a form of process industry (textile) and batch processing (machine tool).

The manufacturing firms were surveyed using a 65 question form covering five broad
areas. The areas are:
> 1. General Information
> 2. Forecasting Practices
> 3. Production Planning and Scheduling
> 4. Shop Floor Information and Control
> 5. Purchasing and Materials Management.

The specific questions covered in this analysis are described where appropriate in the
research explanation. The survey was conducted by mail and phone in western Europe
(34 machine tool and 24 textile firms) and North America (45 machine tool and 50 textile
firms). The survey was carried out by interviewers in a total of 78 companies in Hungary
(19 in machine tool and 17 in textile, with the remainder from companies in other
industries). This research utilizes the data only from the non-fashion textile and machine
tool industries in Hungary.

The field data was collected and coded in dBASE III in both numeric form (e.g., "number of employees" or "finished goods as a percent of inventory") and character form
(e.g., "what department develops the production planning data" or "what is the exposure
to JIT"). The difference in form indicates a difference in the nature of the data. The
character data consists almost exclusively of nominal scale variables, while the numeric

data represents at least interval scale variables. Different methods of analysis were required for the different data forms. In both cases, however, the data was analyzed with multivariate techniques [7]. Exploratory factor analysis was performed to gain overall insights into the data and to avoid having to make any a priori determinations of the importance of any of the variables [8]. This is a different approach than that taken in previous studies of the data. The details of the approach and some of the data preparation steps are presented below.

ANALYTICAL APPROACH

The manufacturing practices data is provided in dBASE III files for each region and industry. There are a total of 35 numeric fields (variables) and 60 character fields (variables) for each company in the original files. We chose to do the statistical analysis with the SYSTAT and SYGRAPH programs [9,10]. This required us to make several changes in the raw data. As a first step we converted the character variables into numeric ones. This conversion did not change the nature of the data (the transformed numeric variables reflected the same nominal scale properties as the original character variables) but the numeric form is required for some of the computations. Secondly, we created some new variables. Some of these were used for grouping the data (i.e., codes for regions or industries), while eight were derived ratio-scale variables (e.g., "sales per employee"). The analysis of the numeric and character variables will be described separately.

THE ANALYSIS OF THE NUMERIC VARIABLES

A principal components analysis was done first. Principal components analysis, which belongs to the factor analysis family, is concerned with the identification of structure within a set variables . It establishes relationships within the data and can serve as a data reduction technique. We used this technique to see if it was possible to reduce (at least) the number of interval scale variables, while maintaining as much of the original information as possible. If reduction is successful, the reduced set of variables can account for most of the variance in the data and provide the advantages of reduced dimensionality. Moreover, the dimensions calculated in the analysis (which are linear combinations of the values of the variables) are used to classify (group) the companies. In this research the classifications are used to determine which practices are grouped by region and which by industry.

The input to the principal components analysis was the matrix of Pearson correlation coefficients among the variables. The analysis combines closely correlated variables into components that explain ever decreasing amounts of the variance (for the unrotated form, we renumbered the rotated factors). To assure ourselves that the variables we added would not affect the results, we carried out this analysis on the complete set of variables and on sets that excluded some of those which we added. The results, in terms of the principal components (factors), were the same in all cases. To determine the number of principal components to retain for further analysis (the data reduction property) we

Table 1. Principal Components (Factors) and Loadings for Numeric Variables

Factor	Variables (loadings)	Factor name
1** 9.17#	LN2: Scheduling horizon (0.786)* IN2: Scheduling increment (0.784) IN3: Dispatch increment (0.713) LN3: Dispatch horizon (0.693) IN1: Planning increment (0.534) LN1: Planning horizon (0.520)	Detail planning parameters
2 8.28	LT: Avg. delivery time (0.779) HI: Max. delivery time (0.763) LOW: Min. Delivery time (0.730) WIP: WIP inventory (0.666)	Delivery leadtime
3 7.93	STK: Make-to-stock % (-0.933) ORD: Make-to-order % (0.930) FIN: Finished inventories (-0.659)	Make-to-stock vs. order
4 7.39	LPROD: Sales/employee (-0.960) PLPROD: Sales/laborer (-0.954) REQ: Sales/equipment value (-0.807)	Productivity ratios
5 5.27	DOC: Written changes (0.902) ORAL: Oral changes (-0.891)	Communication form
6 5.26	LGTHP: Prod. plan. horizon (0.781) INCRP: Prod. plan. increment (0.563) MOD: Forecast changes/year (0.525)	Production planning parameters
7 5.17	INCR: Forecast increment (0.801) LGTH: Forecast horizon (0.771)	Forecast parameters
8 4.26	REV: Prod. changes/year (-0.677) LMT: Demand change limit (0.539)	Production plan Change parameters
9 4.03	XPR: Export sales percent (0.739)	Export sales
10 3.77	UTL: Equipment utilization (0.809) RPW: Direct labor ratio (0.502)	Overhead elements
11 3.67	ERR: Forecast error (0.718) INVREQ: Sales/inv. value (-0.541)	Forecast error and sales to inventory
12 3.34	SFY: Safety stock (0.745)	Safety stock
67.55	*Rotated loadings (correlations with the factor) **The unrotated variables had a different order #Percent of the variance explained by the factor	

applied a criterion of "the root greater than one" (i.e., eigenvalues greater than one), perhaps the most frequently used criterion [7]. A scree test of the excluded components' contribution to the explanation of the variance (very low) confirmed the choice. The total variance explained by the components retained is greater than sixty seven percent.

Table 1 shows the component loadings, principle components retained, variance explained by each, and suggested names for the factors (components). For the most part the variables grouped into quite rational factors. Only factor number eleven combined variables that are not easily related. It takes a great deal of imagination to determine the relationship between the average percentage forecast error and the sales per dollar of inventory.

We grouped the factors into larger combinations we called "metafactors." These are shown in Table 2. The detail, production and forecast parameter factors, under the metabolism metafactor, all relate to the length and increment of the activity (i.e., plans of one year length in increments of one month). The production change factor relates to frequency and conditions for changing the plans. These all reflect notions of reaction time or responsiveness, akin to the heartbeat or "metabolism" of the company. Differences in metabolism certainly could make it difficult for potential partners to merge operating activities.

Table 2. Metafactors from the principal components

Metafactor name	Factor number and name	
Metabolism	1	Detail planning parameters
	6	Production planning parameters
	7	Forecast parameters
	8	Production plan change parameter
External orientation	2	Delivery lead time
	3	Make-to-stock vs order
	9	Export sales
	12	Safety stock
Management (internal) orientation	4	Productivity ratios
	5	Communication form
	10	Overhead elements
	11	Forecast error and sales to inventory

Four factors are grouped under the metafactor called external orientation. The delivery leadtime, make-to-stock vs. order, and safety stock factors all relate to the closeness to the customer (in terms of product delivery). The export sales percentage shows a broader dimension of external orientation, by asking if the firm is looking outside the country. Only the equipment investment per employee variable does not fit well in this metafactor.

The management metafactor contains factors that relate to performance (productivity and forecast error) and practices (oral or written communications and degree of overhead carried). If differences existed in the management approaches of potential partners, they could present significant obstacles to the smooth functioning of a venture.

The metafactors are comprised of the twelve retained factors and the variables which make them up. Many of these variables could be important in establishing viable partnerships between companies. It remains to been seen, however, if they are really significant in discriminating between the regions or industries. In order to determine if there are differences that matter, we carried out an analysis of variance of the factor scores. To help visualize the results of this analysis we developed Chernoff faces using variables from factors that were significantly different. The specific approach is described below.

Analysis of Variance of Factor Scores

The principal components analysis resulted in twelve factors and the coefficients for computing the factor scores for each company in the data set. It remains to be seen, however, which, if any, of these factors differ between regions and/or industries. Individual comparisons of groupings of these factors could be performed, but the number of combinations is large. Even taken just two at a time, there are 66 representations of the results for the regional analysis alone.

In order to (possibly) reduce the number of factors that would need to analyzed, we carried out a one way analysis of variance on the factor scores using first the regions (Hungary, North America and western Europe) and then industries (machine tool and textile) as the grouping variable. Of course, we are fully aware of the assumptions required for using ANOVA and know that some of these assumptions are violated with the factor scores. We used it as a technique of expediency, however, to help reduce the dimensionality of the analysis, counting on the robustness of the procedure to provide helpful insights. Table 2 shows the results, indicating that, indeed, not all factors were "significant" between regions or industries.

One quick conclusion from the results shown in Table 3, is that no significant factors were common between the industry and regional groupings. Thus we can treat our discussion of the two separately. It is surprising that no management factor was significantly different between the three regions. This means that potential partnerships might look first to the metabolism and external orientation areas for potential problems. The variables included in the detail planning parameters factor indicate that Hungarian companies have the "slowest" metabolism. They have the longest horizons and increments of any of the regions. To put that in perspective, for each unit of planning time in western Europe, the North Americans would have two units and the Hungarians ten. The external orientation factors contain variables that indicate North American firms export only about 20% of what either western European or Hungarian firms do. The Hungarian firms carry much less finished goods inventory and make more to order than either of the other regions.

Table 3. Factors that differ significantly
between region and industry

Grouping	Factor	Name	Significance*
	1	Detail planning parameters	0.000
Regional	9	Export sales	0.000
	3	Make-to-stock vs order	0.028
	2	Delivery leadtime	0.000
Industry	10	Overhead elements	0.001
	8	Prod. plan change parameters	0.007
	12	Safety stock	0.026

* Only factors with a probability of less than 0.050 are shown.

The differences between the industries are not surprising and are less important in terms of presenting obstacles to potential partnerships. There is at least one factor from each of the metafactors that is significantly different between industries. In the area of external orientation, the lead times in the machine tool industry are about four times longer than the textile industry and the machine tool firms carry about half the safety stock as the textile firms. For the management area, the textile industry has higher levels of equipment utilization and direct labor. Even the metabolism is different. The machine tool firms change their production plans about half as often as the textile firms.

The results described above provide some feeling of where there were significant differences in the factor scores between the regional or industrial groupings. It is helpful to picture these differences, even though limited to two dimensions. The technique used to do that is Chernoff faces.

Chernoff Faces

In order to help visualize the differences between the regions and industries, Chernoff faces were developed [11]. Since its introduction in 1973, this method has become a useful (but sometimes contentious) tool for representing multivariate data in exploratory data analysis. The use of faces for representing data seems to have potential for, among other things: 1) enhancing the user's ability to detect and comprehend important phenomena, 2) serving as a mnemonic device for remembering major conclusions and 3) communicating major conclusions to others. As Chernoff put it, "This approach is an amusing reversal of a common one in artificial intelligence. Instead of using machines to discriminate between human faces by reducing them to numbers, we discriminate between numbers by using the machine to do the brute labor of drawing faces and leaving the intelligence to the humans, who are still more flexible and clever."

The SYGRAPH program (version 4.1) [10], allows the use of up to 20 variables to represent a single face. The key step in the creation of a face is the assignment of variables to facial features. This assignment may be done randomly or with some procedure. Much of the criticism of Chernoff faces has arisen over the subjectivity of the procedures or over interpretation when the assignment is random. To avoid this criticism, we used the results of the principal components analysis and the analysis of variance to determine which variables to use. We choose variables from the significant factors and assigned related variables to a single facial feature (i.e., mouth or nose). Variables associated with significant differences in factors between regions were assigned to the eyes, pupils, brows and noses of the faces. The industry differences are reflected in variables assigned to the mouth, face, ears and hair. Table 4 shows our assignment of the variables to the facial features.

Table 4. The assignment of variables to
facial features for the Chernoff faces

Feature			Variable	Factor
Eyes	Separation	LN2:	Scheduling horizon	Detail planning parameters
	Height at center	IN2:	Scheduling increment	"
	Slant	IN3:	Dispatch increment	"
	Eccentricity	LN3:	Dispatch horizon	"
	Half-length	IN1:	Planning increment	"
Pupil	Position	LN1:	Planning horizon	"
Brow	Angle	FIN:	Finished inventories	Make-to-stock vs. order
	Height	STK:	Make-to-stock	"
	Length	ORD:	Make-to-order	"
Nose	Width	XPR:	Export sales percent	Export sales
	Length	IEL:	Equipment/employee	-
Mouth	Curvature	LT:	Avg. delivery time	Delivery leadtime
	Length	HI:	Max. delivery time	"
	Height at center	LOW:	Min. Delivery time	"
Face	Height	WIP:	Wip inventory	"
	Upper eccentr.	UTL:	Equipment utilization	Overhead elements
	Lower eccentr.	RPW:	Direct labor ratio	"
Ears	Level	REV:	Prod. changes/year	Product. plan change parameters
	Radius	LMT:	Demand change limit	"
Hair	Length	SFY:	Safety stock	Safety stock

Figure 1 shows the Chernoff faces for the three regions and two industries. All variables were standardized before the faces were drawn. The faces show clearly some of the differences that are disclosed by the statistics. For example, the wider eyes reflect the greater planning horizon in Hungary compared to the other regions. The shape of the head and length of the hair show the differences in equipment utilization and safety stock between the two industries. Clearly, care must be taken in the assignment of variables

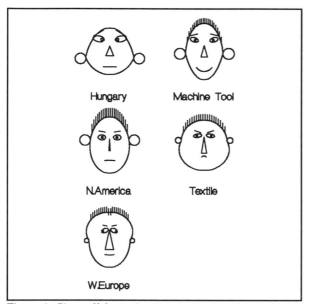

Figure 1. Chernoff faces developed with data for all faces considered together.

and the treatment of the standardization in order to portray differences that might matter.

THE ANALYSIS OF THE CHARACTER VARIABLES

To facilitate the analysis of the character variables from the data base, they were transformed into numeric nominal variables. Still, the set of potential analytical techniques for these variables is very limited due to the nominal scale properties. After transformation, the analysis parallels that of the interval scaled numeric variables. We used a principal components analysis.

To perform the principal components analysis, we used the Goodman-Kruskal Gamma measure of association of the variables [12]. The Gamma association matrix is the character variable equivalent of the Pearson correlation matrix that is the basic input to the principal components analysis. SYSTAT places several restrictions on the computation of this matrix. Only 20 variables could be included in any single matrix and we were limited to 100 companies for the calculation. The practical implication of this is that the industry data had to be analyzed separately from the regional data.

The choice of 20 variables from the 61 possible was guided by previous research on the data for other parts of the world. Variables that had proven to be different between other regions were chosen. Table 5 shows the component loadings, principle components obtained, variance explained by each, and suggested names for the factors (components). Only variables with a component loading greater than 0.5 were selected. The analysis for each industry provided essentially the same principle components.

Of the twenty variables chosen for inclusion in the analysis, fourteen were retained in the principal components. This is some corroboration of the importance of these variables in the previous work. The six groupings are quite reasonable and fall into two of the metafactors that were found for the numeric data. The management (internal) orientation metafactor is made up of the functional groups, sales forecaster and use related factors. The remaining three factors (influence elements, forecast input and leadtime) are included in the external orientation metafactor.

No Chernoff faces were produced for the character data, because of the nature of the variables. Similarly, the analysis necessary to establish whether the differences are significant is beyond the scope of this paper. Never the less, the commonality of variables in the metafactors between the numeric and character data give credence to the claim that they are clearly candidates for practices that make a difference. Knowing what elements influence decision making and who makes decisions is certainly important to communication and coordination in joint ventures.

Table 5. Principal components (factors)
and loadings for character variables

Factors	Variables (loadings)	Factor name
1	RELSE: Group setting priorities (0.765)*	Functional groups
15.52#	STRPU: Group ordering materials (0.737)	
	STRPR: Group starting production (0.718)	
	PLFUN: Group making production plan (0.676)	
	FUSER: Group using production plan (0.564)	
2	BASEPU: Basis for purchase order (0.839)	Influence elements
10.02	BASISP: Basis for production start (0.732)	
	BASISREL: Basis of priority (0.647)	
3	POSIT: Position of sales forecast (0.760)	Sales forecaster
9.05	FUNCT: Group of sales forecaster (0.731)	
4	PUSER: Prod. plan user position (0.708)	Use related
8.54	FRCSTUSE: Uses of the forecast (0.658)	
5	OBJCT: Objective factors in forecast (0.764)	Forecast input
6.92	SUBJT: Subjt. factors in forecast (0.735)	
6	CAUS: Causes of leadtime variation (0.680)	Leadtime
6.73	LAT: Percentage of orders late	
56.78		

* Rotated loadings (correlations with the factor)
# Percent of the variance explained by the factor

CONCLUSIONS

The conclusions from this work fall into three major categories; the usefulness of the techniques of analysis, the work that remains to be done and the insights from the data. One of the first observations to be made is that the use of the multivariate analysis techniques does enrich our understanding of the data and the insights contained therein. We need to make greater application of this body of knowledge. In terms of things to be done, the use of these techniques on the manufacturing practices data base needs to be documented in much more detail for future researchers. In addition, we need to carry the analysis one step further if we want to lay claim to practices that really "make a difference." That is, we need to see which of these factors are really important to the success of joint ventures or other strategic alliances.

The insights gained from this analysis of the data are important in their own right. First, the smaller number of significant factors in the industry comparisons means that we can concentrate on regional differences, an important practical conclusion. The differences between regions involved all three major categories of factors. In the area of "metabolism" we found that Hungary has the slower rate and North America the most rapid. This is reinforced by the character data which gave such prominence to the functions and levels of persons involved in making and executing plans.

Big differences in external orientation also seem to be underscored by the character data on the influences of the planning and execution activities. In terms of responding to market and other external forces, the western Europeans seem to be substantially ahead of the Hungarians and somewhat ahead of the North Americans. Differences were also seen in management practices that could also be reflected in the way that managers allow their operations to be influenced by various forces.

From a practical standpoint, these observations are quite important. Even before additional research establishes those factors which are most important to the success of various international partnerships, the prudent manager should be concerned that the operational practices of any prospective partner be, at least, understood. If this obvious, but often overlooked advice is not taken, it will be difficult to have pity on the failures of otherwise good ideas. Where differences exist, time will be well spent (prior to signing agreements) in working out methods to overcome or reconcile the differences.

REFERENCES

1.* D.C. Whybark and B.H. Rho, "A Worldwide Survey of Manufacturing Practices," Discussion Paper No. 2, Bloomington, IN, Indiana Center for Global Business, Indiana Business School, May 1988.

2.* A. Chikán and D.C. Whybark, "Cross-National Comparison of Production-Inventory Management Practices," Discussion Paper No. 8, Bloomington, IN, Indiana Center for Global Business, Indiana Business School, April, 1989.

3. K.K. Kwong and C. Li, "Sales Forecasting in China, Europe, Japan, Korea and the U.S.," *Proceedings of the Decision Sciences Institute Annual Meeting*, New Orleans, Nov. 1989.

4. K.K. Kwong, W. Fok and C. Hiranyavasit, "Comparing Manufacturing Practices in The People's Republic of China, South Korea and Japan," *Proceedings of the Decision Sciences Institute Annual Meeting*, New Orleans, Nov. 1989.

5.* B.H. Rho and D.C. Whybark, "Comparing Manufacturing Planning and Control Practices in Europe and Korea," *International Journal of Production Research*, 28, No. 12 (Fall 1990), 2393-2404.

6.* B.H. Rho and D.C. Whybark, "Comparing Manufacturing Practices in the People's

Republic of China and South Korea," Discussion Paper No. 4, Bloomington, IN, Indiana Center for Global Business, Indiana Business School, May 1988.

7. W.R. Dillon and M. Goldstein, *Multivariate Analysis: Methods and Applications*, New York, John Wiley and Sons, 1984.

8. I.R. Bernstein, *Applied Multivariate Analysis*, Amsterdam, Springer-Verlag, 1988.

9. L. Wilkinson, *SYSTAT: The System for Statistics*, Evanston, Ill., SYSTAT, Inc., 1988.

10. L. Wilkinson, *SYGRAPH: The System for Graphics*, Evanston, Ill., SYSTAT, Inc., 1988.

11. H. Chernoff, "The Use of Faces to Represent Points in Dimensional Space Graphically," *Journal of the American Statistical Association*, 68, No. 342 (June 1973).

12. Q.A. Goodman and W.H. Kruskal, "Measures of Association for Cross-Comparisons," *Journal of the American Statistical Association*, 40 (1954), pp. 732-764.

* This article is reproduced in this volume.

ACKNOWLEDGEMENTS

A version of this article appeared in the *International Journal of Production Economics*, 23, 1-3 (1991), 251-259, and in R. Grubbström and H. Hinterhuber (eds.), *Issues and Challenges for the 90s*, Amsterdam, Elsevier, 1991.

Compatibility of Manufacturing Practices

Gyula Vastag, Budapest University of Economic Sciences, Hungary
D. Clay Whybark, University of North Carolina-Chapel Hill, USA

ABSTRACT

This paper describes a method for determining the compatibility of manufacturing practices among firms from different regions of the world. The development of successful
joint ventures and other forms of international collaboration depend on the ability to
successfully integrate the production activities. The measure of compatibility developed
in this study is based on the data compiled by the Global Manufacturing Research Group.
Several bi- and multi-lateral comparisons, primarily based on statistical analysis of the
data, have already been performed. The results of these comparisons are helpful in
understanding differences in manufacturing practice, but no simple measure of the degree
of compatibility has yet been developed. This paper proposes a simple model to measure
the compatibility of manufacturing practices of regions relative to a selected base region.
The model is applied to all twelve of the regions in the data base.

INTRODUCTION

Much of the early work of the Global Manufacturing Research Group (GMRG) was built
on the premise that documenting and understanding manufacturing practices was important to successful global alliances. Much more has been done: from establishing a
network of researchers and creating a multiple country data base, to raising questions
about performance as well as practice [1]. The original premise is still an important
aspect of the work, however, and this paper returns to that subject. It is concerned with
the idea that studying the relationship between the similarity of manufacturing practices
and success of international alliances is important. It treats one aspect of that issue, the
similarity of the practices.

International Alliances

There is good cause to be concerned about the success of international alliances. Manufacturing is globalizing at a rapid rate [2]; and firms are establishing alliances in other
countries, often with very different cultures or economic systems [3]. Longevity [4] and
competitive strength [5], among others, have been suggested as measures of success of
such alliances. Success on either of these dimensions, however, seems to be allusive [6].
The lack of success seems to occur even when the alliances are formed between firms
from regions with similar cultures [7]. The premise of the GMRG, that incompatible
manufacturing practices might explain some of the difficulties encountered by firms attempting international collaboration, is quite consistent with these findings.

The concept of a "Martini Merger" was developed to describe alliances that had great
promise from a boardroom standpoint but failed because of operational details [8]. Often,
very clear market, financial, manufacturing, or other advantages persuade top executives

to enter into agreements for collaborating internationally. These agreements must then be made to work at the operational level, and that is where the compatibility of the manufacturing practices comes into play. (These same arguments can be made for within-country agreements as well, but the focus of this paper is on international comparisons.)

The idea of relating manufacturing practice differences to lack of success in international partnerships requires several things beyond the documentation of the practices. Among those are methods for measuring similarities (or differences) in the practices, determination of the criteria of success for the alliances (joint ventures and other forms of international collaboration), and research on the relationship between these measures. This paper presents one way of developing a measure of compatibility of manufacturing practices and relates the results to some of the other approaches.

Measuring Differences

Much of the early work of comparing differences in countries' manufacturing practices relied on parametric and non-parametric bilateral statistical tests [9]. As experience was gained, an analytical framework was developed for these bilateral comparisons and was applied to North America and western Europe [10].

Many multilateral comparisons were also based on statistical tests that explored differences between regions on selected variables [11]. These approaches, both for the bilateral and multilateral comparisons, provide lists of practices that differ significantly between regions. They do not provide an overall picture of the degree or nature of the differences, however.

To provide a comprehensive picture of the differences between two or more regions' manufacturing practices, multivariate statistics can be used. This was the approach taken in describing the manufacturing practice differences in North America, western Europe, and Hungary [8]. These techniques are the basis for determining the major factors (summaries of variables) that describe the differences in manufacturing practice between regions. This provides a comprehensive, though technically complex, way to assess manufacturing practice compatibility. For this paper, multivariate statistics were applied to a larger set of countries. When this was done, the factors became more diverse and the descriptive clarity of the technique diminished.

The complexity of the multivariate techniques and the difficulty of interpretation, as more countries are added to the analysis, provide sufficient reason to explore simpler measures of compatibility. One approach for developing such a measure is based on "scoring" each firm on selected practices. The individual scores are then used for developing overall measures for the country and/or industry. These in turn are used for constructing an index of compatibility of practices between countries.

That is the basic approach used for this paper. It is very straightforward, but must be considered preliminary since many of the steps involve subjective choices and the impact of alternative choices is not yet known. In the first step, the variables that are considered important for compatibility (with respect to international alliances) are chosen. The second step is to score the companies on those variables. Finally the compatibility index is developed for each of the regions.

In making the choices of the variables, some theory was used; but individual interest (from firms considering joint ventures, for example) or other theories could lead to different choices and different outcomes. The same is true of the scoring. Different weights or scoring schemes could change the compatibility indexes. This work must be considered preliminary and exploratory in this regard. Until some feeling for the sensitivity to these changes is developed and the importance of the index for follow-on research is better understood, firmer guidelines to the choices can't be made.

The development of the compatibility index facilitates the testing of very broad hypotheses about differences between regions and/or industries. It also allows for comparisons with the findings of the other approaches already used for analyzing the GMRG data base. Similarities to outcomes of other studies would add to the credibility of the index. The driving force behind the development of a compatibility index was not to rank companies or regions according to their manufacturing performance but to describe their existing practices from the point of view of a potential cooperation with other regions.

MULTIVARIATE ANALYSIS

Multivariate analysis of the data base has involved the use of principal components, clustering, and Chernoff faces (for a general description of these techniques see Bernstein) [12]. When these techniques were applied to the data from Hungary, western Europe, and North America, the principal components analysis of the ratio scale data resulted in 12 factors comprised of 31 variables that explain more than 65% of the variance. Clear differences between the regions were easily visualized with the Chernoff faces and cluster analysis.

These 12 factors were further combined into three meta factors: metabolism, external orientation, and internal orientation. The first of these meta factors has to do with the detail, length, and frequency of change of the planning and forecasting activities. It is an indication of the nimbleness of the organization. The external orientation meta factor has to do with the firm's orientation to customers, both domestic and export. The final meta factor describes some of the internal management activities in the firm. Each of these captures elements of practice that, if sufficiently different, could lead to operational problems in any attempt to integrate activities.

Sixteen character variables, accounting for more than 55% of the variance, were combined into six factors using factor analysis. These six factors fell into two of the meta factors: the external and internal orientation. The combining of variables into factors and factors into meta factors, for both the numeric and the character variables, resulted in useful meta factors. The meta factors, in turn, are helpful in describing practice differences between firms in the regions.

In another set of comparisons, six countries were analyzed; and the factors were nearly the same, but more variability was introduced [13]. For this paper, 10 countries were analyzed, using the principal components analysis. The results showed even more variation than with six countries and the factor assignments were considerably more ambiguous, even though the same primary factors were present. That variation provides an important motivation for this study, the search for a simpler procedure.

While multivariate analysis can provide useful insights, it is very time consuming and not transparent. The nature of the differences can be well described and, to a certain extent, the degree of differences assessed. The complexity of the procedure, however, suggests it is worthwhile to pursue something simpler that would be just as informative and just as valuable. That is the objective of the development of the compatibility index described in the next section.

COMPATIBILITY INDEX

In this section, the construction of the compatibility index is presented. The reasons for the choice of the variables, their scoring and the combination of scores into an index are all described. Even before variable selection was started, however, some principles were developed to guide the choices and the scoring. First, all sections of the questionnaire were to be represented. Secondly, there were to be a limited number of variables in total and per section (something like 2-5 questions for each of the five sections or 15-25 overall). Thirdly, only variables that reflect manufacturing practices were to be used. Finally, no special weighting of any of the variables or sections of the questionnaire was to be done. Different variables, weightings or scoring schemes would have to wait for new theory, some sensitivity analysis and/or a specific requirement.

Selection of the Variables

Three different approaches were taken to identify candidate variables for scoring. The first thing considered is the need for international partnerships to integrate their manufacturing activities. The entire questionnaire was reviewed for those questions which were felt to indicate areas where differences could frustrate integrating manufacturing practices. This provided more than 30 potential candidates. Some of the 65 questions were dropped because of overlap with other questions, some because there were too many missing responses, and others because it was felt they would not be important to the integration of manufacturing practices.

The other two approaches to nominating candidates for scoring were based on previous studies. First, bilateral statistical comparisons of regions were used to determine variables that were consistently mentioned as being different between regions. Special attention was paid to regions that had quite different practices (China and western Europe) and those with similar practices (western Europe and North America). Second, the multivariate analysis results were used to identify variables that were included in the factors describing the differences between regions.

Table 1. Variables and modal values

Variables	Modal Value (USA and Canada)
Company profile	
Average capacity utilization (%)	75-85
Days of training per employee per year	≤5
Sales forecasting	
Position of forecaster	Vice president
Length of forecast horizon (months)	12
Increment of forecast (days)	30
Number of forecast modifications per year	4-12
Most important uses of the forecast	Budget preparation and material planning;
	Material planning and production planning;
	Budget preparation and production planning
Production planning and scheduling	
Most important factors for production plans	Actual orders and production capacity;
	Actual orders and forecast
Length of production plan horizon (months)	12
Increment of production plan (days)	30
Most important uses of the production plan	Operations scheduling and material planning
Number of production plan revisions per year	12
Shop floor control	
Basis of the order for starting production	Actual customer order
Way of delivery date determination	Determined by the company
Average customer delivery time promise (days)	60
Purchasing and materials management	
Work-in-process inventory (days)	≤40
Finished goods inventory (days)	≤20
Most frequent method for purchasing raw materials	Purchase according to production plan or schedule
Number of vendors per purchased part	2-3
Most important reasons for subcontracting	Production load and production difficulty

Scoring the Variables

For the scoring scheme used in this study, the variables were scored relative to a base of comparison. The North American (USA and Canada) data were selected as the base. Thus, the index measures the degree of compatibility with USA and Canadian manufacturing practices. In all cases, a score of one or zero was given to each of the variables for each firm. If the practices matched those of USA and Canada, the score was zero; if not, the score was one. A practice matched USA and Canada if it was the same as the modal practice in the USA and Canada.

For categorical data, the determination of the mode is straightforward. It is simply the most frequently chosen response. In the case of continuous variables, the mode depends upon the method used for describing the data. For this study, histograms were used, and the interval that contained the largest number of responses was the modal range. To develop the histogram, the SYSTAT routine (based on Scott [14]) was used. The resulting modal values are summarized in Table 1 for each of the variables chosen.

Developing the Index

A total of twenty variables were selected for scoring. Ideally, there would be a value for each of them for every company surveyed. Unfortunately, however, for any individual company, there may be missing responses on one or more of these variables. If only those firms for which there was data for every variable was chosen for scoring, the sample sizes in each region would be very small. Therefore, each company was used and all variables for which there were responses were scored. This increased the sample size, but means that there is not strict comparability between the scores of the firms in the region.

The score on each variable is a one (if the practice matches USA and Canadian modal practice) or zero (if the practice is different). To develop the total score for a region, the individual variable scores were summed for every company and then all company scores were summed for the region. The average score for the region was then computed by dividing the region's total by the total number of individual variable scores (i.e., the non-missing responses). This average varies between zero (no compatibility with USA and Canadian modal practices) and one (complete compatibility).

To compute the compatibility index, the average scores were divided by the USA and Canadian score. This standardizes the scores of the other regions with the USA and Canadian score. Other approaches to computing the index can lead to different results as pointed out by Vastag [15]. The average scores and compatibility indexes for each of the twelve regions are shown in Table 2.

Table 2. Average scores and compatibility indexes for the regions

Region	Average score	Index
USA and Canada	0.526	1.000
Australia	0.494	0.939
Bulgaria	0.469	0.892
Finland	0.450	0.856
Western Europe	0.435	0.827
South Korea	0.431	0.819
Hungary	0.427	0.812
USSR	0.420	0.798
Japan	0.407	0.774
Chile	0.389	0.740
Mexico	0.364	0.692
China	0.363	0.690

DISCUSSION AND CONCLUSIONS

It is encouraging to see that the compatibility index reflects the relationships between regions that has been reported in other GMRG studies. China appears as one of the countries most distant from the USA and Canadian practices, for example, something that is ratified in many other studies. There are a couple of surprises though. The position of western Europe, which has always been shown to be quite similar to the USA and Canada, is further removed than was expected. Bulgaria, which is still dominated by central planning influenced practices, is much closer than the differences in the economic systems would suggest. Several countries have been included in this comparison which have not been involved in any of the other comparisons. Mexico and Finland are examples. Finland's position is not surprising, but Mexico's is.

This application of the compatibility index does show that it has some promise as a research tool. It is simple to compute and does seem to bear out the results of other studies. It looks promising as a measure that could be used for studies of international alliances or other performance oriented research. Moreover, the variables that are used in scoring can be selected to focus on the practices relevant to a particular study. There are some issues with the concept, however.

The properties of the index lead to some anomalies. Since a base of comparison is selected, all index values are relative to that base, but the indexes are not necessarily symmetrical. That is, the index for China compared to the USA and Canada may not (probably won't) be the same as for USA and Canada with China as the base.

The individual variability of the companies in a given region means that a region may not be as compatible with itself as another region since not all the firms in a region will have the modal practice values. Although that didn't occur in this study, the average score for USA and Canada in Table 2 was only 0.526. This means that only slightly more than half of the responses in the region matched the modal practice value.

Missing values mean that the comparability of the scores between firms may not be as high as desired. The problem of missing values is one that continually arises in empirical research. In addition, the index itself, as constructed for this study, is not very discriminating. The values of the index are quite close together and individual company or variable average scores must be computed to get some idea of the significant differences between regions (see Vastag [15]). This all suggests that there is considerable room for future work with the index.

One of the first areas of future work is on the index itself. Clearly, some notion of the sensitivity to other means of computing the index is needed. Some arguments can be made for using averages of average company or variable scores to get some measure of variability. Another variation would be to construct the index using a modal basis as was done for the scoring. A modal compatibility score could come from a histogram of the individual company compatibility scores, for example. Many other combinations can be devised for computing the index and such work would be of value.

The other area where additional work is needed is in relating the index to other measures. The GMRG started with the question of the relationship to the success of mergers, but the performance on other dimensions are also important to study. Such studies will provide some insight into whether it is important be concerned about the how the index is computed when relating the index to performance results. Again work on this will be of value. Finally, it is important to say that the index is still too broad a measure for a micro level decision. A company that is interested in forming an alliance can use the concepts presented here, but needs to focus on individual firms rather than broad geographical results.

REFERENCES

1.   D.C. Whybark and G. Vastag (eds.), *Global Manufacturing Practices: A Worldwide Survey of Practices in Production Planning and Control*. Amsterdam, Elsevier, 1993.

2.   M.M. McGrath and R.W. Hoole, "Manufacturing's New Economies of Scale," *Harvard Business Review*, 70, 3 (May/June 1992), 94-102.

3.   E. Anderson, "Two firms, One frontier: On assessing joint venture performance," *Sloan Management Review*, 31, 2, (1990), 19-30.

4.   M.J. Geringer and L. Hebert, "Measuring Performance of International Joint Ventures," *Journal of International Business Studies*, 23, 2 (Second Quarter 1991), 249-264.

5.   G. Hamel, Y.L. Dos and C.K. Prahalad, "Collaborate With Your Competitors--And Win," *Harvard Business Review*, 67, 1 (Jan/Feb 1989), 133-139.

6.   S. Sherman, "Are Strategic Alliances Working?" *Fortune*, 126, 6 (Sept. 21, 1992), 77-78.

7.   D. Savona, "When Companies Divorce," *International Business*, Nov. 1992, 48-51.

8.*   D.C. Whybark and G. Vastag, "Manufacturing Practices: Differences that Matter," *International Journal of Production Economics*, 23 (1991), 251-259.

9.*   B.H. Rho and D.C. Whybark, "Comparing Manufacturing and Control Practices in Europe and Korea," *International Journal of Production Research*, 28, 12 (Fall 1990), 2393-2404.

10.* G. Vastag and D.C. Whybark, "Comparing Manufacturing Practices in North America and Western Europe: Are There Any Surprises?" in D.C. Whybark and G. Vastag (eds.) *Global Manufacturing Research Group Collected Papers, Vol. 1*, Chapel Hill, NC, Kenan Institute, University of North Carolina, 1992.

11.* R.B. Handfield and B. Withers, "A Comparison of Logistics Management in Hungary, China, Korea and Japan," *Journal of Business Logistics*, Fall 1992.

12. I.R. Bernstein, *Applied Multivariate Analysis*, Amsterdam, Springer-Verlag, 1988.

13. G. Vastag and D.C. Whybark, "A Global View of Local Manufacturing Practices," Proceedings of the First International Conference of the Decision Sciences Institute, Brussels, June, 1991, 157-160.

14. D.W. Scott, "Optimal and Data-based Histograms," *Biometrika*, Vol. a (1979), pp. 605-610.

15.* G. Vastag, "Technical Considerations in Analyzing the Manufacturing Practices Data."

* This article is reproduced in this volume.

SECTION SEVEN
FUTURE ACTIVITIES

This section contains material quite different in form and content from that of the preceding sections. It has an introductory paper on future GMRG activities. The other material relates to the conduct of the second round survey. There is a set of suggestions for conducting the survey and a complete copy of the survey itself. Some information on the data entry program that has been developed to produce consistent data files has also been included.

D. Clay Whybark
Future Activities for the Global Manufacturing Research Group

This short article describes some of the projects that are planned by the GMRG. In addition to continuing work on the initial data base, the work on the second survey has already started. New work in more focused projects and pedagogical developments are also possible.

Suggestions for GMRG Survey Research
Guidelines for the Second Round Survey

The suggestions that are provided here have been sent to researchers beginning to use the new survey. They are based on the experience developed by the GMRG in the first round of data gathering.

Global Manufacturing Research Group
Manufacturing Practices Survey

This is the full survey that was developed during 1991 and 1992 for future data gathering activities on manufacturing practices.

Michael Bohlig and Gyula Vastag
Data Entry Program

This paper intoduces a data entry program created to accompany the GMRG second round survey.

Future Activities for the Global Manufacturing Research Group

D. Clay Whybark, University of North Carolina-Chapel Hill, USA

ABSTRACT

This brief paper describes several possible future activities for the Global Manufacturing Research Group and others interested in international research and education. The activities include carrying out additional analyses of the data base from the first survey, conducting the second survey, doing other international studies, and undertaking the development of some educational material.

INTRODUCTION

Members of the Global Manufacturing Research Group have been engaged in international research since 1985. The group itself was formally founded in 1990 at an international meeting in Seoul, Korea. Since that time, many other people have become associated with the group, although it has remained an informal, loose federation of interested (and interesting) researchers, concerned about global manufacturing issues.

With the publication of this book, a milestone of sorts has been reached. The broader circulation of the data base, made possible by its inclusion with this volume, increases the potential for new researchers to join in the research activities. The development of the second survey brings with it opportunities for answering different types of research questions and, again, increases possibilities for new persons to get involved. This all signifies the potential for increasing the scope and variety of research activities in the future.

It also may be time for the group to undertake activities that would help make manufacturing a positive career alternative for students of management and technical programs. Such activities would include developing teaching material and classroom delivery techniques, creating joint international seminars, and forging closer associations with manufacturing executives from around the world. Manufacturing is too important to our global economy for us not to make every effort to increase our understanding of manufacturing practices and to effectively disseminate this information.

Several of the potential future activities will be discussed below. They range from items related to the survey to other projects that the group might undertake.

INITIAL SURVEY

A copy of the questionnaire used for the initial survey is included in this book. The results of analyzing the data from that survey form the basis for most of the articles, as well. Some of the research that has been done on this data is not included here. Part of the research was not in English, some was presented at meetings in abbreviated form, and some very likely has simply not been brought to the attention of the editors. Nevertheless, a broad array of analyses has been performed on the data.

The data base that was compiled from the initial data gathering effort is also included here. The improved access to the data should entice other researchers to develop their own analyses of the data. It is quite clear that the more people who get involved with the effort, the greater the variety and quality of the research that is accomplished. The data already have been used to address questions in a highly creative manner by people who were not initially associated with the project.

There is more to be mined from the data and this paper is an invitation to join in the activity. The broad pictures of differences between regions, which seem to outweigh the distinctions between the two industries, can be refined with sharper analysis. The differences between practices that arise because of factors other than size, country, economy or industry have scarcely been addressed in the work to date. Some possible new factors could include degree of external orientation, amount of vertical integration, make-to-stock orientation, and many others that are possible by slicing the data differently. All are untapped research potential using the data included here. Studies providing insights into how these factors influence practices would help in building a better understanding of global manufacturing activities.

The creation of the second survey opens up many new opportunities for comparisons by adding a time dimension. The results of the first survey can now be compared to the practices at a later time to evaluate some of the dynamics of manufacturing and determine whether the dynamics vary in different parts of the world. Some of these questions are particularly important in the reforming economies of eastern and central Europe. They also may shed light on what might be expected from China in the future.

SURVEY OF 1992

Though much remains to be gained from additional analysis of the first survey data, the second survey has opened up even more opportunities. Several members of the group worked hard to produce a questionnaire that would be easy to answer, is consistent with the first survey, and provides for some analyses that were not possible with the first one. Three principles guided the group's work. The first was to maintain comparability with the first survey, even when a question needed to be changed to facilitate responding to it or analyzing the response. The second was to have a minimum of questions that required financial information, especially that which would be subject to the accounting regulations of a specific country. The third was to drop questions if new ones were added, in order to maintain some limit on the total length of the questionnaire.

A copy of the new survey is included in this book. The format is quite different from that of the first survey. It is the result of our experience in gathering and analyzing the data from the first survey. Some questions were added and several were dropped. Those for which the research to date had indicated little gain or which were felt less important than the new questions, were deleted. The focus, however, has not changed. The survey is still largely concerned with manufacturing practices.

Although some questions on performance have been added, there are not many. The argument against including more performance oriented questions was concern with comparability between countries and a possible loss of sample size because such questions increase the number of non-responses. The net result is a survey that is fairly easy to administer (even though it is long), does not require a lot of sensitive information, and can be analyzed quite easily.

In addition to the new survey, a coding program has been developed that will assure data uniformity. This program is used to facilitate entering the data from hard copy of a completed new questionnaire and create data files for analysis. The coding program follows the format of the survey and provides reasonableness tests for the responses to many of the questions in order to trap some of the obvious data entry errors. It creates files of the data that will be identical for each country in which data is gathered and compiled. The files produced by the program will assure a common format of the data base for use by the analyst.

OTHER STUDIES

The documentation of manufacturing practices and studies of their similarities and differences is an important undertaking. Many researchers want to do much more than this, however. They raise concerns about improving manufacturing performance. They would like to conduct studies that disclose practices resulting in improved performance in a multitude of settings. To do this effectively, however, requires a different survey and/or different forms of research and analysis.

The network of global researchers that form the GMRG provides an obvious platform for conducting such studies. In addition, the group members provide some overlap with other organizations having experience and contacts to help perform such studies. A natural future activity, therefore, is to expand the breadth of research undertaken in order to cover issues not contemplated in the manufacturing practices survey. These might include more targeted studies, which could then use the network to help conduct them.

As operations management matures as a field, more experience with empirical research will be gained and the insights provided will improve. This provides both a challenge and an opportunity for the GMRG. The group should be on the leading edge in building this experience, both for survey research and for other forms of empiricism that provide useful knowledge to the field. The GMRG network and its contacts can facilitate achieving this goal.

PEDAGOGICAL ACTIVITIES

The conduct of international manufacturing research remains a primary aim of the GMRG, but the broader objective of the research is disseminating the knowledge thus gained and changing manufacturing practices. Consequently, the GMRG might undertake programs to improve pedagogical material and delivery. Cases on manufacturing practices in different parts of the world would be a welcome addition to the teaching literature. Also useful would be exercises and other forms of enhancing understanding based on observations of manufacturing in a variety of settings.

Manufacturing managers all over the world thirst for knowledge. They don't usually get out to talk to people in other companies, a job reserved for the marketing/sales people and occasionally a purchasing person. The need for contact among manufacturing personnel provides an opportunity to put together seminars that would bring to executives some of the knowledge gained from the research. By communicating the diversity of experience, a better case could be made for management and technical students to consider manufacturing as a career path.

SUMMARY

The development of the new survey provides the basis for expanded activities in the future, but these activities certainly are not limited to survey research. Use of the GMRG network for other, more targeted, studies, is an important opportunity for the group. The need to improve pedagogy in manufacturing instruction continues around the world and presents another great opportunity for the GMRG. The basis exists for meaningful future growth in the activities of the group. It is up to all of us to seize the opportunity.

Suggestions for GMRG Survey Research

Guidelines for the Second Round Survey

These suggestions for the second round of the GMRG survey research were compiled from observations made during the administration of the first survey, from insights provided by experts in survey research and from the requirements needed to share and use the data. The suggestions are for your consideration in making the survey data as valid and widely useful as possible. Several areas will be covered; from the translation of the survey (if required) to incorporating the data into the data base. We hope you find these pages helpful, and we are anxious to learn of any new inventions or discoveries you make, so that we can pass them on.

Some procedures are required for us to share and archive the data. These are indicated at the end of the document, but we will mention some procedures throughout. Most of the required procedures are found in the latter steps of the data gathering process. We recommend, however, that you become familiar with them before starting to gather the data. Some of the information needed is most easily gathered as the survey is being conducted.

TRANSLATION

The survey was developed in English, so we have no experience regarding translation of this version to pass on. We have found from the first survey that if you need to translate, a forward and backward translation is useful. Such a translation works this way: Translate the questionnaire first into your language, using a person familiar with the field rather than a translation specialist. Then have a different person translate the survey back into English. Comparing the two English versions will be embarrassing and entertaining, and will disclose those questions requiring editing. Finally, we will want to archive a copy of the translated survey that was used for data gathering.

Two other points to keep in mind if you create a different version of the questionnaire. First, try to keep the sequence of questions the same, both for data coding purposes and to prevent any bias from question order differences. Second, try to preserve the amount of "white space" that is found in the English version. The amount of white space was approved by an experienced social science survey researcher.

PILOT TEST

Using an interviewer to conduct a pilot test of the survey in a couple of firms has proven to be very useful. Besides learning about difficulties with the response scales, about questions that are still unclear, or about other problems with the survey, you will learn other things that will be helpful later on. Note, for example, where new difficulties arise, where someone else has to be consulted to get the data, how long it takes to fill out the survey. If you relay this information to respondents when you are gathering data in other firms, they will appreciate it.

The pilot test may reveal that it is necessary to reword some questions, to provide better introductions to some of the sections, or to add explanations for how some of the questions are to be answered. These changes should be made in the final version of the survey.

Substantial changes need not mean the loss of the pilot study data. That data can often be added to the rest of the data, although you may need to go back to the company that provided the pilot data for some corrections to the original entries.

CHOOSING THE FIRMS TO SURVEY

We certainly want to continue gathering data on our two original industries--non-fashion textiles (bed sheets, pillow case, industrial cloth, curtains, rugs, underwear, etc.) and small machine tools (lathes, milling machines, drills, grinders, etc., for industrial use). If you want to survey another industry, however, that is fine, but you will need to inform us when you submit the data.

The first choices for data sources will be the firms from the original survey. Using the same firm code number as before will make it easy to study the changes that have occurred in those companies. Where possible, use systematic sampling procedures (random, stratified, clustered, etc.). This will help us in publication and is consistent with the procedures of other groups doing survey research (e.g., Boston University's Global Manufacturing Futures Project and IMD's Project 2000). The use of convenience samples is certainly acceptable when systematic sampling is not possible.

Whatever the method used for choosing the firms to be surveyed, let us know what it was (even if it was a bunch of your friends, i.e., a convenience sample). Also keep track of the response rates (i.e., how many of the firms contacted completed the survey?). Documenting the sampling procedures and results is important to the research community and we will want this data for the archives here.

DATA GATHERING METHODS

Several different methods of gathering data have been tried by the GMRG. Some of these have been very creative, and we expect there will be more new ideas from this round. What is important is gathering the sample data efficiently, have few missing values, get high response rates, and preserve the relationships with the companies that have cooperated. Here are some of the methods with pros and cons.

Mail Survey

Mail surveys have been used by many researchers for all kinds of studies. They are fairly easy to administer, simply mail the survey to the selected firms. The mailing can be preceded by a personal or telephone contact to solicit an agreement to participate. A follow-up visit or call can be made to answer questions (and encourage responses). Many variations of precontact and follow up have been used. Example contact and follow-up letters have been developed if you would like some ideas.

The disadvantage of mail surveys is very low response rates. In countries where survey research is not very well known, a simple mailing of the survey has resulted in response rates of much less than 5%. In North America, less than half the firms that agreed beforehand to participate in a mail survey did so. These are not very encouraging numbers, especially if the total number of firms in the industry is small to begin with.

Telephone

Telephone surveys work well if they can be timed to meet the respondents' schedules. The response rate is usually much higher for a telephone interview than for a direct mail survey. A telephone interview can be scheduled in advance and continued later, if necessary.

A disadvantage of phone interviews is that the respondent may not have all the answers, and the interview may have to be broken up. In addition, it is almost always necessary to hold the interview during working hours for both parties, which takes some of the flexibility out of the process and occupies both people.

During a phone interview, the respondent should have a copy of the survey in front of him or her, even if the interviewer is going to record the data, so that the respondent can create a copy of the responses for later reference.

Personal Interview

This is one of most effective ways to gather data. Personal interviews permit questions of interpretation or other difficulties to be answered immediately. They can be set up ahead of time and dates can be confirmed, which improves the response rate. The contact also helps to maintain the relationships with the companies involved.

The problem, of course, is that personal interviews are expensive and time-consuming to conduct. Some researchers have been able to persuade trade associations to help with the interviews, arguing that the value of the study to the association was worth the cost of helping out. One researcher hired sociologists trained in interviewing and gave them additional training in the nature of the data needed. This was effective, but still costly.

Group Interviews

A clever variation on the personal interview is gathering together a group of responding firms (either at a plant or at the researcher's location) and surveying them together. This allows good contact with the respondents and offers a chance to discuss the work with them. The response rate for those firms that agree to come to participate in such a session is very high, and when the respondents are gathered, the interviewer has a captive audience.

A disadvantage of this method can be the need for some respondents to telephone their firm, or actually return to it, in order to get some of the data. Another drawback is that a mutual time for several people from the companies as well as for the researcher must be found. The company representatives do expect some value added while they are being surveyed, so you will need to plan for this.

PREPARING THE PLAYERS

It is necessary to prepare both the people gathering the data and the people providing the data in order to get good results.

Data Gatherers

The data gatherers (whether they work on the phone, in the field, or in a meeting room) need to be trained. They need to know how to interpret the questions and understand something about the research objectives. They can do a lot to make a respondent feel comfortable if they sound confident in what they are doing and understand why.

Survey Respondents

In order to get the appropriate responses on the survey, the respondent should be made to understand and feel comfortable about several things. These matters can be communicated to them in person, through the mail or by phone, or in the group meetings.

You may have other ideas, but here are some of the things that we have learned are important to communicate to the respondents:
1. Confidentiality of the data, individuals, and the firm's identity. Copies of research papers based on previous surveys will help demonstrate that the data are truly confidential (and are used).
2. The survey not a test. There are no right or wrong answers.
3. Approximate answers are much better than no answers at all.
4. The response formats may not be familiar to some people, and may look intimidating. Assure them that once they have established a rhythm, it will go quickly.
5. The data should all be for the same unit, whether it is a company, plant, or group of plants. Consistency is key.
6. The survey is long and the respondents should be aware how long it will take to complete it. (If anything, overstate the time required.)

RESPONDENT RELATIONS

Several researchers have found it useful to provide sample research articles when requesting data and to follow data gathering with some form of feedback to the respondents (even when the interviews are done in a group). One useful form of feedback is to provide each respondent with the mean responses (for the appropriate industry and country) on a copy of the survey. If you do this, however, remind the respondents to keep a copy of their results to compare to the industry averages. This is a good form of thanks for the time spent in filling out the survey and provides respondents with some value.

You want to be able to identify an individual respondent in case there are questions when the data are coded. Consequently, you will need to assign a unique code number to each respondent and we will ask for this code with your data. This must be done carefully, however, so there is no question about the data confidentiality. The code should not contain letters or groups of letters that appear in the responding firm's name.

SURVEY FORM STORAGE

We suggest that you store the original survey forms where you can get to them easily. It should be possible to retrieve an individual form to verify a response if necessary. We will keep a copy of your translated survey here in order to facilitate answering any such questions.

DATA ENTRY

A data entry program has been developed for this version of the survey and it is necessary that you use it to prepare data for the data base. A complete set of instructions on how to use the program is included with the coding program. Use of the program does not guarantee accuracy, although it will help. We urge you to have the data verified (by a different person than the coder) after it has been coded. Running off a hard copy and scanning it can disclose problems. Data errors are an embarrassment for all of us and can hurt our research credibility.

SENDING IN THE DATA

The data should be provided on disks (both 3.5" and 5.25") in the format produced by the coding program. The program will ask you to specify some general information including the survey methodology, your name and address (in case there are questions about the data), information on new industries surveyed, and the exchange rates to use in converting to US dollars. Individual firm data will require assigning company codes and describing any special meaning the code might have. Along with the data you should include a copy of the survey that was used to gather the data. We will also ask you to sign the data release form, shown on the next page, to provide the data to the group for research purposes.

GMRG DATA RELEASE FORM

The enclosed disks contain the data gathered and coded for the following country(s):

_____

The data is in the format produced by the coding program provided by the Global Manufacturing Research Group and has been checked for accuracy. I hereby release the data for distribution to other members of the Global Manufacturing Research Group for academic research purposes. In so doing, I expected to receive data from the Group for my own research purposes.

Name: _____

Institution: _____

Address: _____

_____

_____

Telephone: _____

Fax: _____

Bitnet: _____

Signature: _____

GLOBAL MANUFACTURING RESEARCH GROUP

MANUFACTURING PRACTICES SURVEY
(September 19, 1992)

1. Company Description and Background Information

The information in this section of the survey will be useful to researchers in studying rela-tionships between company characteristics and manufacturing practices. As with the answers to questions in subsequent sections of the survey, the information that you provide will not be used to identify individual companies. Please feel comfortable giving **approximate** responses; in most cases, our research has shown that it is more important to have approximate answers than none at all.

> In this survey, we have used the word "company" to represent the unit for which you are answering questions. We ask that you be consistent throughout the survey and report the sales, employment, practices, etc. for the same unit, whether it is a company or a plant.

1.01 **Approximately** how many employees work for the company? _____ employees

1.02 How many of these employees are production workers? _____ production workers

1.03 How many production shifts are there per day? _____ production shifts per day

1.04 In this company, how many hours per year does a production employee work?

_____ hours per year

1.05 About what percent of the company's total manufacturing cost is for labor?

_____% of cost

© Global Manufacturing Research Group, 1992

1.06   What were the company's sales last year?

(State in currency units.)

_____ in domestic sales

_____ in export sales

1.07   How many product lines or product families are produced by the company?

_____ product lines or product families

1.08   What percentage of company sales is represented by the company's largest selling product line?

_____ % of sales

1.09*   About what percentage of the company's products are make-to-stock?
        About what percentage of the company's products are make-to-order?

(These should sum to 100%.)

_____ % make-to-stock

_____ % make-to-order

1.10   What percentage of the ownership of the company is represented by each of the following?

(These should sum to 100%.)

_____ % domestic (within the country)

_____ % foreign (outside the country)

1.11   What percentage of the machines in the company are grouped as follows?

_____ % of machines are grouped by machine type (e.g., all lathes together)

_____ % of machines are grouped in combinations for products or product families

---

* For centrally planned economies, the category "make-to-state plan" was added.

1.12   What is the company's average capacity utilization rate for plant machinery or equipment?

_____ % capacity utilization rate

1.13   Over the last two years, about what percentage of **annual** sales has the company spent on new production equipment?

_____% of annual sales

1.14   What is the company's approximate total investment in production equipment?

(State in currency units.)

_____ investment in equipment

1.15   What is the approximate average age of the company's production equipment?

_____ years

1.16   About what percent of sales is the total manufacturing cost? _____% of sales

1.17   To what extent does the company use computers for the following?

	not at all				to a great extent
(Circle a number for each use.)					
sales forecasting	1 . .	2 . .	3 . .	4 . .	5
production planning . .	1 . .	2 . .	3 . .	4 . .	5
production scheduling .	1 . .	2 . .	3 . .	4 . .	5
inventory management	1 . .	2 . .	3 . .	4 . .	5
purchasing.	1 . .	2 . .	3 . .	4 . .	5
product design	1 . .	2 . .	3 . .	4 . .	5

1.18   For each of the characteristics listed below, how does the company compare with its
       competitors?   (Circle a number from 1 to 5 for each characteristic.)

	far worse than competitors	about the same as competitors	far better than competitors
unit cost of manufacturing.	1 . . 2 . .	3 . .	4 . . 5
quality of products	1 . . 2 . .	3 . .	4 . . 5
manufacturing throughput time (speed).	1 . . 2 . .	3 . .	4 . . 5
delivery speed	1 . . 2 . .	3 . .	4 . . 5
delivery as promised	1 . . 2 . .	3 . .	4 . . 5
flexibility to change product	1 . . 2 . .	3 . .	4 . . 5
flexibility to change output volume. . .	1 . . 2 . .	3 . .	4 . . 5
product design time	1 . . 2 . .	3 . .	4 . . 5

1.19   In the last two years, to what extent has the company invested resources (money, time
       and/or people) in programs in the following areas?
                                                          (Circle a number for each program.)

	not at all			to a great extent
cellular manufacturing	1 . .	2 . .	3 . .	4 . . 5
computer hardware/software	1 . .	2 . .	3 . .	4 . . 5
employee participation programs . . .	1 . .	2 . .	3 . .	4 . . 5
factory automation	1 . .	2 . .	3 . .	4 . . 5
just-in-time systems	1 . .	2 . .	3 . .	4 . . 5
manufacturing time reduction	1 . .	2 . .	3 . .	4 . . 5
material requirements planning	1 . .	2 . .	3 . .	4 . . 5
productivity improvement.	1 . .	2 . .	3 . .	4 . . 5
setup time reduction	1 . .	2 . .	3 . .	4 . . 5
process analysis	1 . .	2 . .	3 . .	4 . . 5
statistical process control	1 . .	2 . .	3 . .	4 . . 5
supplier partnerships	1 . .	2 . .	3 . .	4 . . 5
total quality management	1 . .	2 . .	3 . .	4 . . 5
recycling of materials	1 . .	2 . .	3 . .	4 . . 5

2. Sales Forecasting

This section of the survey is about the methods that the company uses to anticipate demand for its products. We have found that a wide range of methods are used by manufacturing firms, and that both formal and informal approaches are effective. Thus, as with the other sections, your answers will provide us with insights about actual company practices, and there are no right or wrong answers. Please remember, also, that for questions that ask for numerical answers, your answers may be **approximate**.

2.01 Which of the following best describes the **position** (level) of the person who has **primary authority** for producing the company's sales forecasts?

Position (Mark only one.)
- ☐ president/CEO/managing director
- ☐ vice president/director
- ☐ department/division head
- ☐ group/section manager
- ☐ owner

2.02 Which of the following best describes the **function** of the person who has **primary authority** for producing the company's sales forecasts?

Function (Mark only one.)
- ☐ administration
- ☐ planning
- ☐ production
- ☐ engineering
- ☐ sales
- ☐ marketing
- ☐ finance
- ☐ accounting

2.03 To what extent does the company use each of the following for sales forecasting?

(Circle a number for each method.)

	not at all				to a great extent
time series models (e.g., moving average)	1	2	3	4	5
causal models (e.g., regression)	1	2	3	4	5
qualitative methods (e.g., delphi)	1	2	3	4	5
past experience	1	2	3	4	5

2.04   To what extent is each of the following factors considered in the company's forecast?

(Circle a number for each factor.)

	not at all				to a great extent
current economic conditions . .	1 . .	2 . .	3 . .	4 . .	5
current political conditions. . .	1 . .	2 . .	3 . .	4 . .	5
general company situation . . .	1 . .	2 . .	3 . .	4 . .	5
general industry situation . . .	1 . .	2 . .	3 . .	4 . .	5
customer information	1 . .	2 . .	3 . .	4 . .	5
supplier information.	1 . .	2 . .	3 . .	4 . .	5
results of market research . . .	1 . .	2 . .	3 . .	4 . .	5
past sales (demand) history . .	1 . .	2 . .	3 . .	4 . .	5
current order backlog	1 . .	2 . .	3 . .	4 . .	5

2.05   Is the company sales forecast developed for individual products or product lines?

(Mark one **or** both.)

☐   individual products

☐   product lines

2.06   For about how many individual products or product lines does the company develop forecasts?

_____ products are forecast

_____ product lines are forecast

2.07   How is the company sales forecast expressed?

(Mark one **or** both.)

☐   monetary value of products

☐   physical units of products

2.08   How far into the future does the company's sales forecast extend? _____ months

2.09   What is the **smallest** time increment into which the company's sales forecast is divided?

(Mark only one.)
☐ days
☐ weeks
☐ months
☐ years

2.10   About how many times per year is the company sales forecast modified?

_____ times per year

2.11   What has been the **approximate** average percent forecast error over the past two years?

_____ percent average forecast error

2.12   To what extent is the company's sales forecast used for the following purposes?

(Circle a number for each purpose.)

	not at all				to a great extent
budget preparation	1	2	3	4	5
production planning	1	2	3	4	5
subcontracting decisions	1	2	3	4	5
material/inventory planning	1	2	3	4	5
sales planning	1	2	3	4	5
human resource planning	1	2	3	4	5
new product development plans	1	2	3	4	5
facilities planning	1	2	3	4	5
equipment purchase planning	1	2	3	4	5

2.13   How often is the company's sales forecast changed by production personnel for production planning purposes?

(Circle a number.)

	never				very often
forecast changes by production	1	2	3	4	5

3. Production Planning and Scheduling

This section is about your company's practices in the areas of production planning and detailed production scheduling. **Planning** refers to activities which express units of production and inventory in aggregated terms for the purpose of specifying overall output and capacity requirements. **Scheduling** refers to detailed scheduling (often known as master production scheduling) that is used to determine the timing and output levels for specific products or components.

Production Planning Questions

3.01   Is the company production **plan** developed for individual products or product lines?

(Mark one or both.)
☐   individual products
☐   product lines

3.02   For about how many individual products or product lines does the company develop production **plans**?

_____ products in production **plan**
_____ product lines in production **plan**

3.03   How is the company's overall production **plan** expressed?

(Mark all that are appropriate.)
☐   monetary value of products
☐   physical units of products
☐   labor time (e.g., hours, days)
☐   machine time (e.g., hours, days)

3.04   How far into the future does the company's production **plan** extend? _____ months

3.05   What is the **smallest** increment into which the company's production **plan** is divided?

(Mark only one.)
☐   days
☐   weeks
☐   months
☐   years

3.06 About how many times per year is the company's production **plan** revised?

_____ times per year

3.07 To what extent is each of the following factors considered in the development of the company's production **plan**?

(Circle a number for each factor.)

	not at all				to a great extent
customer order backlogs	1	2	3	4	5
previous sales	1	2	3	4	5
machine capacity	1	2	3	4	5
labor capacity	1	2	3	4	5
customers' future plans	1	2	3	4	5
inventory levels	1	2	3	4	5
the forecast	1	2	3	4	5

3.08 When **demand exceeds capacity**, how often does the company respond in each of the following ways?

(Circle a number for each alternative.)

	never				very often
hire more workers	1	2	3	4	5
use overtime	1	2	3	4	5
add shifts	1	2	3	4	5
subcontract production work	1	2	3	4	5
backlog customer orders	1	2	3	4	5
lease temporary capacity	1	2	3	4	5

3.09 When **demand is less than capacity**, how often does the company respond in each of the following ways?

(Circle a number for each alternative.)

	never				very often
lay off workers	1	2	3	4	5
allow idle capacity	1	2	3	4	5
eliminate shifts	1	2	3	4	5
reduce work day or week	1	2	3	4	5
build inventory	1	2	3	4	5
lease capacity to others	1	2	3	4	5

3.10   To what extent is the company's production **plan** used for the following purposes?

(Circle a number for each purpose.)

	not at all				to a great extent
budget preparation.	1 . .	2 . .	3 . .	4 . .	5
production scheduling	1 . .	2 . .	3 . .	4 . .	5
subcontracting decisions.	1 . .	2 . .	3 . .	4 . .	5
material/inventory planning.	1 . .	2 . .	3 . .	4 . .	5
sales planning	1 . .	2 . .	3 . .	4 . .	5
human resource planning.	1 . .	2 . .	3 . .	4 . .	5
facilities planning	1 . .	2 . .	3 . .	4 . .	5
equipment purchase planning.	1 . .	2 . .	3 . .	4 . .	5

3.11   How often does the company consider the following when subcontracting work?

(Circle a number for each.)

	never				very often
excess production load at your company .	1 . .	2 . .	3 . .	4 . .	5
production difficulty at your company . .	1 . .	2 . .	3 . .	4 . .	5
top management directive	1 . .	2 . .	3 . .	4 . .	5
subcontracting allows earlier delivery date	1 . .	2 . .	3 . .	4 . .	5
subcontractor's costs are lower.	1 . .	2 . .	3 . .	4 . .	5
subcontractor's quality is higher	1 . .	2 . .	3 . .	4 . .	5

Production <u>Scheduling</u> Questions

3.12   How is the company's overall production **schedule** expressed?

(Mark only one.)
- ☐   monetary value of products
- ☐   physical units of products
- ☐   units of product lines
- ☐   labor hours of output
- ☐   machine hours of output

3.13   How far into the future does the company's production **schedule** extend? _____ weeks

3.14   What is the **smallest** time increment of the company's production **schedule**?

(Mark only one.)
☐ days
☐ weeks
☐ months
☐ years

3.15   Is part of the company's production **schedule** frozen or otherwise made difficult to change in the near future?

(Mark yes or no.)
Yes   No
☐   ☐

3.16   Does the company have formal measurements of the following?

(Mark yes or no for each one.)
Yes   No

	Yes	No
inventory record accuracy	☐	☐
bill of material accuracy	☐	☐
shop floor routing accuracy . . .	☐	☐ .
time standard accuracy	☐	☐
forecast accuracy.	☐	☐
late deliveries	☐	☐
delivery speed.	☐	☐
manufacturing throughput time . .	☐	☐
product design time	☐	☐
raw material quality	☐	☐
work in process quality	☐	☐
finished goods quality	☐	☐
customer satisfaction.	☐	☐
productivity	☐	☐
inventory levels	☐	☐
setup times	☐	☐
employee turnover	☐	☐
employee absenteeism	☐	☐
number of employee suggestions	☐	☐

4. Shop Floor Control

Shop floor control refers to the set of operating-level activities associated with the implementation of detailed production schedules. This includes decisions about lot sizes, when to start a production order, sequencing at work centers, and when to make changes in the schedule. Our research has shown that a wide range of practices are used. As with the other sections, your answers to these questions will assist us in understanding actual practices.

4.01   Which of the following best describes the **position** of the person in the company who authorizes the plant to start work on an order?

Position (Mark only one.)
☐   president/CEO/managing director
☐   vice president/director
☐   department/division head
☐   group/section manager
☐   owner

4.02   Which of the following best describes the **functional group** in the company that authorizes the plant to start work on an order?

Function (Mark only one.)
☐   administration
☐   planning
☐   production
☐   engineering
☐   sales
☐   marketing
☐   finance
☐   accounting

4.03   To what extent is each of the following considered in company decisions to authorize start of work on an order?   (Circle a number for each factor.)

	not at all				to a great extent
actual customer order	1	2	3	4	5
production plan	1	2	3	4	5
detailed production schedule	1	2	3	4	5
parts shortage list	1	2	3	4	5
inventory level	1	2	3	4	5
importance of the customer	1	2	3	4	5

4.04 How often do individuals in the following groups decide on the **sequence** in which jobs will be processed at machines or work centers?

(Circle a number for each.)

	never				very often
high level company managers.	1	2	3	4	5
production control personnel .	1	2	3	4	5
work center supervisor.	1	2	3	4	5
operators	1	2	3	4	5

4.05 How often is the processing sequence at machines or work centers in the company established by the following criteria?

(Circle a number for each criterion.)

	never				very often
order in which jobs arrive	1	2	3	4	5
customer order due date	1	2	3	4	5
processing time of the job	1	2	3	4	5
work remaining at subsequent stations .	1	2	3	4	5
minimize number of set ups	1	2	3	4	5
top management directive	1	2	3	4	5
how easy or difficult the job is.	1	2	3	4	5
past experience	1	2	3	4	5

4.06 How often does each of the following factors **change** the company's production schedule priorities **after** the plant has started an order?

(Circle a number for each factor.)

	never				very often
pressure from marketing	1	2	3	4	5
pressure from customer	1	2	3	4	5
labor shortage	1	2	3	4	5
material shortage	1	2	3	4	5
energy shortage	1	2	3	4	5
manufacturing problem	1	2	3	4	5
insufficient machine capacity .	1	2	3	4	5
machine breakdown	1	2	3	4	5
change in sales plan	1	2	3	4	5
change in delivery due date . .	1	2	3	4	5
change in demand	1	2	3	4	5
engineering design change . . .	1	2	3	4	5
top management directive . . .	1	2	3	4	5

4.07   How often are the due dates that are promised to customers determined by each of the following?

(Circle a number for each.)

	never			very often
the customer	1 . .	2 . .	3 . .	4 . .	5
the company	1 . .	2 . .	3 . .	4 . .	5
negotiation with customer.	1 . .	2 . .	3 . .	4 . .	5
state (government) plan . . .	1 . .	2 . .	3 . .	4 . .	5

4.08   When a customer places an order, approximately how many days into the future does the company promise delivery?

_____ **minimum** days to delivery promise date
_____ **maximum** days to delivery promise date
_____ **usual** days to delivery promise date

4.09   On average, what percentage of the company's orders are delivered to customers **after** the promised delivery date?

_____% of orders are delivered **after** promised date

4.10   For customer orders that are **not delivered on time**, what is the average number of days late?

_____ days late

4.11   When the company's finished goods are delivered late to customers, how often does each of the following occur as the cause of lateness?

(Circle a number for each possible cause.)

	never			very often
insufficient machine capacity.	1 . .	2 . .	3 . .	4 . .	5
machine breakdown	1 . .	2 . .	3 . .	4 . .	5
material shortage	1 . .	2 . .	3 . .	4 . .	5
energy shortage	1 . .	2 . .	3 . .	4 . .	5
insufficient labor capacity	1 . .	2 . .	3 . .	4 . .	5
material quality problem	1 . .	2 . .	3 . .	4 . .	5
production quality problem.	1 . .	2 . .	3 . .	4 . .	5
production bottleneck	1 . .	2 . .	3 . .	4 . .	5
scheduling error	1 . .	2 . .	3 . .	4 . .	5
change of schedule priorities	1 . .	2 . .	3 . .	4 . .	5
finished goods transportation problem	1 . .	2 . .	3 . .	4 . .	5
change in customer due date	1 . .	2 . .	3 . .	4 . .	5

4.12   On approximately what percent of its orders do **customer schedule changes** occur after
the start of production?

_____ % of orders

4.13   On approximately what portion of the company's orders do **engineering or design
changes** occur after the start of production?

_____ % of orders

4.14   To what extent does each of the following determine the size of the company's production
lots?

(Circle a number for each.)

	not at all				to a great extent
setup time	1 . .	2 . .	3 . .	4 . .	5
setup cost	1 . .	2 . .	3 . .	4 . .	5
processing time	1 . .	2 . .	3 . .	4 . .	5
size of the order	1 . .	2 . .	3 . .	4 . .	5
storage cost	1 . .	2 . .	3 . .	4 . .	5
storage space.	1 . .	2 . .	3 . .	4 . .	5
material handling capacity.	1 . .	2 . .	3 . .	4 . .	5
cost per unit	1 . .	2 . .	3 . .	4 . .	5
past experience	1 . .	2 . .	3 . .	4 . .	5

4.15   What are the company's approximate reject or return percentages at each of the following
stages?

(Provide an answer for each one.)

_____% rejects of incoming material

_____% rejects during processing in factory (scrap rate)

_____% rejects at final inspection

_____% returns from the customer

4.16   About how much time typically elapses from the **start** of the first operation until a batch of
the company's products is **finished** ?

_____ days **minimum** time from start to completion
_____ days **maximum** time from start to completion
_____ days **usual** time from start to completion

4.17   Of the elapsed factory time described in the previous question, approximately what
percentage is spent in actual processing operations (i.e., non waiting time)?

_____ % of elapsed factory time is spent in actual operations

4.18   About how much time typically elapses from the receipt of a customer order until it is
shipped?

_____ days **minimum** time from order to shipment
_____ days **maximum** time from order to shipment
_____ days **usual** time from order to shipment

4.19   What proportion of the company's orders is completed on or before the time specified by
the production schedule?

_____% of orders completed on time or early

4.20   For approximately what proportion of its manufacturing operations does the company have
time standards?

_____% of operations have time standards

4.21   About what percentage of the company's time standards are **accurate**?

_____ % of time standards are accurate

4.22 Over the past two years, what has been the **annual** percent of change in the following? For each of these measures, please indicate increases with a "**+**" sign and decreases with a "**—**" sign. If there has been no change, use a zero (0).

	sign **+** or **—**	approximate annual change
physical output (units, meters, etc.) .	( )	_____ % per year
productivity	( )	_____ % per year
product design time.	( )	_____ % per year
cost of manufacturing.	( )	_____ % per year
quality of products	( )	_____ % per year
manufacturing throughput time .	( )	_____ % per year
delivery speed	( )	_____ % per year
on-time deliveries	( )	_____ % per year

5. Materials Management

Materials management includes a wide range of activities associated with purchasing, managing, distributing, and controlling inventories within the plant. Inventory includes raw materials, component parts, work in process, and finished goods. As with the other sections of the survey, we are interested in the practices employed in your plant; a variety of approaches to materials management have been shown to be effective, so there are no right or wrong answers.

5.01   To what extent does the company consider each of the following in determining purchase **quantities**?

(Circle a number for each.)

	not at all				to a great extent
cost to place an order	1	2	3	4	5
difficulty of placing an order.	1	2	3	4	5
quantity discounts.	1	2	3	4	5
amount needed for customer order	1	2	3	4	5
transportation cost.	1	2	3	4	5
transportation distance.	1	2	3	4	5
storage cost	1	2	3	4	5
storage space	1	2	3	4	5
cost per unit.	1	2	3	4	5
size of transport equipment.	1	2	3	4	5
expected quality	1	2	3	4	5
delivery reliability	1	2	3	4	5
amount specified by supplier	1	2	3	4	5
past experience	1	2	3	4	5

5.02   Approximately what percent of the parts and components that comprise the company's products are fabricated within the plant?

_____% fabricated in plant

5.03   What percent of the company's total manufacturing cost is for purchased material?

_____% of cost

5.04   How often does the company use each of the following policies when initiating purchase
        orders?

(Circle a number for each policy.)

	never				very often
order at periodic interval (e.g., monthly). .	1 . .	2 . .	3 . .	4 . .	5
order based on inventory level	1 . .	2 . .	3 . .	4 . .	5
order based on production **plan**	1 . .	2 . .	3 . .	4 . .	5
order based on production **schedule** . .	1 . .	2 . .	3 . .	4 . .	5
order based on material shortage list. .	1 . .	2 . .	3 . .	4 . .	5
order for actual customer order	1 . .	2 . .	3 . .	4 . .	5
order based on past experience	1 . .	2 . .	3 . .	4 . .	5

5.05   What proportion of the company's purchase orders are delivered by suppliers as follows?

(These should sum to 100%)

_____% delivered **early**

_____% delivered **on time**

_____% delivered **late**

5.06   Of the purchase orders that are delivered **late** by suppliers, what is the approximate average
        lateness?

_____ days late

5.07   About how many suppliers does the company have, on average, per part?

_____ suppliers per part

5.08   What is the approximate total number of part numbers in each segment of the company's
        inventory system?

_____ raw material part numbers

_____ component part numbers

_____ finished goods part numbers

5.09   What is the value of the company's total inventory? _____
                                                            (Please state in currency units.)

5.10   What is the percentage distribution of the company's inventory values?

(These should sum to 100%)

_____ % of value in purchased materials and parts

_____ % of value in work-in-process

_____ % of value in finished goods

Comments

Thank you for your help with this survey. Your participation will contribute to a better world-wide understanding of manufacturing practices. Are there any important issues that you feel have been left out? If so, please comment in the space provided on this page.
Comments and Additional Remarks:

GLOBAL MANUFACTURING RESEARCH SURVEY
(This page will be separated from the data.)

NAME OF COMPANY:

_____

EXAMPLES OF COMPANY PRODUCTS:

_____
_____
_____

Please indicate, below, the name and the address of the person responsible for coordinating the completion of the GMRG survey in your company.

SURVEY COORDINATOR

Name _____

Title/Function _____

Mailing Address _____

_____

_____

_____

_____

Phone Number _____

How many years have you been with the company? _____ years

How many years have you held your current position in the company? _____ years

DATE _____

Data Entry Program

Michael Bohlig, Pacific Institute for Research and Evaluation, USA
Gyula Vastag, Budapest University of Economic Sciences, Hungary

ABSTRACT

This paper introduces a computer program that was developed to enter the data from the Global Manufacturing Research Group questionnaire. The program assures that the data are encoded electronically on a uniform basis from country to country. It also helps to catch obvious errors in the data.

INTRODUCTION

Great care was taken in preparing the data from the first GMRG survey for analysis to make sure that the data were strictly comparable from country to country. The names of the variables, the lengths of the data fields, and the conventions for indicating missing variables were all addressed.

The second round survey was designed to achieve the required consistency through a computer program that facilitates data entry and assures a common format for each variable. The program also screens out some incorrect entries and provides a signal if other entries are out of range or are beyond expected bounds. Some of the concepts that were built into the program are presented below.

DATA ENTRY PROGRAM

The data entry program mirrors the questionnaire question-by-question, helping assure that the data are entered for the correct field. The program assigns the variable name to the field, thereby ensuring that it is the same for every country. The data are then encoded into dBASE IV files for reading into data analysis programs. Three basic aspects of the program are described below.

Nature of the field

The program distinguishes between character and numeric fields and makes the proper assignment of both the field type and variable name to the variable. This is important in ensuring that the analytical technique can be applied uniformally to that variable, regardless of the country from which it comes.

A second aspect of the nature of the field has to do with missing variables. When missing values are encountered, as they inevitably are in survey research, they must be indicated in some way. Moreover, they need to be indicated the same way for every country from which data are collected. The data entry program helps provide this consistency by checking for appropriate indications of missing values (e.g., -99 in numeric fields and not blank characters). The importance of this for uniformity of analysis is clear.

Data errors

While the program cannot prevent incorrect data entry, it can help catch some obvious errors. Incorrect field types (e.g., trying to put a character into a numeric field), missing values, and other potential errors are flagged by comparing the entered values with a range of expected values. If an entered value is not in the acceptable range, the program generates an audible tone and displays an error message indicating that the value is not in the expected range and asks for the item to be reentered. This feature does not prevent the unexpected response from being entered, but simply asks for verification that the entry is what is wanted.

Two types of variable ranges are evaluated. The first deals with questions asking for a set of responses that should add up to 100%. Any sum that doesn't equal 100 is questioned. The second is based on expected values derived from the first GMRG survey. The range comes from previously observed values, such as high and low values of sales per employee or inventory turns per year.

Data files

The program produces data files that contain the values of the variables (the responses to the survey) on a company-by-company basis. These data files are identical regardless of the source of the data. This set of common files facilitates analysis.

In addition to the data files, another file is produced for each country in which data are collected. This file contains information on the conduct of the survey, the researcher involved, and other information particular to the country and time of the survey. These data are helpful in identifying aspects of the survey that might be different from one location to the next.

SUMMARY

A users' manual has been prepared to go with the coding program. It provides considerably more detail on the technical aspects of the ranges, use of the program, and the files that are produced. Use of the coding program does not guarantee that incorrect entry of the data will not occur, so it is important to verify data entry and to make a visual check of the data. Use of the program, however, will greatly reduce detail data preparation and will assure identical files for each set of data.

APPENDIX

Global Manufacturing Research Group Data Base

ABSTRACT

This appendix contains a brief description of the Global Manufacturing Research Group (GMRG) data base. It was developed from the data gathered using the questionnaire presented in the first section of this book. It was the initial survey of manufacturing practices developed by the group. An introduction to the files and some information on the data from each country is included here. In addition to the files, the disk distributed with this book also contains a data base utility program that allows the user to access the GMRG files. A brief description of that program is presented here also.

INTRODUCTION

The GMRG data base contains descriptions of manufacturing practices in the small machine tool and non-fashion textile industries from twelve different regions of the world. The twelve regions incorporate data from more than twenty countries. The data were compiled from surveys administered between 1985 and 1990 using the questionnaire presented in the first section of this book. The data are distributed here in non-indexed dBASE III files, with a file for each region-industry combination.

The coding of the data follows the format presented in the first section of this book. The data structure and variable names for every file follow that format. (An exception is the NOTES (memo) variable which has been deleted. Key information from that variable is included in the files as described below.) A substantial amount of effort has gone into the coding of the data to assure that the formats are consistent. In addition, an effort was made to make sure that the treatment of missing variables is consistent throughout. The files have been checked and rechecked, but some of the differences may have been missed.

Any time this much data is pulled together, especially from around the world, there are some inconsistencies that arise. The coding program that will be used for the second round of data gathering will help solve this problem in the future. For now, however, there may still be the occasional misspelled variable, slightly different field length, or a non-standard treatment of missing variables in the data. As always, therefore, it is a good idea to do a quick visual check of the data before performing any analysis. If there are questions on the data from any region that are not answered in these files, we have provided the names and locations of the principal researcher(s) for each region.

In addition to checking data accuracy and consistency, this disk has been checked for viruses. However, neither the publisher, the editors, the Global Manufacturing Research Group, nor the authors makes any warranty or representation, express or implied, with respect to the usefulness of the data contained in these files or assumes any liability with respect to the use of, or for damages arising from the use of, the data contained herein.

CONTENTS OF THE DISK

There are three types of files on the data disk. One of these is a READ.ME file both in
ASCII format, and as a .DBF file. This file contains much of the information that appears
here in a format that can be read on the screen or be printed out. The second category
is the data files. These are the dBASE III files containing the data from each region and
industry. The final set of files is a database utility program, DBFVIEW.EXE and its
related help file, HELP.DBF. The utility program provides easy access to the data. The
data files and utility program are described below.

Data files

The data file names identify the country (or region) and industry in which the data was
gathered. All the data file names and their meaning are listed below. With only a couple
of exceptions there is a textile and machine tool industry file for each region. The exceptions include both missing and added industries for some regions. Since all the data
files are in dBASE III, each data file has a DBF extension.

File name	Country or region	Industry	Firms
AUSTRALT.DBF	Australia	Textile	46
BULGARIM.DBF	Bulgaria	Machine tool	15
BULGARIT.DBF	Bulgaria	Textile	13
CHILEM.DBF	Chile	Machine tool	10
CHILET.DBF	Chile	Textile	11
CHINAM.DBF	People's Republic of China	Machine tool	44
CHINAT.DBF	People's Republic of China	Textile	56
EUROPEM.DBF	Western Europe	Machine tool	34
EUROPET.DBF	Western Europe	Textile	24
FINLANDE.DBF	Finland	Electronics	13
FINLANDM.DBF	Finland	Machine tool	21
HUNGARYM.DBF	Hungary	Machine tool	19
HUNGARYO.DBF	Hungary	Other*	42
HUNGARYT.DBF	Hungary	Textile	17
JAPANM.DBF	Japan	Machine tool	18
JAPANT.DBF	Japan	Textile	36
MEXICOM.DBF	Mexico	Machine tool	14
MEXICOT.DBF	Mexico	Textile	22
NAMERM.DBF	North America	Machine tool	45
NAMERT.DBF	North America	Textile	50
SKOREAM.DBF	South Korea	Machine tool	89
SKOREAT.DBF	South Korea	Textile	33
USSRM.DBF	USSR (before the breakup)	Machine tool	18
USSRT.DBF	USSR (before the breakup)	Textile	32

* The specific products are listed in the file

The data base utility program

A data base utility program is available on the disk. It is the DBFVIEW.EXE file. Typing
DBFVIEW at the system prompt will open the program. All the dBASE III files

(including HELP.DBF and README.DBF) are accessible through the program. It will enable the user to look at the data in each file in a matrix format. To use the program, "OPEN" the file of interest and "BROWSE" through the data. The variables appear as column headings and the rows are the company responses for eacg variable.

The utility program also permits the user to do some modifications of the data: changes can be made, and even new records can be added. (In that case, please copy the files to another disk or to your hard disk first, in order to maintain integrity of the original data.) The HELP.DBF file can be printed out with (Shift-)Print Screen if hard copy detail is desired.

DATA DESCRIPTION

The development of the data in dBASE III format provided some advantages, but also incurred some costs. The advantages are related to the compactness of the files and the discipline necessary to create them. By having to specify the name, format and character of each variable exactly, data consistency was enhanced. On the other hand, many researchers prefer to use other programs for their data analysis work.

The dBASE III format can be imported directly into many popular spreadsheet and statistical analysis programs. This is the first step to try in attempting to communicate with other programs for data analysis. If the analysis program does not read the dBASE III files directly, then the dBASE III program can be used to produce ASCII files.

The principal researcher(s) that were responsible for data gathering and coding in each region are identified below. Also, certain files contain additional information. For example, in some cases the code numbers used to identify the company also identifies the location of the firm more specifically. Other files have additional explanatory information. Where such information is present, it is described below. If nothing appears except the name of the researcher, then there is no special information in the file(s). In general, information on company products, if any, will be found under PRODUCTS in the file.

Australia

The data from Australia came only from the textile industry. The principal researchers were:
> *Prof. Danny Samson, University of Melbourne, Victoria, Australia.*
> *Prof. Amrik Sohal, Monash University, Victoria, Australia.*

Bulgaria

The principal researcher from Bulgaria was:
> *Prof. Pavel Dimitrov, University of National and World Economy, Sofia, Bulgaria.*

Chile

The PRODUCTS field in the Chilean data base contains the number of products the firm produces. The principal researchers were:

Profs. Antonio E. Kovacevic and Juan Claudio Lopez, Catholic University of Chile, Santiago, Chile.

China

The data from the People's Republic of China were mostly gathered in the Shanghai region, but companies north and south of the Shanghai area were also included. The first digit of the code number used for each firm indicates the town or area where the firm is located. The coding used is presented in Table 1.

Table 1. Code numbers for regions in China

Code	Town, province or area
100	Shanghai
200	Jiang-Shu and Zhe-Jiang provinces
300	South China
400	North China
500	Chang-Zhou
600	Outside Shanghai

The principal researchers were:

Prof. Xiao Cheng Zhong, Shanghai Institute of Mechanical Engineering, Shanghai, China.

Prof. D. Clay Whybark, University of North Carolina, Chapel Hill, North Carolina, USA.

Western Europe

The term western Europe is used here to distinguish it from eastern or central Europe. In fact, the data were collected only in northern European countries. The first (or first two) digits of the code number of each company indicates the country from which the data comes. The coding used is shown in Table 2.

The principal researcher was:

Prof. D. Clay Whybark, University of North Carolina, Chapel Hill, North Carolina, USA.

Finland

The files from Finland contain data on the machine tool industry and electronics industry. There is no textile industry data. The principal researcher was:

Prof. Allan Lehtimäki, University of Oulu, Oulu, Finland.

Table 2. Code numbers for countries in Western Europe

Code	Country
100	England
140	Scotland
160	Ireland
200	Sweden
300	France
400	West Germany (before reunification)
500	Switzerland
550	Austria
700	Belgium
800	The Netherlands

Hungary

In addition to the files on the textile and machine tool industries in Hungary, there is an additional file that contains the same data on a variety of other companies. The description of the products produced by those companies are contained in the file. Also, the textile file contains information on companies that produce textiles and "clothes." They are firms 3,5,11,12,14,59,75, and 76. They are identified, also in the PRODUCTS variable. The principal researcher was:

Prof. Attila Chikán, Budapest University of Economic Sciences,
   Budapest, Hungary.

Japan

The principal researcher from Japan was:
   Prof. Toshihiro Murakoshi, Waseda University, Tokyo, Japan.

Mexico

These files include the name of each company surveyed and substantial information on the products produced by them. This data is included at the end of each record in two additional variables. The principal researchers were:

Profs. Arturo Macias and Felipe Burgos, University of the Americas,
   Puebla, Mexico.
Prof. Benito E. Flores, Texas A&M University, College Station, Texas,
   USA.

North America

The North American data comes from Canada and several regions of the United States. The location of the company is included in the first digit of the code number assigned to the firm. The coding used is shown in Table 3.

Table 3. Code numbers for locations of North American firms

Code	Location of firm
100	Northeastern U.S.
200	Southeastern U.S.
300	Midwestern U.S.
400	Western U.S.
500	Canada

The principal researcher was:

> *Prof. D. Clay Whybark, University of North Carolina, Chapel Hill, North Carolina, USA.*

South Korea

The principal researcher for South Korea was:

> *Prof. Boo-Ho Rho, Sogang University, Seoul, Korea.*

USSR

There are notes on the safety stock values contained in a variable at the end of each record. The principal researchers were:

> *Prof. Alexander Ardishvili, IMEMO, Academy of Sciences, Moscow, Russia.*
> *Prof. Arthur V. Hill, The University of Minnesota, Minneapolis, Minnesota, USA.*

ACKNOWLEDGEMENTS

We would like to thank the following publishers for allowing us to reproduce the articles listed below:

pages 121-138: **MCB University Press Ltd.**
Manufacturing Practices in the Soviet Union
by Alexander Ardishvili and Arthur V. Hill
Scheduled for publication in the *International Journal of Operations and Production Management*

pages 173-184: **Taylor & Francis Ltd.**
Comparing Manufacturing Planning and Control Practices in Europe and Korea
by Boo-Ho Rho and D. Clay Whybark
Previously published in the *International Journal of Production Research*, Vol.28, No.12, pp.2393-2404, (Fall, 1990)

pages 213-232: **Council of Logistics Management**
A Comparison of Logistics Management in Hungary, China, Korea and Japan
by Robert B. Handfield and Barbara Withers
Previously published in the *Journal of Business Logistics*, (Winter, 1992)

pages 233-246: **MCB University Press Ltd.**
Global Manufacturing Practices and Strategies: A Study of Two Industries
by Scott T. Young, K. Kern Kwong, Cheng Li and Wing Fok
Previously published in the *International Journal of Operations and Production Management*, Vol.12, No.9, pp.5-17, (1991)